INVENTORY MODELS

THEORY AND DECISION LIBRARY

General Editors: W. Leinfellner and G. Eberlein

Series A: Philosophy and Methodology of the Social Sciences
Editors: W. Leinfellner (Technical University of Vienna)
G. Eberlein (Technical University of Munich)

Series B: Mathematical and Statistical Methods
Editor: H. J. Skala (University of Paderborn)

Series C: Game Theory, Mathematical Programming and
Operations Research
Editor: S. H. Tijs (University of Nijmegen)

Series D: System Theory, Knowledge Engineering and Problem
Solving
Editor: W. Janko (University of Economics, Vienna)

SERIES B: MATHEMATICAL AND STATISTICAL METHODS

Volume 16

Scope

The series focuses on the application of methods and ideas of logic, mathematics and statistics to the social sciences. In particular, formal treatment of social phenomena, the analysis of decision making, information theory and problems of inference will be central themes of this part of the library. Besides theoretical results, empirical investigations and the testing of theoretical models of real world problems will be subjects of interest. In addition to emphasizing interdisciplinary communication, the series will seek to support the rapid dissemination of recent results.

The titles published in this series are listed at the end of this volume.

Table of contents

Distributors

---

for the United States and Canada: Kluwer Academic Publishers, 101 Philip Drive, Norwell, MA 02061, USA

for Hungary, Albania, Bulgaria, China, Cuba, Czechoslovakia, German Democratic Republic, Democratic People's Republic of Korea, Mongolia, Poland, Romania, Soviet Union, Democratic Republic of Vietnam and Yugoslavia: Akadémiai Kiadó, P. O. Box 24, H—1363, Budapest, Hungary

for all other countries: Kluwer Academic Publishers Group, Distribution Center, P. O. Box 322, 3300 AH Dordrecht, The Netherlands

Translated by

Mrs B. Szőnyi (Part I)
Peter Kelle (Part II)

Library of Congress Cataloging-in-Publication Data

Inventory models / Éva Barancsi ... [et al.]; editor, Attila Chikán.
    p.   cm. — (Theory and decision library. Series B,
Mathematical and statistical methods)
   Includes bibliographical references.
   ISBN   0792304942
   1. Inventory control—Mathematical models. 2. Inventories—Mathematical models.
I. Barancsi, Éva. II. Chikán, Attila.
III. Series.
TS160.I56   1990
658.7'87'015118—dc20

                                 89-24423
                                     CIP

ISBN 0—7923—0494—2

Joint edition published by: Kluwer Academic Publishers, Dordrecht, The Netherlands, and Akadémiai Kiadó, Budapest, Hungary

© 1990 by Akadémiai Kiadó — Attila Chikán, Budapest, Hungary

© English translation: Mrs B. Szőnyi and Peter Kelle

Printed in Hungary by Szegedi Nyomda Vállalat, Szeged

INVENTORY MODELS

by

**ÉVA BARANCSI, GÉZA BÁNKI, RUDOLF BORLÓI,
ATTILA CHIKÁN, PÉTER KELLE, TAMÁS KULCSÁR
and GYÖRGY MESZÉNA**

editor
ATTILA CHIKÁN
Budapest University of Economics
Hungary

KLUWER ACADEMIC PUBLISHERS
DORDRECHT/BOSTON/LONDON

Introduction

Inventory modelling is one of the most developed fields of operations research. Most of the basic theoretical principles now considered to be "classical" came into being some thirty years ago (ignoring some early precursors) and more than two decades have passed since the publication of the fundamental general works. Papers on inventory studies still appear regularly in scientific journals, and new books are often published, some of which are of a very high standard. Publications on inventory models could fill a small library.

Why one more book then? Of course because we feel (just like any other author would) that the results of our research can contribute to a better understanding of the inventory problem. We have used an approach which is certainly new—and the interest it has received, during presentations we have given for various international audiences during the preparation of this book, has been encouraging.

When carrying out our research we have always had in mind a double objective: we wanted to obtain theoretical results and find ways to encourage practical application. We did not whish to go into a detailed analysis of the discrepancy between theory and practice; there are a great number of publications analysing this aspect. In this Introduction, we can only state that we hope that our work will be of interest to both academics and practicioners.

Our research has been oriented to existing inventory models. We have collected and studied 336 models from the literature, and our analysis and conclusions are based on this sample of models. We have restricted our interest to the most traditional types of models, and have not dealt with many related areas (such as water storage, manpower stocks, production planning, etc.). Of course, to draw borderlines is not easy and this cannot be considered strict.

The train of thought of the book is as follows:

Part One is analytical in character, while Part Two gives a description of the individual models. In the first, introductory, chapter, the background of inventory models and modelling is dealt with. The second chapter gives the analysis of item-level inventory systems; here we have confronted the connection between the real problems and the models. Even in the most complex system, the inventory problem is, after all, concerned with an actual item; therefore, the reference system of traditional operation research models is the item-level system. The systems analysis provides at the same time the economic background for model analysis.

Chapter Three is an historical review of the development of inventory modelling. The purpose here is to outline the temporal progress of the modelling issues in order to acquaint the reader with the "environment" of the individual models and, simultaneously, with what further development may be reasonably expected. This chapter is mainly addressed to readers having a theoretical interest.

Chapter Four summarizes the logic and the methods of research we have used. This chapter is recommended for careful study by those readers who wish to obtain a thorough understanding of inventory models, but an understanding of the logic presented here is essential for any application of our inventory model system.

Chapter Five gives a comprehensive description of the statistics characterizing the model system. Through this the structure, and the characteristic features, of the inventory models will hopefully become clear. Besides that, it indicates the "blank areas" (problems for which the modelling has not yet been developed) of interest to the theoreticians, while the application experts will be able to determine to what extent existing models can help solve their problems.

The second part of the book contains the description of the 336 inventory models we have considered, not only giving the background to the analyses but also providing experts with the opportunity to obtain very specific information.

In the descriptions, we have not emphasized the mathematical aspects of the models (this is given only with some basic models described in full) but our main purpose has been to demonstrate the assumptions on which the models are built.

We strongly advise not to consider the descriptions given as being sufficient for becoming fully acquainted with the models: reference to the original source (exact references are given) is recommended.

The classification we have applied to the sample models is one of the theoretical results of our research. Its basis is the economic, systems analytical and mathematical–statistical analyses, reported in the first part of this book.

This book is a somewhat modified version of the original edition published in 1983, in Hungarian. Even though many new results have been achieved from that time on in inventory research, and our views have also changed to some extent, (and also we continued the research, collected and studied about 300 further models) we believe that the main contents and message of the book can still be of interest for those studying inventory systems.

Acknowledgements

Our research has been supported by the Hungarian Academy of Sciences for over six years, within the framework of a long-term research project devoted to theoretical and practical questions of the operation of socialist enterprise. The changing supervisors of this project have always shown interest in our work, and without their moral and financial assistance this book could never have been written.

This book was first published in Hungarian. The readers of the Hungarian edition, Aurél Hajtó and Zoltán László, have been extremely helpful in the compilation of the book from the large amount of material it has drawn upon.

Special thanks must be expressed to János Pintér for his very thorough over-viewing of the English edition and for the many suggestions he has made, which have certainly improved the quality of the book.

It is our pleasant task also to mention here Professor Robert W. Grubbström who has encouraged and promoted the publication of the English edition.

THE INVENTORY SYSTEM
AND ITS MODELLING

I. Models and Modelling

I.1. The Concept of Model and the Modelling Process

"Model" is a category which applies in almost every field of science. A joint characteristic of the concept of model as applied in science generally is that it is taken to be an instrument of cognition and explained as a system relating the real system to the person studying it. Real systems are, in general, so complex that they cannot be directly observed and described. In addition, the limitations of the individual (e.g., the lack of capability, opportunity, or subjective interest), leads to further constraints on the cognitive process, making full knowledge impossible. As a consequence, models must be objectively used in the study of any scientific problem.

What is a model? From among the large number of definitions which exist let us take one, that of V. Stoff: "... the model is a qualitatively special category of scientific cognition which cannot entirely be deduced to other instruments and forms of recognition though it is closely interlinked with them. This property may most completely be expressed in the following definition: the model is a materially realized or theoretically created system which replaces (represents) the object of research in the process of recognition, and is in a specifically expressed relationship of similarity with the latter (isomorphic relation, analogy, physical similarity etc.), and as a consequence of this, the studying of the model and the operations carried out with it make possible the acquisition of information about the real object of research." (V. Stoff 1973)

Thus, the model originally is a general methodical scientific concept which, in principle, can be applied in any domain of scientific cognition. Although, in many cases, it is not simply the recognition aspect but the possibility of intervention that is the main element in the relation between the model and reality: a very important role is attributed to models in designing and regulating the real phenomena.

The situation is similar to the inventory models examined by us: the above definition (with necessary specifications) is valid for this type of models as well.

As a matter of course, the model can only be used for studying the properties of a real system, or for designing and/or directing the operation of the said system, if it correctly reflects reality. Thus, the elaboration and application of models (in brief: the modelling process) has rather strict rules.

Modelling is a process which may be divided in stages in different ways. We accept the following main stages of the modelling process given by András Kocsondi (1976):

— recognising the necessity of modelling,
— theoretical preparation of modelling,

- creation (selection) of the model,
- analysis of the model,
- transmission of knowledge from the model to reality,
- verification and checking of the new knowledge,
- implementing the results in practice or inserting the new knowledge into scientific theory.

No detailed analysis of these stages will be given here. Nevertheless, it is advisable—with reference to the forthcoming analysis of the inventory system—to survey briefly how models, which appropriately conform with the real system, can be created.

As mentioned before, real systems are usually so complex that it is normally impossible to identify their basic elements and to take into account all the parameters determining their state—in most cases, this would not even be reasonable. It would be impossible because the basic elements and parameters constitute a very large numerical set in general, and it would not be reasonable because from the point of view of modelling, a considerable number of the elements of this set play a negligible role. That is why we can disregard some of the individual elements of the real system (abstraction), while other elements will be considered in aggregation. This leads to a new "abstract" system called the model of the original system. The most significant property of this new system (the model) is that its variables respond to the changes in the controlling or influencing parameters in the same way as the real system would react. This property is called isofunctionality, and the plotting itself leading to the model is called homomorphic transformation. Homomorphic transformation is operation constant but several operandi may give the same picture, i.e., the transformation is unambiguous in one direction only. In other words: the model will be created from the real system but, even knowing the model, due to the abstraction and aggregation mentioned previously the original system cannot be recognized unambiguously.

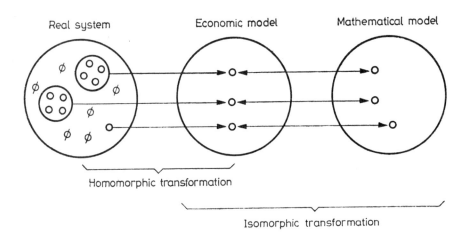

Fig. 1. Modelling as transformation

4

In the case of economic systems, the abstract system created in this way is considered to be the economic model of the original system. If we want to describe or solve any economic problem by means of mathematical modelling, we have to go further: the economic model must be "translated" into a mathematical language, i.e., the elements of our abstract system (economic model) are to be brought into mutual and unambiguous correspondence with the elements of a new system: the mathematical model. This type of plotting is called isomorphic: with knowing one of the systems and the compilation code, the other system can be deemed known, too.

The transformations referred to are indicated in Fig. 1 by a simplified scheme. For simplification, only the elements are shown, without their interrelations; the logic of plotting may be followed also in this way. The step of abstraction is indicated by crossing out the irrelevant "elements", and aggregation is indicated by circling elements.

I.2. Classification of the Models

The basis for the classification of models can be very different; any significant properties may serve as a starting point. Only some aspects of classification—the most important ones from our point of view—will be discussed here.

A trivial mode of classification has already been mentioned implicitly: classification by the object to be modelled. According to this, our interest is the modelling of inventory problems.

Differentiation between two major families of models, the "material" and the "ideal" "intellectual" ones, can be considered as generally accepted. As stated by A. Kocsondi (1976), "the most characteristic feature of the material models is reflected in the fact that they objectively exist and act in accordance with the rules of the external world. Their elements are the various objects and phenomena of organic and inorganic nature, as well as the material–objective products of human activity. The speciality of the ideal models lies in that they only exist in the mind of the recognizing person (subjectum) consisting of elementary depictions and symbols, and they function only in consequence of speculative operations which the subjectum carries out in the course of constructing and modifying them".

Both described groups of models can be analysed further (both Kocsondi and Stoff do that in their cited works), however, it should be sufficient here to know that economic models are members of the larger family of ideal models (with the exception of several experiments where economic phenomena have been demonstrated by material models, e.g., equilibrium problems by the principle of pipelines and reservoirs). Thus, inventory models are also ideal models.

Two further classification aspects are based on principles also widely used in the literature. Here we will describe these aspects according to Miller and Starr who have given a very detailed analysis (Miller and Starr 1969).

According to the first aspect normative and descriptive models can be distinguished. Descriptive models describe facts and experiences without evaluating the system. They normally answer the question: "what if...". These models may be excellent tools for the learning and recognition process, especially if we manage to avoid unintentionally introducing normative elements among them (e.g., in compliance with individual expectations and hypotheses of the designer or user of the model) because such elements may be the source of considerable deformations. The construction of descriptive models must precede the construction and application of normative models for both logical and practical reasons: it would hardly be correct to speak about "what should be" before we know "what is".

The construction of normative models involves certain norms (standards) or rules. These models describe "what should be", and they, of necessity, contain evaluation criteria for selecting from the elements of the set of decision alternatives. Normative models also contain methods and means for the selection of the decision alternatives themselves.

According to this grouping, inventory models of operations research are mostly normative models, with the exception of a few descriptive models.

A final classification aspect is to distinguish between qualitative and quantitative models. In the general consciousness, the "model" is often identical with the concept of quantitative models, although, as seen in the general description of models, even considerations of reality take shape as qualitative modelling; and nor does the normative approach necessarily require a quantitative form—which means that numerical definition of the outputs is not absolutely indispensable. The statement "the operation of the system will improve, if..." may be an acceptable normative description. "Will improve" is not a numerical requirement, but useful information. In many cases, the quantitative observation or description is impossible as a matter of course, whereas qualitative results can be written in a quasi-quantitative form (e.g., with ranking).

According to this latter classification aspect, inventory models are mostly quantitative models.

I.3. Decision Models

Let us focus our attention to the family of models most important to us: the decision models. In general. several different types of models are considered to be "decision models". According to the terminology we use here, these are normative and quantitative models entering the decision process where the analysis and evaluation of decision alternatives takes place. We test the various alternatives on the operation of the system and choose the version which, according to the model, leads the most efficiently to the realization of the objective of the model.

It is an extremely important, though trivial, fact that decision models can only provide guidance but not actually make the decisions. That is the responsibility of the management.

According to the most common considerations, decision models comprise four fundamental parts:

— the decision variables,
— the system of assumptions,
— the objective function, and
— the methods applied for evaluation.

6

The decision variables contain decision parameters, the value of which is to be determined by means of the model taking into account the interest (decision criteria) of the decision maker. In this way, the closest correlation exists between the questions to be answered by the model and the decision variables.

As a matter of course, each economic decision is influenced by only a subset of all the possible variables. The situation often encountered is that the system described by the model has some variables which will not be taken into consideration in the set of the decision variables, though their actual value affects the operation of the system. These variables will be put in the system of assumptions either with a fixed value or, if their value changes as a result of modifications of the decision variables, they will be regarded as state variables of the system.

In addition to specifying the problem, another basic function of the decision variables is that their character and contents determine the structure of the model as a whole. The most important classification aspects of the decision models can be fixed simply on the basis of the character of these decision variables: for example, if the decision variables are time-dependent, the model is dynamic, otherwise it is a static model; whereas stochastic models will result if the variables have a random character, otherwise the model will be deterministic.

The other major part of the models is *the system of assumptions* which is essentially the set of the relations postulated in the set of the variables. This is the part which decisively sets a limit for the scope of validity of the model. Obviously, in different systems of assumptions we get a different "optimal" value of the same decision variables; or the adaptability of the model to different real situations as well as the mode and the possibility of answering the questions put through the decision variables are determined just by the system of assumptions. The homomorphic feature mentioned, as well as the examination of the modelling process, is ensured essentially by means of the system of assumptions. It is clear, in this context, that the structure of the system of assumptions plays a decisive role from the viewpoint of the complexity of the model: the consideration of more assumptions increases the accuracy of the model but, at the same time, restricts its general validity.

The system of assumptions can be divided into two different parts. The first are the assumptions quantitatively given in the model (e.g., various capacity limitations); the other are the assumptions qualitatively given in the model (a great part of which could be described by mathematical formulations, too, as, for example, the assumption referring to the constant character of specific quantities, or the linearity of interrelations).

The objective function is a function of the decision variables, some extreme value of which expresses the desirable status of the system under consideration from the viewpoint of the decision maker. In general, the purpose of the decision models is to identify the values of the decision variables which maximize the effectiveness of the system. What should be the measure of efficiency? This depends basically always on the interest of the decision maker. Its actual form should, of course, bear a close relationship to the remaining parts of the model. (For example, when stochastic variables are present among the decision variables, an expected value should be chosen as the efficiency measure; in the case of deterministic decision variables, this measure itself may be deterministic as well.)

The fourth part of the decision models consists of those *methods applied for evaluation* of the model, which analyse the effect of the possible values of decision variables on the objective function in the system of assumptions of the model.

In this way, the mathematical aspects are logically determined by other characteristics of the model—at least as far as the main features are concerned. It must be emphasized that, although every algorithm satisfying the assumptions and other criteria, which leads to the optimal solution, is involved in the concept of the evaluation methods, and this concept is not limited to, for example, some given computer program or analytical method, those solutions which cannot really be computed are not suitable for a decision model. Therefore, for such models we must not be satisfied with the theorem verifying the existence of the solution, or by an iterative procedure which cannot be accomplished by existing computer techniques. The best result during the optimization process is if a procedure can be found which leads, in the set of solutions, to the determination of the optimal parameters. This is not always possible. In other cases, it is possible, though not economical. Finally, and most importantly, the decision maker himself does not necessarily strive to seek the optimum solution in many cases: procedures which lead to a mathematical suboptimum can, nevertheless, be quite appropriate taking into account the economic considerations of the decision maker.

I.4. Fields of Inventory Modelling:
Models on Macroeconomic, Company and Item-Level

Speaking of inventory modelling, we have to distinguish between the modelling of three levels of inventory systems having a hierarchic relationship with each other:

— models on a national economic (macro-) level,
— models on a company (micro-) level and
— models on an item (submicro-) level.

The reference system of macroeconomic models on the national economy level is the economy as a whole. Of course, it does not mean that these models comprise the total economy but according to what has previously been said on modelling and its rules, the real reference system here is the national economy.

These models constitute an integral part of macroeconomic theory. They can be grouped in different ways. The most characteristic feature is maybe that most of these models have no normative or optimizing character, but, rather, a descriptive character investigating the mechanisms of operation in the economy. The aggregate inventory in the economy is the result of autonomous company decisions, thus macro-level optimizing models can rarely be realistic. The situation is different in some socialist countries where, due to the planning directives and the centrally-planned distribution of items, stockpiling is a macroeconomic category and, therefore, an optimal inventory volume can be spoken of at a macro-level as well (it is a quite different question, of course, how one can find the instruments which are needed to realize these optimums).

Microeconomic inventory mostly belong to the family of behavioural models of the companies and are in close connection with the macro-models. A summary of the classical approach is given in the famous book of Holt, Modigliani, Muth and Simon (1960) in which the authors give decision rules for production, labour force and inventory accumulation, postulating quadratic cost functions and profit-maximizing companies. An essential element of these models is the assumption of profit maximization although one can find some different models, oriented, for example, in particular towards the magnitude of inventory investments or containing alternative criteria as well.

Microeconomic models mostly deal with the aggregate inventory of a company, based on the economically reasonable fact that the principal question for management is, how much from the total capital should be devoted to holding inventories? It is obvious that the inventory of each individual item "optimal" is in vain if the total of these inventories is so big that its realization is not feasible, because it is in conflict with other, maybe similarly "optimal" requirements originating from other partial functions of company operation.

These models usually treat the company as a homogeneous system, paying attention to external effects only, and having no regard to the structure of the inventories, except in some cases when input, work-in-process and output inventories are distinguished.

Obviously, this approach is rather biased, which does not question its justification, but attention needs to be paid to a multidimensional approach. The final purpose of inventory holding is satisfying demand, and demands always emerge at a given specific location and time for a given product. Thus, the other side of the coin is that though the aggregated inventory stock of the company may be as sufficient as possible in relation to other major parameters of the company, it is impossible to conduct a successful inventory management and, in this way, to obtain the desired profit, when the structure of inventories is not sufficient, i.e., when the volume of inventories of the individual items are not appropriate. This gives the reason for item-level modelling which will be discussed in detail in the following chapter.

One has to see clearly that the inventory models of the three system levels discussed above logically correspond to each other. If the national economy is defined as a system of material flows and inventories (Chikán 1980b), the taxonomic structure of the system can be determined in two fundamental dimensions: either the companies (the domain of microeconomic models) or the items (the domain of item-level models) may be treated as elements of the system. Obviously, modelling objectives will decide which dimension we should rely on in macro-modelling: Should we start from any of the two approaches, in the course of disaggregating we arrive at the question of the connection between the two dimensions in any case.

These two main sections of the possible breakdowns starting from the macro-level will meet in the item-level system of a given company. This system belongs to the classical areas of operation research investigations; from among the above-described three levels this, from the economic viewpoint, submicro level (i.e., item-level) has up to now been the most intensive subject of research work. We have devoted this book to this system and its modelling.

9

II. The Subject of Operations Research Modelling: The Item-Level Inventory System

In this chapter we analyse the item-level inventory system. The conceptional and methodological apparatus of systems theory have been applied to the analysis; we have relied particularly on the works of Churchman (1968), Hajnal (1976), Nemény (1973), Pawelzig (1974) and Sadowsky (1976). In the course of the *in merito* exposition, we have used our earlier works (Chikán 1973, 1977) to a considerable extent.

The system to be examined can be defined as follows: the *item-level inventory system* is a system for satisfying the *demands* for a given item, wherein the source of the *output process* generated by the demand is a certain *stock* of the given item, the replenishment of which is provided by the input process regulated through ordering.

We do not go into the detailed interpretation of this definition now; the analysis of the system given below is intended to provide the explanation. To avoid misunderstanding, we add only that the expression "item" is used here in a quite general sense, independent of its degree of processing; thus semi-products, machine parts, tools, finished products, etc., are all included.

We relate our system to the usual system classifications. According to the well-known classification of Bertalanffy (1951), the item-level inventory system is an open system; it belongs to the third of nine system levels (control mechanisms or cybernetic systems) while with regard to the dual-aspect classification of Stafford Beer (1959), it may belong to the complex stochastic systems.

As a starting point of the analysis of the system, we use the thesis of Churchman (1968) that the following "must be kept in mind when thinking about the meaning of the system:

1. the total system objectives and, more specifically, the performance measures of the whole system;
2. the system's environment: the fixed constraints;
3. the resources of the system;
4. the components of the system: their activities, goals and measures of performance;
5. the management of the system."

We do not agree in every respect with Churchman in the actual explanation of the factors listed, and particularly not in his detailed exposition not cited here. Nevertheless, the five basic aspects (objective—environment—resource—structure—management) have already proved to be a very useful principle for systematic ordering in the course of our other analyses. Thus, we will use this logical system further on as a starting point for the item-level inventory management system, with the difference that as a consequence of the properties of the system, to be detailed later, we change the sequence of the single aspects. Furthermore, we discuss points 4 and 5 together.

It also should be mentioned that there arise substantial differences in connection with most of the investigation motives of the system, depending on the particular type of elements of the inventory (the subject of the item-level inventory system, e.g., raw materials, finished products, merchandise, etc.). In the exposition, we try to present a general view but with reference to the special features of the item deriving from its character, where this is inevitably necessary.

II.1. The Objective of the System

Our basic point is that no entirely autonomous objective can be set for the item-level inventory system. Its objective can only be deduced from the objective of the systems on a higher hierarchical level. This initial theorem comes from the fact that the item-level inventory system, due to its economic gist, is a material system, the motivations and goals of which cannot be explained by themselves; only such objectives can be spoken of which may be deduced from an external organization being in possession of some self-reliant motivations.

Where can this objective be deduced from? The item-level system is located at the lowest level of the hierarchical structure. In this structure, the higher system level having the greatest effect on the objective is the enterprise. Namely, if we distinguish the levels of national economy—enterprise—inventory subsystem—item-level inventory system according to the general subdivision usual in the hierarchical structure, then from among these the enterprise is the very level which possesses—through its separated means and, in connection with this, through its own interest and motivations—independent objectives which can be further divided. We do not enter into details here. In a previous work (Chikán 1980a) it has been specified how the overall objective of the company implies the partial objectives of stockholding. We will survey only the most important aspects here.

We have already stated as a definition that the goal of stockholding is to satisfy demand for the items stored. This necessity emerges from the fact that it is neither physically possible nor economically reasonable to produce every single item, when and where the demand for them arises. For this reason we have to hold inventories, i.e., to withdraw provisionally resources from the reproduction process. The word "provisionally" has to be stressed here for—according to the intention of the person or organization responsible for the decision of holding inventories—the items held on inventories, i.e., kept inactive, will sooner or later become active participants of the reproduction process.

The objective after all is manifested in covering demand in general, but more exactly it has to be interpreted as the economical covering of demands. As in every economic problem, a comparison of costs and benefits will decide what is to be considered as an economical satisfying of demands. Instead of a detailed exposition we refer again to the relevant literature (Nagy 1975b, Chikán, Fábri and Nagy 1978), but, in advance, it is clear that both satisfying and non-satisfying of demand implies some costs and benefits as well. Thus, the measure of value of the economical (or optimal) satisfaction of demands according to Churchman is the magnitude of costs (i.e., the minimization of costs is the optimum criterion).

II.2. The Resources of the System

As for the resources of the system, "these are inside the system. They are the means that the system uses to do its job." (Churchman 1968, p. 37)

The resources for the item-level inventory system are the inventory itself: the stock of items held as inventory both in a physical sense as actual items which can be used for satisfying demand and in an economic sense as a means of making profit.

In order to highlight the role of inventories as resources perhaps it is sufficient to refer to the fact that for a smooth production (and distribution) not only items (materials, spare parts, finished products, etc.) are needed, but also the presence of their inventories, since inventories are indispensable for smoothing out discontinuities of the production (distribution) process in time and space, due to objective and subjective reasons alike. And if this is true, then inventories as conditions of the production process can be defined as production factors, i.e., resources.

To the explanation that the inventory is a resource pertains the theorem that the item on stock has a value in an economic sense because a demand exists for it. The demand mobilizes the inventory as a resource which has been inactive for a certain time and so it becomes the promoter of the whole system. (This statement exactly corresponds to the explanation given above from another point of view, namely, that the objective of holding inventories is to satisfy the emerging demands for the item.)

As a consequence of what has been said above, the analysis of the properties of the item on stock is of primary importance when examining the resources of the given item-level inventory system. The question has to be put here: in which phase of the production process is the inventory a resource (i.e., in the usual terminology, what is the type of the inventory: raw material, semifinished goods, spare parts, etc.)? Numerous additional characteristics of the system depend on this fact (e.g., whether the end of the input or that of the output—maybe neither of them or both—falls outside the given enterprise, i.e., outside the sphere of control of the enterprise).

Investigations related to the value, as well as the stability of value, the storage and management possibilities, storability, the manageability and other properties of the item in connection with stockpiling, constitute a part of resource analysis.

Another aspect has to be mentioned, too: it is obvious that the time-level system would not be able to operate without human activity "pumping life" into it. Nevertheless, in our conception, this human activity belongs to a higher hierarchical system level, i.e., to the enterprise level inventory management system, thus, it is not subject to our investigations here (we will return to it when discussing environment).

II.3. The Environment of the System

We begin the discussion of environment-related problems with the statement that the item-level inventory system is an open system in a systems theory sense. This means that the exchange of materials, energy and information with the environment belongs to its essential features. The relationship between the system and its environment will be examined on this basis.

The first stage in the investigation is the most exact description of the system and its environment. By all means, this is not a trivial step for most of the systems, neither with the item-level inventory system. We refer to the definition of this system and give a description in this sense.

The environment is—says Churchman (1968)—that which falls outside the sphere of control of the system, but has a partial influence on how the system operates.

According to Pawelzig (1974, p. 52), the investigation of the relationship between a system and its environment may start from two fundamental viewpoints. One of them is that "the environment of the system will be subdivided on the basis of those factors and groups of factors, of which the relations to the system constitute the subject of the investigation, whereas in the other approach "we may discuss the environment based on the so-called system-hierarchic considerations, too", i.e., we can identify subsystems which are parts and elements of the system investigated. Both approaches are justified and both have their limitations. Since these two methods of investigation may, in our opinion, successfully complement each other, let us look over the item-level system of stockholding from both aspects, combining the two approaches.

We shall begin with the second approach, for this one is simpler and we have already dealt with it, and particularly because our method of investigation will be just to survey the characteristics of the connection of the item-level inventory system to the individual systems at the higher hierarchical levels.

Regarding the hierarchical structure, the closer environment of the item-level inventory system is the inventory management system of the enterprise comprising all the functions as well as the partial units for executing them, which are in direct connection with stockholding (e.g., material management, spare parts management and the handling of other kinds of inventories). In addition to the enterprise inventory system, all the systems which involve this as a subsystem constitute at the same time the environment. Of these, two levels are important to us: the whole of the enterprise and the national economy as a system. (An essential remark has to be made here: we did not mention, among the elements of the item-level inventory system, the decision maker himself, who directs the system. In this way, we maintained the property of the system that the human factor does not have a direct role in it, only in its effects, i.e., as an element of the environment. In our interpretation, the decision maker is a part of the enterprise's inventory system.)

It is also worthwhile mentioning that these further system levels are themselves diversely structured, but we do not deal with this structured character here. Accordingly, when speaking of the relations to the enterprise, we do not discuss through which functional unit (production, research and development, sales, etc.) these relations actually appear. Furthermore, we do not address the question of which other enterprises or authorities play a role in some material or information relation within the national economy.

We consider two of the, logically emerging, three groups of relevant factors: material, energy, and information exchange. The problem of energy exchange is neglected in this chapter, since this is an inessential question as far as the whole of our investigation is concerned.

When discussing the material and information aspects, we rely on the system structuring aspects of J. Kornai, according to which economic systems can be divided into two spheres: the control and the real sphere (Kornai 1971, p. 61—64). According to Kornai's approach, information processes occur in the control sphere, whereas material processes occur in the real sphere. The significance of this interpretation may be confirmed by more detailed considerations.

The relation between the item-level inventory system and the environment may be investigated by symbolically completing the following table:

System level / Characteristics of relations	Enterprise-level inventory system	The whole of the enterprise	National economy
Information relation (affecting the control sphere)			
Material relation (affecting the real sphere)			

The environmental effects influence every component of the inventory system (they even determine the total system in many cases). As a matter of course, even a detailed analysis is unable to provide a complete picture; that is why the actual operation of a seemingly similar inventory system will be different under various economic, political, cultural conditions. This is discussed in many textbooks, and so we shall not consider this in detail here.

II.4. The Structure of the System

What we have explained so far regarding the aspects of system analysis according to Churchman—"the components, activities, objectives and performance measures of the system"—may be described, in short, as the structure of the system. From the viewpoint of the main purpose of our investigations (modelling) this is perhaps the most important part of the analysis.

In order to investigate the system structure, a selection must be made of those aspects which form the basis of our analysis, since a given system can be structured in many different ways. In revealing the structure of the item-level inventory system, we will follows the train of thought of A. Hajnal (1976), who considers three structures which can be examined in a hierarchical manner:

— taxonomic structure,
— static relations structure and
— dynamic relations structure.

It is worth noting that, as a matter of course, this is not the only possible approach to structure analysis. The system described by V. Nemény (1973), for example, can be applied successfully as well; particularly in the case of more sophisticated systems than the one being investigated here. Nemény gives five aspects for structure analysis: objectional, criterial, functional, hierarchical and decisional structures. It is clear that these two systems of analysis do not contradict each other but, rather, complement each other.

II.4.1. The Taxonomic Structure

Obviously more than one taxonomic structure can be outlined for a given system. We have chosen a structuring based on an activity (functional) basis, because this seems to be the most convenient for our purposes. It should be pointed out that structuring should be considered from as many aspects as possible for a complete investigation of the structure of the system, especially when treating more compound systems—but in the case of our simple system the complexity of the functional investigation is sufficient.

The exploration of the taxonomic structure calls for the determination of the elements of the system and their classification. Nemény (1973, p. 53), Pawelzig (1974, p. 47) and Churchman point out that the elements are the carriers of the internal functions of the system, i.e., we can attain the determination of the elements of the system by breaking down the given principal function of the system into perhaps multilevel hierarchical partial functions. Following this logical approach, we give the taxonomic structure of the item-level inventory system as follows.

1. The first step is—as mentioned before—that the system should be divided into a control and a real sphere. The task of the control sphere is the reception, processing and forwarding of the information coming from the external world (the environment) and from the real sphere itself.

Since we do not treat the decision maker as an element of the item-level system, it is possible to identify, as part of the control sphere of the system, the management of the system, thus it is reasonable to discuss Churchman's examination steps 4 and 5 together. We will return to this problem when explaining the control sphere in detail. The task of the real sphere is the implementation of the stocking activity as a material process: withdrawal from the inventory, given demand, and replenishing the inventory upon receiving fresh supplies.

2. The control and the real sphere can both be divided into further elements. In the control sphere, the relevant functions (reviewing of inventory—operating of the inventory control mechanism—ordering) appear in compliance with the control system's elements (measuring—comparison—intervention), whereas in

15

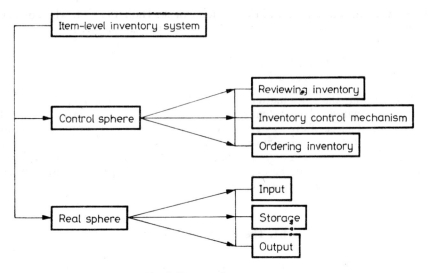

Fig. 2. Taxonomic structure

the real sphere the inflow (input)—carrying of inventory (storage)—outflow (output) and related activities take place.

3. Each of the listed elements can be divided further into partial activities. Consider: how many operations, which can further be hierarchically divided, have to be accomplished in order to realize any of the above elements of activity? Thus, an appropriate means of recording inventories is necessary for an inventory review, as well establishing suitable means of handling and relaying information. Fulfilment of the storing function is covered extensively in the literature. Further subdivisions would, however, not serve the main purpose of our investigations and, therefore, the taxonomic structure most suitable for our requirements is shown in Fig. 2.

II.4.2. The Static Relations Structure

The static and the dynamic relations structures express the relations existing between the components of the system. By elucidating the static relations structure (the possible relations between the elements of the system), the various states of the system can be demonstrated, whereas the dynamic relations structure shows the changes of these states, i.e., the operation of the system. While in the taxonomic structure, hierarchy can be explained by the expression "it comprises" (for instance, the control sphere "comprises" the review of inventories). The relation structure describes hierarchy by causal relations: the cause of an effect is of a higher level than the affected element. Thus, the static relations structure of the item-level inventory system can be indicated by describing the connections between the elements listed in the taxonomic structure as well as the states of the system

16

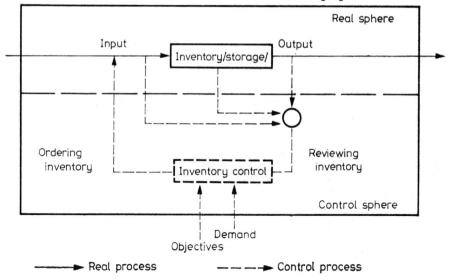

Fig. 3. Static relations structure

resulting from these states, whereas, the dynamic relations structure expresses, how (by what kinds of motions of the relations) the system proceeds from one state to another. The static relations structure is described in its simplest form by Fig. 3.

The static relations structure can be simply explained in the following way.

Logically, the first step in establishing the connections between the elements is the appearance of a demand. Demand comes from outside the item-level system, similarly to the objectives posed in expectations of the item-level system. This is followed by ordering, which influences the input process. The other factor influencing the input process is the source of supply (placed also outside the system). All these are control relations. The next step occurs in the real sphere: a shipment is received which, of course, may not necessarily be identical with the order, and increases the volume of inventories. Outflow takes place from this stock which may not be identical with the demand. At this stage, we return again to the control sphere insofar as information concerning the meeting of demands is relayed again to the element controlling the inventory, where it meets the new demand which has arisen in the meantime as well as the information concerning modifications of the objectives and starts the described process again.

The process described above is a logical system of relations connecting the elements. Though this description refers to a process, it is still of a static character because it does not show how these processes take place but only indicates their direction. If, in addition, the state characteristics of the system is given, then the possible actual state of the system can be defined which—due to the present static approach—is characteristic of the system at a given moment.

The actual appearance of the state characteristics in the various systems can be very different. Major characteristics generally appearing are the following:

17

— inventory status (inventory at a given time or average inventory during a given period);
— the level of supply (ratio between planned and actual inventories);
— intensity of demand (demand in unit time, actual demand, or in the case of random process, expected demand);
— intensity of input (similar to that of demand);
— the service level (ratio between actually occurring, and satisfied, demand);
— the level of input supply (relation between ordered, and received, quantity).

These characteristics develop according to the actual realization of the connections indicated in the static relations structure at a given date (period) and jointly describe the state of the system. Other secondary characteristics can be deduced from these (e.g., relating to excess or to shortage).

We should point out that these characteristics—though they may be used for estimating the effectiveness of the system's operation—are not efficiency indices. It was stated in the discussion of the objective of the system that the objectives are derived from outside the system. Accordingly, the estimation of efficiency occurs outside the system as well. As explained previously, cost assessment for the fundamental processes of the system—which have been described here by the state characteristics—can be carried out by means of specific cost factors determined from outside, and applied ex ante for choosing the operational parameters of the system (i.e., for establishing the general objectives), and ex post for evaluating the efficiency of the system.

II.4.3. The Dynamic Relations Structure

For an analysis of the dynamic relations structure, let us survey the interrelations of the processes taking place in the system.

The demand mobilizing the system arises from outside, and, as an environmental effect, influences directly the regulation sphere and within this the regulating element which we call the inventory mechanism. Information related to the objectives of the system, which—as seen before—will be conveyed towards the item-level subsystem immediately by a part of the environment which is called "the inventory management subsystem of the enterprise", have the same direction. Here, it must be mentioned that "demand" can arise from not only outside the item-level system, but also outside the inventory subsystem, and even outside the enterprise: in order to realize the objectives of the enterprise, demand for the products belonging to its profile must be satisfied. This "external" demand is for the finished product and leads to an "internal" demand for all the components of the inventory (raw materials, secondary materials, spares, etc.). This internal demand will be conveyed by the inventory subsystem to the item-level system. Consequently, the characteristics of the demand significantly depend on the type of inventory (raw material, finished product, etc.).

Stemming from the above, there appear five demand categories (in correlation with each other) in the item-level inventory system:

— the abstract demand which will be generated from outside the company and

18

means the demand which may be induced for the given item in principle in the operation sphere of the enterprise;

— the demand recognized by the enterprise which, due to reasons arising from the enterprise itself, e.g., lack of market knowledge, and/or to external reasons, e.g., customer information deviates from actual demand, may differ from the abstract demand. This demand category comes from the enterprise level (i.e., from outside the inventory system), too;

— the demand considered satisfactory—this is the volume of demand which the enterprise wants to meet in a given period and for which the enterprise is prepared to satisfy (and which corresponds with its own aims and capabilities, e.g., resource limitations). The determination of the volume of this demand begins at the enterprise level and is ultimately concretized within the inventory subsystem, and, as will be seen, is a factor affecting the management and operation of the item-level subsystem;

— the demand actually arising which, in contrast to the demand categories listed above, can be recognized only ex post, after decisions which concern the item-level inventory system have been made, and which may, depending on the sense, differ from the various categories discussed above;

— the actual demand satisfied, which, in contrast to the above described concepts of demand, is a phenomenon of the real sphere: this is the quantity of items which has been despatched from the store in a given period.

The correct interpretation and treatment of the categories given above is fundamental to understanding and analyzing inventory control processes and, as such, is the starting point of an analysis of the dynamic relation structure.

The operation of the system begins so that the information concerning the demand (which is to be satisfied) comes into a control element known as "inventory management", where it meets the inventory mechanisms and norms which express the "expectations" of the item-level system and which derive directly from the inventory subsystem, and this mobilizes the system.

How do the elements and relations, as encompassed by the static relations structure, correlate, i.e., how can the dynamic relation structure be defined?

Our starting point is that the item-level inventory system is a control system in the classical sense, i.e., the management of the system proceeds on the basis of information derived from the system itself, by negative feedback. If we deal with inventory control as a control problem, in addition to the economic, systems-theoretical and operations-research considerations: we will also be involved in control theory or cybernetics.

In order to discuss the control of the system, some well-defined conditions concerning the factors and processes of the system must be met:

— the system must possess a controlled variable, the value of which is to be kept at a planned level (such as the target value of an objective function);

— certain perturbations must exist, which change the value of the controlled variable from its target value;

— the actual value of the controlled variable must be directly or indirectly comparable with the target value;

— there has to be a manipulated variable inside the system which changes according to the influence of the decision maker, the changes of which have a determinable effect on the value of the controlled variable.

These conditions are completely fulfilled by the item-level inventory system as we have defined it. In addition to the above concepts, we have to explain the concept of the regulating variable in the control sphere, the task of which is to establish the value of the controlled variable to be maintained by the control process depending on the intention of the decision maker, whose position is outside the system according to the definition given. We add here that the control sphere, as defined above, is capable of making only routine decisions.

As long as the norms (i.e., the parameters of the inventory mechanisms to be defined shortly) in the regulating variable of the control sphere, based on the conditions dictated by the decision maker, are unchanged, the control system can be "left alone", since it is able to operate as expected by the decision maker. As soon as the decision maker realizes (from outside the system) that the conditions have been changed, he modifies the norms (new values are introduced into the control sphere through the regulating variable), and the system is ready to operate again. (The inventory system operates like this in reality, too. The decision maker, e.g., a material manager, endeavours to form an inventory volume in accordance with the inventory norms i.e., he orders "automatically" as long as the norm is valid.)

The first step in determining the control sphere of the inventory system is the identification of the variables:

Controlled variable —
— the volume of inventory which the decision maker shall keep according to his intentions as indicated by the regulating variable;
Disturbance —
— this covers all the effects which divert the level of inventory from the norm. Its most important element is the actually-arising demand, but the vagaries of procurement (e.g., the length of time between ordering and delivery) also belong here, along with any external disturbing effect (e.g., the incidental diminishing of stocks, natural disaster, theft, etc.);
Manipulated variable —
— the order placed by the decision maker;
Regulating variable —
— the intentions of the decision maker reflecting the factors of the system as well as their changes, which appear after all in the parameters of the inventory mechanisms and in their changes, respectively. In addition to other, less important factors, the demand to be satisfied is also a part of this variable.
Without going into details, it is useful to comment here on the control theoretical explanation of the inventory system:
— It is a property and not the essential of the inventory system defined by us that it is a control system. Depending on the intentions of the decision maker, other types of regulation could also be applied, but, in our opinion, control, in the above sense, is the proper instrument being in compliance with a wide range of properties of a system.
— It is obvious from comparing the above control-theoretical concept with the

general description given in the Introduction that the material processes and information flows of the system progress in opposite directions when considering the important elements of the system.

A question of primary importance when analyzing the dynamic relations structure concerns the form in which control actually takes place. Since the specific control steps are: measurement, comparison and intervention, we must examine to which factors of the inventory system these steps are linked and how.

a) In defining the term "measurement" (which takes place in the "inventory review" element of the taxonomic structure), of fundamental significance in the inventory system is the statement of control theory according to which the required value of the controlled variable can be maintained so that some other unambiguously related variable will be measured instead. Consistent with this, we may state that control in the inventory system can be carried out directly, based either on a measurement of demand or its satisfaction. We will return to this matter later on.

b) Comparison (which may pertain to "inventory management") occurs by means of the parameters given by the regulating variable. Since we defined the item-level inventory system as a control system, where the decision maker wants to compensate for any disturbing effects on the basis of information originating from the system itself, and ordering takes the place of the manipulated variable, the specifics of the control process in accepted terminology means the "ordering rule", or the "mechanism" of inventory contol. This is determined by answering the following two questions: (1) when, and (2) how much should the decision maker order? In inventory theory (according to practical inventory management) two basic types of answers can be given to each of these questions. As to the time of ordering:

— orders can be made at fixed intervals (in what follows this is denoted by t);
— the decision concerning inventory replenishment is made when the level of inventory has decreased below a certain minimal value (the order point, s).

Two possibilities arise concerning the volume of the order:

— the volume (amount) to be ordered is fixed (the order quantity, q);
— the volume shall be so determined that after delivery the inventory reaches a fixed maximal level (the order level, S).

In this sense, inventory control mechanisms are possible combinations of the above answers. In principle, therefore, we may speak of inventory management mechanisms (t, q) (t, S) (s, q) and (s, S) denoted by the controlled parameters. For more details of this classification of inventory control systems, see Naddor (1966).

Figure 4 demonstrates thematerial processes in the given inventory control mechanisms. We use the following basic assumptions:

— replenishment time can be neglected,
— the review of the inventory is undertaken periodically, the magnitude of demand (and its probability distribution function) is given for the reviewing period (τ),

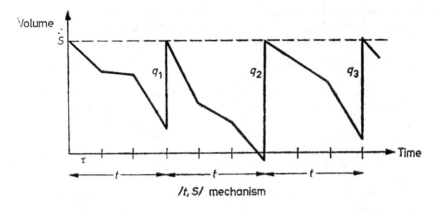

/t, S/ mechanism

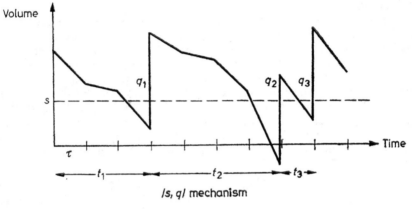

/s, q/ mechanism

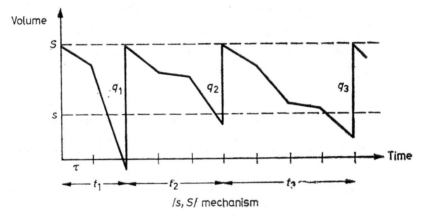

/s, S/ mechanism

Fig. 4. Inventory control mechanism

— at the delivery of shipments, the demand backlogged from the previous period will immediately be satisfied (this is the case of "back ordered demands"),
— demand arises continuously and uniformly in the reviewing period,
— the process of delivery is deterministic.

(We note that these assumptions imply the simplest possible case and serve only for demonstration. This simplification can be allowed, as the application of a more complex system of assumptions is logically the same but would make more difficult the demonstration of the basic ideas. Concerning an interpretation of the conditions, see the study of J. Berács (1973).)

In the case of the assumptions above, the operation of the individual mechanisms is shown in Fig. 4. In the case of stochastic processes, the figures can be regarded as a realization for each of these processes. It is important to note here that of the control mechanisms which are actually possible, the mechanism (t, q) has been excluded because it is only suitable for control under very special (primarily deterministic) conditions, and in these cases it can be considered as a special case of any other mechanism. (Nevertheless, for the sake of simplicity it is often reasonable to describe a problem with the mechanism (t, q), but in these cases the equivalence to one of the other mechanisms is fairly obvious.)) Each of the remaining three possibilities (t, S) (s, q) (s, S) are connected—in the sense of the internal logics of the control system—to some inventory level. It is to be pointed out that there is an important difference of content between the parameters t and q as well as a and S concerning the answers to be given to the "basic questions": the former answer the questions directly, whereas the latter provide the answer by involving some of the inventory levels and taking into consideration the actual processes occurring in the system. This latter circumstance enables controlling the system.

c) Intervention takes place through ordering the purpose of which is to replenish the inventory, when the decision maker is able—by involving the regulating variable—to introduce his own objectives and expectations into the system, as aspects. In the case of deterministic replenishment, this intervention is free of disturbances; in a system with a stochastic input process, some of these elements are to be treated as components of the disturbing effects.

In the control process of the item-level inventory system, the (objectively) continuous time will be made discontinuous by the decisions of the decision maker concerning ordering. A consequence of this is that considerations of quantity—as seen before—are more important aspects of the demand process than the aspect of time. The latter can be estimated from the replenishment process and not from the demand process even if the specific date of placing orders always depends on the actual situation regarding demands. On the other hand, in the case of the input process, the aspect of time becomes apparent: the fundamental parameter of the control process is the reviewing period—the basic unit of the order decision. This is the time interval between successive reviews of the state of the system. In the case of a continuous review this interval is zero, otherwise it has a positive, usually previously-fixed value.

Upon investigating the state of the system, the decision maker can decide whether intervention, i.e., ordering, is required or not. The ordering period is the time between two decisions (i.e., two orders) for replenishing the inventory.

The features of this depend on the inventory mechanism of the given system. If the answer to the question "when?" is t (interval), then the ordering period is a constant and coincides with the review period. But if the date of ordering corresponds to some minimal level of inventory (s), then the ordering period is a variable (in the case of a periodic inventory review, it still remains an integer multiple of the review period).

Another essential time interval must be defined for the analysis of the input process. This is the period of replenishment, i.e., the interval between ordering and arrival. (The corresponding period in the demand process is the time for the demand to be satisfied; nevertheless, this does not have the same importance as the replenishment period as far as the input process is concerned: its value is non-zero usually because of administrative and technical reasons only, i.e., if a stock of the required item is available, delivery follows ordering immediately. If there is a difference between the date a particular demand arises and when this is satisfied, this is due to the intention of the decision maker without disturbing effects in merito on the given item-level system. This is not true, of course, for the customer as the source of the demand: this may exert a very negative effect on him but the demand of the customer is the input problem of another item-level system, and thus we have returned to the problem of replenishment.)

If the replenishment interval is positive (and this is the usual situation in practice), the decision maker has to take into account two inventory levels: (1) the stock actually available in the store, and (2) the quantity of items already ordered but not yet received. How this alters the operation of the inventory system is shown in Fig. 5 which demonstrates a simple case of the (t, S) mechanism with lead-time. In this figure, an example can be seen where the lead-time (L) is 4/3-times more than the ordering period (t), i.e., if, for instance, the order is placed four months after placing the order. Thus, the "inventory on the books" (S) comprises three lots in each period t_i:

— the actual stock in hand (the part below the continuous line in Fig. 5);
— the ordering q_{i-1} placed in the preceding period t_{i-1}, which will be received after the first third of the reference period (t_i);
— the order q_i placed at the beginning of period t_i, which will be received in the period t_{i+1}. This quantity q_i has to be determined so that the total amount of the three components listed should be just S'.

If the length of the replenishment period is fixed, it does not disturb the operation of the system, but if this varies randomly (i.e., by causes independent of the decision maker, by factors not known to him), the operation of the whole system is essentially influenced, since we cannot estimate the actual effect of the ordering, i.e., what the manipulated variable of the system will display.

In our previous analyses and those which fo lows, it is most important to verify that the defined inventory mechanisms are really in compliance with the control process to be realized in the inventory system outlined, i.e., knowing the initial state of the system, and the actual values of the decision variables, the system is able to control the processes taking place in it. Namely, in this case it is true that the parameters of the inventory mechanism responsible for the control will lead to the optimal operation of according to the objectives of the decision maker.

24

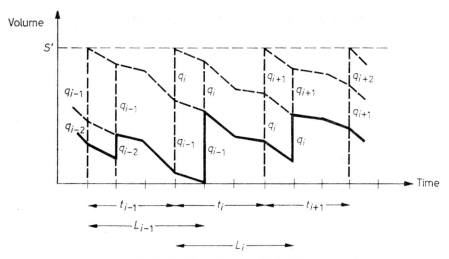

Fig. 5. *(t, S)* **mechanism with leadtime**

We have already emphasized several times that from both economic and technical point of views the "driving force" of the system is the demand. It is therefore self-evident that demand must play a significant role in the control process, too.

In the course of analyzing the regulation process of the inventory system we have pointed out (for basic cases see Chikán (1973), for certain modified conditions see Berács (1973)) that control can be performed with the inventory mechanisms defined earlier, so that with a given initial stock and decision parameters, the measured magnitude of demand alone determined the ordering, i.e., demand really plays a decisive role in the operation of the system.

The selection of the operation mechanism for the inventory system is for the decision maker to determine, and is based on the specific way he wants to achieve his objectives.

The inventory mechanisms have been defined by different parameters. On the basis of what has been described so far, we can see that the regulating variable manifests itself in these parameters. These regulating variables are called "norms". These norms indicate either, directly, some inventory level (s and S), or variables which have only an indirect connection with the size of the inventory but yet have a fundamental influence (t and q).

We do not consider inventory norming. For the purpose what follows, however, it should be noted that the goal of inventory optimization models is simply to determine the optimal value of these parameters and norms, or those derived from them.

III. A Historical Review of Inventory Modelling

The aim of this chapter is to present an historical overview of inventory modelling as a classical field of operations research. The large number of inventory models and the hundreds of papers considering many different versions of these models make it difficult to give an exhaustive historical review. The results achieved in inventory modelling will not be summarized here in strict chronological order; rather, typical groups of models and methodologies which characterize subsequent research periods will be emphasized. Through the development of modelling and solution techniques, the historical review of the most important theoretical and practical results in inventory control will be briefly summarized in this chapter. The more detailed description of a large number of inventory models is the subject of the second part of the book.

Even though we have tried to cover a substantial part of the literature, there are certainly many missing references, for which we apologize.

III.1. The Economic Order Quantity Model and Its Generalized Versions

The first classical mode of inventory control is the economic order quantity model (EOQ model) which was first published in the book of F. Harris (1915). A known demand of constant rate must be satisfied. No shortage is allowed and there is a fixed leadtime of deliveries. A fixed ordering cost and an inventory holding cost proportional to the amount and time of carrying inventories is considered in this model. The economic order quantity (EOQ) is the order amount which minimizes the total holding and ordering costs.

The formula which expresses the value of the EOQ is often referred to as the Wilson formula, since it was derived by R. H. Wilson in 1934. In the German literature, it is often called the Andler formula after the book of K. Andler published in 1929, but this model is described also in the books of B. Margansky (1933) and K. Steffanic-Allmayer (1927).

The assumptions of the EOQ model are very rigorous and they are rarely fulfilled in practice. In contrast, the Wilson formula is still the model most cited and used for an approximate solution even in the computer packages of inventory control. What is the secret of the EOQ model? One advantage is that on the basis of few data the calculations involved are simple. In addition, the simplicity of the model makes it easily understandable and applicable for people not well trained in mathematics and operations research. However, the most important reason why it is still so often used is its insensitivity to the input parameters. The order amount given by the Wilson formula does not result in a considerable increase in cost as compared with the optimal cost in the case of an inaccurate but not very bad estimation of the input parameters.

Many authors have worked on generalizations of the EOQ model. A summary of the early results was given by Whitin (1957). Through the formulation of the different assumptions different model-variants have been constructed and solved. Here, we summarize the deterministic versions of the generalized EOQ models, for which all the parameters are assumed to be known. Research into stochastic inventory models can also be considered as generalizations of the above models for those cases where some of the parameters can only be statistically estimated. If these parameters are considered as random quantities for which we have statistical observations or hypotheses, then a better solution can be achieved. Research into stochastic models started later than that into deterministic ones, but, subsequently, both fields developed side by side. The development of stochastic models can be connected with other goups of models which will be treated later.

Many different model variants deal with the case when *shortage is permitted*. The first contributions were published by Churchman, Ackoff and Arnoff (1957), and by Sasieni, Yaspan and Friedman (1959). A group of these models assumes that in the case of a shortage the customer is waiting for the delivery of the next order, at which time his demand will be fulfilled. This is the so-called backorders case, while, in the other case, the customer is not prepared to wait if there is a shortage; this is the so-called lost sales case.

In the backorders case, three different cost factors may occur when there is a shortage. The shortage cost may depend on the quantity or time, or shortage, or both, or it may be a fixed cost. All these cases may happen in practice. The equivalent models and their solutions were first detailed in the books of Hadley and Whitin (1963) and Naddor (1966). The lost-sales case is considered in these books, too. Here, the shortage cost also includes the loss of profit resulting from the lost sales. The appropriate mathematical models enable the optimal order ouantity and reorder level to be determined.

A lot of different versions of *constraints and bounds* have been built into the EOQ model. From the practical point of view the most important ones are the storage capacity constraint, the investment (budget) constraint and the limited total number of the annual orders. In all of these cases, the constrained optimum can be found using the Lagrange multiplier method. A detailed description of these models has been given in the books of Hadley and Whitin (1963), Naddor (1966), Popp (1968), and Klemm and Mikut (1972).

The EOQ model has also been generalised for different versions of the delivery process. In the original EOQ model, an immediate delivery was assumed, which can be easily extended to the case of constant leadtime. Only the reorder level must be transformed, it must be equal to the demand during the leadtime. In the case of continuous delivery, both the order quantity and reorder level have other optimum values, but they can also be easily expressed explicitly. These models are detailed along with others in the book of Hadley and Whitin (1963).

The *time-dependent demand* is first considered by Naddor (1966). The power-demand pattern is defined where a parameter characterizes the way in which quantities are taken out of inventory. Here the total demand for a given period does not change but during the period there are changes in the demand intensity.

Linearly increasing (or decreasing) demand was first considered also by Naddor (1966), but he could give only an approximate solution for the optimal ordering policy. The exact optimum was derived by Ryshikow (1969). The quantity of demand was expressed by Barbosa and Friedman (1979) as a power function of time and a general solution method was given for this case.

Different approximate solution methods have been derived for when the demand cannot be expressed as a simple function of time. These methods are based on the following two specific aspects of the EOQ model, which are, however, not always valid for a general demand process:

(1) The minimum point of the average annual cost coincides with the minimum point of the average cost concerning a unit of item.

(2) The average annual cost of ordering and that of inventory holding are equal.

The least-unit-cost model based on (1) and the cost equilibrium model based on (2) give in general not the optimal solution, but an approximation which is usually suitable for practical use. Both methods result in very simple algorithmic solutions in the computer packages of inventory control.

The optimal ordering policy can be derived by the method of *dynamic programming* in the case when demand changes from period to period in a known way. The first dynamic-type lot-size model was published by Wagner and Whitin (1958). The general dynamic programming method was specified for the inventory problem in the sense that the number of candidate solutions for the optimum could be decreased drastically. This results in an economic computing time. Further improvements of this method are due to Hadley and Within (1963) and Popp (1968). Here, a positive leadtime is also allowed when this is an integer multiple of the length of the period.

The requirement that no shortage is allowed was first considered by Zangwill (1966). This model was formulated as a production planning model in which the production lot-sizes are equivalent to the amounts ordered in the usual inventory control model formulation. In the similar production planning model of Popp (1968) the production capacity is bounded, and this can only be increased by surplus cost resulting from overtime. Taking this cost into consideration, the optimal production plan was derived using an improved dynamic programming algorithm.

An investment and production planning model was published by Ryshikow (1969) in which the total cost of inventory holding, investment, production and un-utilized capacity was minimized. The optimal solution was given in the form of equations based on dynamic programming.

Changes in prices and costs result in structure changes in the cost function. In addition to the fixed ordering cost, the purchasing cost may also influence the optimal ordering cost. The purchasing price may depend on the purchase quantity in different ways. In the most common case, discounts are offered for the purchase of large quantities. The discount may apply to every unit purchased or only to the incremental quantity. Both cases have been discussed in detail by Hadley and Whitin (1963), but there are earlier results in this field published, among others, by Churchman, Ackoff and Arnoff (1957) and Sasiani, Yaspan and Friedman

28

(1959). The purchasing price is assumed by Naddor (1966) to be a continuous function of the amount ordered. Price changes in time were also considered first by Naddor (1966). For a known price increase at a given, or probable, time Naddor gave the conditions which underline the increase of the lot-size. The holding cost may also be a nonlinear function of the time, or of the quantity carried: this also is discussed by Naddor (1966).

The *time discounting* of costs usually influences the optimal decision. A production-inventory system is considered by Schussel (1968) in which the unit set-up cost is a monotonously decreasing function of the lot-size. The classical EOQ model for the case of time-discounted costs was first solved by Hadley (1974).

Research into multi-item deterministic models started at the beginning of the sixties. The first multi-item version of the EOQ model considered the case for which *joint ordering* of items brings cost saving in ordering cost. The optimal joint interval of ordering and the optimal amounts of ordering for each item were expressed in simple explicit formulas by Naddor (1966).

The optimal cycle-length of lot releases in production was derived by Bomberger (1966) using the technique of dynamic programming.

Another connection between items of an inventory may be the *joint constraints*. The most common being the inventory capacity constraint and the financial (investment) constraint. These may also figure together. The solution can be easily derived in these cases on the basis of the Lagrange multiplier method as has been done, for example, by Naddor (1966), and Klemm and Mikut (1972).

A production-inventory system, under constrained resources, has been considered by Kleindorfer and Newson (1975). A production plan was developed which minimized the sum of the setup, production, and holding, costs. In the generalized version of this model manpower and overtime-management has also been included.

The inventory policy of *substituting items* is a rather difficult problem. An algebraic solution was derived by Wolfson (1965) for such a problem. For a given product there is a demand for different sizes. Only some specific sizes are produced, the demand for other sizes is satisfied by the next larger size, thus reducing the surplus amount. The optimal stocking policy minimizes the total losses, including those resulting by the cuts.

III.2. Single-Period Stochastic Models

Research into stochastic inventory models started in the forties with the so-called "Christmas tree model" or "newspaperboy problem". The first summary was published by Morse and Kimbel (1951). The models referred to above deal with the classical problem, when a merchant orders a quantity of a seasonal item for which no further ordering possibility exists. The demand for the item is not known at the time of ordering, and is a random amount. The surplus stock at the end of the season cannot be sold. The purchasing cost is smaller than the selling price. The optimal order quantity is that for which the expected value of the total profit is maximal.

In the generalised form of the above model given by Hadley and Whitin (1963) the surplus stock at the end of the season can be sold at a decreased price. It was also extended to multi-items with joint capacity of investment constraint.

Some model versions have been described in the book of Buchan and Koenigsberg (1963) for the stocking of spare parts with a single ordering possibility. The sum of the expected holding and shortage costs is minimized.

A systematic review of the single-period stochastic models has been given by Hochstädter (1969). The case of the fixed and the non-fixed ordering cost is examined separately for immediate delivery. The backorders case and the lost sales case are distinguished for single period models with leadtime. The method of solution is given for all the above models.

Optimal ordering for two consecutive periods has been investigated dy Naddor (1966), and three periods have been considered by Whitin (1973). These are straightforward generalizations of the single period model. They have, however, more complex solutions based on dynamic programming methods.

III.3. Stochastic Order-Level Systems

III.3.1. The (s, S) Inventory Models

The models described in this section are based on the (s, S) inventory policy which was defined in Section II. 4. 3. The first exact formulation of such an optimization problem was given in the book of Arrow, Harris and Marschak (1951). This model is the precise mathematical formulation of an inventory policy which has been applied in practice. Two critical inventory levels are to be fixed, s means the reorder point and S the order level ($s \leqq S$). The inventory level is reviewed periodically (there is also a continuous-review version which will be described in the next section). As the inventory level decreases under the reorder levels, an order is given which increases this level up to the order level S. In general, the task is to determine the values of s and S which minimize the total expected cost of the system. Another interesting question is under which circumstances does this policy guarantee the overall optimal inventory policy, i.e., the optimal among all possible periodic review policies. The search for efficient solution algorithms and for the proof of overall optimality of the (s, S) policy proceeded in parallel fashion between 1958 and 1968. In the seventies, algorithmic questions came more into consideration.

The optimality of the (s, S) policy was first analysed in the papers of Dvoretzky, Kiefer and Wolfowitz (1952, 1953). Dynamic programming aspects were studied by Bellman, Glicksberg and Gross (1955). Bellman (1957) was first to give a sufficient condition for the overall optimality of the (s, S) inventory policy. Arrow, Karlin and Scarf (1958) compared different ordering policies and proved the optimality of the (s, S) policy under rather strong conditions. The first general result concerning its optimality was published by Scarf (1959). The proof is valid for an arbitrary demand distribution, convex cost function and finite time horizon. The definition of K-convexity was introduced here, and further general results are based on this concept. In the context, amongst others, the papers of Zabel (1962) and Iglehart (1963), should be mentioned. The proof was extended to the case of an infinite time horizon. For the optimal parameter values, new bounds were given leading to more effective solution procedures. The optimality (s, S) policy was proved for stationary demand distribution by Iglehart, also. This result was published in the book edited by Scarf, Gilford and Shelly (1963). Veinott (1966) made further relaxations concerning the optimality conditions and, for a unimodular cost function, proved the optimality of the (s, S) policy, even for a non-stationary demand distribution. For the probability density function of the demand distribution, a sufficient condition was given by Limberg (1968) in the form of a differential equation. Further results in this field are summarized in the book of Hochstädter (1969).

The theory of Markovian decision processes and the renewal theory have been applied to prove further specific results. Demand depending on the stock level was considered by Tijms (1972) who gave the optimality proof, also for an undiscounted cost function, and derived conditions for the unicity of the optimal solution. Optimality criteria were given by Tur (1972) under an inventory capacity constraint and by Wijngaard (1973) under an order quantity constraint.

For an *exact solution* of the optimal (s, S) policy problem, even nowadays we have no generally effective numerical method; however, many publications are concerned with this question.

In the paper of Arrow, Harris and Marschak (1951) mentioned in connection with the first exact formulation of the (s, S) model, a method of solution for the optimal decision parameter was derived, too. This method is, however, not effective. It involves the numerical solution of a functional equation which contains the cost function and is built up on the basis of Bellman's dynamic optimality principle.

The *value iteration* method of dynamic programming was first applied to the solution of the above-mentioned functional equation. This is a successive iteration process derived by Bellman, Glicksberg and Gross (1955). The *decision iteration* method of dynamic programming was sketched already in the book of Bellman (1957) and a detailed description has been given by Howard (1960). This method may usefully be applied to the value iteration concerning the theoretical speed of convergence and the stopping criteria. However, the investigations of Beckman and Hochstädter (1968) showed that, in practice, value iteration is still more effective, since this needs considerably less computing time for each iteration step, than that required for decision iteration.

The above two general methods of dynamic programming have proved to be unefficient in practice. Methods based on specific aspects of the inventory control model are efficient.

The dynamic programming algorithm can be considerably simplified in the case of some special demand distributions. Of the large number of such investigations, the results of Naddor (1966) and Ryshikow (1969) merit mentioning, due to their effectiveness.

The steady-state probabilities of the inventory positions were determined by Wagner (1962) using the theory of Markov processes, and by Greenberg (1964) using a Laplace transformation, generalizing the result of Arrow, Karlin and Scarf (1958) (which was valid for exponentially distributed demand). Knowing the steady-state probabilities, the expected cost function can be expressed and minimized by a numerical procedure. The main difficulty is that the cost function is usually not convex, therefore it may have many local minima.

The most powerful of the exact solution methods is the Wagner—Veinott algorithm (1965). Here the upper and lower bounds are first calculated for the optimal values of s and S, then in the range of the possible optimality domain the cost function is calculated for each possible (discretized) pair of (s, S). Often a rather narrow range can be determined for the optimality domain, in which case the method is effective.

The practical importance of the (s, S) policy and its overall optimality (as compared with other policies under fairly general conditions) underline the necessity of effective numerical procedures. The demand of practical applications is for quick algorithms to be constructed which do not necessarily serve the exact optimum but ensure a good approximation with little computation. Investigations proceeded side-by-side in two different fields.

The first approach is to simplify the algorithm, while the other is to simplify the model. In the first case, the model is considered in its original form and the simplified algorithm, requiring few calculation steps, serves as an acceptable approximation of the optimal solution. In the second case, the simplified model is solved exactly and this solution is considered as the approximate solution of the original model.

The *approximate solution* of the (s, S) model is of great practical importance. The first such method is the Roberts approximation (1962) which is still being applied today. It is based on an asymptotic theorem of the renewal theory which is used for approximating the optimal average length of the order cycle. If the value of $D = S - s$ is a large number when related to s, it approximately equals the value of the EOQ, calculated by the Wilson formula, in which the mean demand rate is taken to be the demand rate parameter. The value of s can be easily determined using the density function of the demand distribution. If this value of s is not much smaller than the value of D calculated by the Wilson formula, then a correction factor suggested by Wilson must be applied.

The exact optimum calculated by the dynamic programming algorithm has been compared with the approximate value given by the Roberts approximation. Such investigations were published by Wagner, O'Hagan, and Lundh (1965) and by Beckman and Hochstädter (1968). In both papers, new correction formulas are suggested.

Upper and lower bounds for the optimal values of s and S were first given by Iglehart (1963) then by Veinott and Wagner (1965). These bounds can also be used as approximate solutions for the optimal parameter values. The calculations are very simple in the case of special demand distributions, such as a normal distribution. For the approximation, only heuristic error estimations are achieved.

The original form of the Roberts approximation is valid for continuous demand distributions. It was extended to discrete distributions by Girlich (1973). Here, the optimal value of s is determined by iteration. Tijms (1972) derived a simple iteration procedure also for the optimal value of S, for which convergence was proved under general conditions.

The method of stochastic approximation is used for the determination of the optimal parameters in the paper of Griessbach (1975). Although, it can be applied under general conditions, it is often numerically inefficient. Methods have been described which increase the speed of convergence based on the results of stochastic dynamic programming. The papers of Bartmann (1975) and Van Numen (1976) should be mentioned here.

III.3.2. The (t, S) and (s, q) Policies as Simplified Versions of the (s, S) Policy

Fixing the length of the order period yields a simplified version of the (s, S) policy since at each review an order is placed, independently of the inventory level. The fixed length of the order period is denoted by t_p, where the subscript p means "prescribed". Hence, here we have only one variable to consider. This is the amount of order (q) in the case of (t_p, q) policy or the order level (S) in the case of (t_p, S) policy. Both policies approximate the (s, S) policy when the fixed order cost is low relative to the inventory holding cost.

The (t_p, S) policy is often referred to as "order up to level S" since the inventory level is increased by an order to the level S at the beginning of each order period. Methods of solution for the optimal value of S have been given by many authors under different assumptions. Buchan and Koenigsberg (1963), Hadley and Whitin (1963) have considered the case of constant and random lead-time, the backorders and the lost sales case, and have derived simple formulas for different types of demand distributions. Naddor (1966) introduced the so-called "inventory bank system" which is a generalized version of the (t_p, S) policy, since the value of the order level may change from period to period depending on the changes in demand in the previous periods. Time-dependence during the order period is taken into account in the model of Prékopa (1972). A further generalisation is given there, in which the delivery of an order is assumed to arrive not in a single lot, but delivery may occur in random lots at random times during the order period. This kind of delivery process is called the "interval-type delivery process". Gerencsér (1972) compared the optimal costs for different unit cost factors and carried out a sensitivity analysis for the (t_p, S) model.

The dynamic lot size model was investigated by Hadley and Whitin (1963) for the case of stochastic demand. The optimal value of the orders was derived, which may be different in different order periods. Both backorders and lost sales cases are considered. The lead-time may also change from one period to another, it may also be random variable, but orders placed earlier must be delivered earlier. The model can be solved using dynamic programming methods. Ryshikow (1969) considered the case of urgent ordering, with immediate delivery and extra charge. Control theoretical results are applied by Bessler and Zehna (1967) to the determination of optimal ordering quantities. The deviation of the expected inventory level, from the ideal level prescribed, is minimized. The demand distributions of different periods may be different. The leadtime is supposed to be an integer multiple of the value t_p. The results of stochastic programming were applied by Prékopa (1973) to give the optimal amounts of order for multiple periods. The joint distribution of demand in subsequent periods is considered, thus the stochastic interdependences in the demand process are taken into account, in contrast to the usual models which assume independent demands in the different periods. The solution of the stochastic programming model is reduced to a convex programming problem, for which many numerical procedures are available.

The length of the order period may be also subject to control. This case is detailed in the books of Hadley and Whitin (1963), and Naddor (1966). The cost function under stochastic demand and random leadtime was derived by Hadley and Whitin (1963). For this model, a simple approximate solution was achieved by Naddor (1966) for the case of a deterministic leadtime. The case of stochastic leadtime was solved by Agin (1966) in the form of a minimax procedure.

Hadley and Whitin (1963) also investigated the more general type of model (t, s, S), where the order period t, the reorder point s, and the order level S, are all subject to control. The cost function was formulated for an arbitrary demand distribution, but no efficient algorithm was suggested. Many local minima may exist, thus the usual numerical optimization procedures do not guarantee a global optimum. This is also valid for the algorithm of Teicholz and Lundh, which has been derived directly for the optimization of the above cost function, and yields a local optimum usually more quickly than the above general methods. For the case of Poisson, or normal, distributed demand, special algorithms have been derived by Hadley and Whitin (1963). The length of the periods are assumed to form a Markov process in the model of Denardo (1968). The inventory capacity is bounded and delivery occurs immediately after ordering. For this case, a solution procedure based on a fixpoint method is derived which serves the optimal value of all the three parameters under control.

The prescription of the reorder point was considered first by Naddor (1966). The maximal demand of an inventory review period is assumed to be a known final value. No shortage is allowed, thus the reorder level is fixed in advance by this constraint. The optimal value of the order amount may be easily expressed.

In the inventory system described by Ryshikow (1969) a fixed amount q is ordered in each period. This is the only variable subject to control. The demand is random. If the inventory level decreases below a certain level, then an urgent order has to be placed since no shortage is permitted. By an increase in the inventory level the capacity bound of the inventory may be exceeded. The surplus stock can be sold only at a loss. This complex system is described with the help of a Markov process.

In many practical inventory systems, the amount which may be ordered is determined by some external conditions, e.g., package size. In the model of Naddor (1966) the optimal reorder point is given. When the inventory falls below the reorder point s, an integer multiple of the package size is ordered. This integer is determined in such a way that the order has to increase the inventory level above the order level S which is also determined by the optimization of the cost function.

The total cost of n subsequent periods is minimized by Wijngaard (1973) under a restricted order amount. The dynamic programming method applied yields a single optimal reorder point for each period under rather general conditions. The optimal order amount coincides with the upper bound.

The optimal policy for slow moving items is the so-called $(S-1, S)$ ordering rule, where the inventory level coincides usually with the optimal order level S. In this kind of model a discrete demand distribution is usually supposed. The customers arrive according to a Poisson distribution in the model of Freeney and Sherbrooke (1966). The amounts demanded have a discrete distribution. An algorithmic solution for the optimal S is derived for a geometric probability distribution. By the assumption of Croston (1974), a normally distributed amount is demanded on average every p periods. In this case, the optimal order level is expressed in a simple explicit form. For a demand characterized by a Poisson process an explicit solution was derived in the paper of Smith (1977).

34

In the model of Higa, Feyerhorn and Machado (1975) the customer waiting time is random. The expectation of the waiting time must be below a specified level: this is the prescribed customer service level. The leadtime is random and has an exponential distribution. The same model was considered by Sherbrooke (1968) assuming a constant leadtime, and by Das (1977) assuming a constant waiting time and the case of lost sales. In both cases, geometric Poisson-distributed demands were assumed, as in the above-mentioned model of Feeney and Sherbrooke (1966).

III.3.3. Generalizations of the (s, S) Policy

Order policies depending on the *time of decision* are based on dynamic models. Seasonal demand fluctuations which can be described by a discrete Markov process have been considered by Riis (1965). Kao (1975) formulated an order policy where the order amount depends not only on the actual inventory level but also on the date of occurrence of the last demand. The function of the model is described by a Markovian renewal process, and for the optimal ordering policy a linear programming procedure or another iterative scheme is derived.

In another direction of the development of generalized models, production and investments are directly included together with the inventories. Some characteristic models of this type will be mentioned here.

A *production-inventory* system has been considered by Suddenth (1965), where stochastic demand occurs at the end of each period in one lot. Unsatisfied demand awaits production. The total cost of production, inventory holding and shortage is minimized in the optimal production schedule, which is determined by an iterative procedure in a more simple way than the dynamic programming solution requires.

The *production capacity* may be increased according to the model of Rao (1976). The demand may change from period to period but is supposed to be known in advance. It can be satisfied by production or from inventory. The optimal production plan and the amount of capacity increase (if this is necessary) is to be determined, and, for this, a dynamic programming procedure is suggested.

A model of *production for stock* has been published by Liberatore (1977). Known amounts are demanded at specific, though irregular, times. There are specific demands which must be satisfied individually. The problem is to determine the optimal setup times which minimize production and inventory costs. Cost savings can be achieved by satisfying various demand lots toghether. The time of producing a given lot is random. The optimal strategy can be derived using a dynamic programming method.

The *urgent ordering* possibility was first investigated in a single-period model by Barankin (1961) under the restriction that the amount of urgent ordering is fixed. For a bounded amount and for n periods a generalized form of this model was derived by Daniel (1963). The inventory holding cost function is linear. Optimal ordering is characterized by two critical levels changing from period to period. Taking these values into account, the order amounts for the normal and for the urgent ordering may be expressed as a function of the actual inventory level. The order cost is assumed to be proportional to the amount of orders with a higher unit cost of urgent ordering. No constant ordering cost is considered. A backorders case is assumed. This model was generalized for a convex inventory holding cost and the lost sales case by Neuts (1964).

A constant ordering cost has also been considered by Fukuda (1961) as well as the normal and urgent ordering cost factors. The cost-optimal ordering strategy is given by a critical value together with two critical numbers (s_n, S_n) depending on the period. The critical value is the reorder point of the urgent ordering. For normal ordering, the (s, S) policy is implemented with the parameter values s_n, S_n. Urgent ordering is satisfied immediately, while normal ordering suffers a period of delay. The case of two different urgent orderings with different leadtimes and costs has been examined, too. Some special models with different urgent-ordering strategies have been described in the paper of Bulinskaya (1964). The papers of Kaminsky (1966), Allen and d'Esopo (1968) also contain useful results.

III.4. Models with Continuous Review

For an *arbitrary demand distribution*, the case of continuous review (and ordering possibility) was first examined using the method of dynamic programming. If, in the periodic (s, S) model, described previously, the length of the review period t tends to zero, then, in the limit, the continuous review case is obtained. The results concerning the (s, S) policy obtained by dynamic programming were adapted first by Beckman (1961) for continuous review in this way. The optimality question of continuous review policies can be similarly treated.

Most of the methods of solution derived for (s, S) models can be adapted for continuous review models; however, models which take into account the pecularities of continuous reviewing are more efficient in practice. (Methods based on dynamic programming techniques are usually not fast enough.)

The assumption of a stationary demand distribution has been generally used. The first approximate solution under this assumption was given by Galliher, Morse and Simond (1959).

The first systematic survey of continuous review models appears in the book of Hadley and Whitin (1963). For the optimal lot-size and reorder level, only an approximate solution is given there for the case of an arbitrary demand distribution. The shortage is not taken into account by a calculation of the expected annual stock level based on the supposition that the amount of shortage is negligible relative to the amount of stock. Herron (1967) and Ryshikow (1969) made also the assumption that the duration of shortage is so small that the mean stock level can be well approximated without the need to take into account stockout situations. A fast numerical and graphical solution has been given by Herron (1967) and an iterative method is given by Ryshikow (1969) for the approximation of the optimal solution. For the convexity of the cost function, sufficient conditions have been derived by Gerencsér (1972) which also ensure that the local optimum is global at the same time. The stability of the solution was also examined.

The optimal inventory policy was approximated by Wagner (1969) who considered a fixed leadtime. A similar iterative procedure was given by Das (1975) together with an examination of stability. Cowdery (1976) considered the case where units are demanded, and he derived the optimal policy as the solution of a system of equations with two variables.

Demands occurring at *random times* have also been the subject of a number of investigations. In the model of Sivazlian (1974) units are demanded according to a renewal process. There is no leadtime. The optimal value of the reorder level can be determined analytically. In the deterministic case, this coincides with the solution of the discrete lot-size model. In the paper of Snyder (1974), the amount demanded may also be random with an arbitrary distribution. After every demand occasion, the stock level is reviewed. The leadtime is supposed to be negligible and no shortage is permitted. This result was later generalized also for positive leadtime (Snyder, 1975). In the first case, the optimal ordering policy was determined using the theory of Markovian renewal processes, while, in the second case, the solution method is based on dynamic programming.

The consideration of a *specified demand distribution* is characteristic in continuous review model investigations. The reason for this is that solutions derived for an arbitrary demand distribution are either too sophisticated, or do not guarantee optimality (and even, no estimation exists for the deviation from the optimum). These problems may also frequently appear by using specific demand distributions.

Some specific demand distributions were considered by Hadley and Whitir (1963). The leadtime demand was described by a Poisson process. It is a good model for when units are demanded at random times. The optimal solution can be calculated only under simplifying assumptions for the backorders and lost-sales cases.

Gamma distributed demand and partial backorders are considered in the model of Burgin (1970). The loss of profit due to lost sales is also included in the cost function. The optimal policy can be calculated by an iterative process. A similar solution was given by Ryshikow (1972). Beside the normal- and Poisson-distributed demand, the special case was also investigated by Gebhardt (1973) for when the demand arriving at random times a geometric distribution, while the time between consecutive demands is exponentially distributed. In the paper of Snyder (1974), the amount of demands is characterized by an exponential distribution and the interarrival times of demands may have an arbitrary distribution. In the paper of Lavratshenko (1973) the amount of demands may be arbitrarily distributed, while the interarrival times of demands have exponential distribution. The cost belonging to the annual inventory level is minimized instead of the expected annual cost, as is usual. The optimal parameters are the solutions of cubic equations.

The leadtime demand was approximated by a normal distribution by Psoinos (1974) and the method of Hadley and Whitin (1963) was extended. Nomograms and tables were presented which increase the efficiency of the procedure. The Weibull distribution has been shown to characterize the leadtime demand in many practical cases. For the estimation of the parameters and for the optimization of the reorder level, some simple procedures have also been derived. In this model, the probability or expectation of shortage may also be constrained.

Queuing theory may also be applied for the construction of different solution procedures. The following equivalent of a queuing model is described by Buchan and Koenigsberg (1963): the service corresponds to the ordering and the customers represent the demand. A service channel is busy, if an order has been placed. The beginning of a service is equivalent to the placement of an order, while the conclusion of a service is equivalent to the delivery of an order. The number of channels M is the ordering amount which is to be optimized. In accordance with the simplest queuing model, the demands arrive one by one and the lead-time has an exponential distribution. The cost function can be derived from the stationary solution of the state equations. The optimum can be easily calculated.

In another model of Buchan and Koenigsberg (1963), the stock level is the equivalent of the queue length waiting for service. This can be controlled by changing the input (order) rate. Optimization can be carried out also using the stationary solution of the state equations.

An all-embracing application of queuing theory for many different inventory models is described in the book of Ryshikow (1969). Here the following equivalence is applied: the possible maximal stock is the total number of the service channels, the actual stock is the number of the free channels, the shortage is the length of the queue. In some of the models, the amount of shortage is constrained, while, in others, no shortage is allowed. The different usual input and service processes of queuing theory are adapted for inventory models. In most of these cases, only the state probabilities are derived. The expected costs can then be expressed (based on the state probabilities) and minimized using some general numerical procedure of function minimization. For some models, iterative procedures are given. The difficulties of the numerical computation stem from the fact that often several local minima may exist, among which the global minimum is to be selected (while most optimization algorithms find only local optima).

III.5. Reliability Constraints and Reliability-Type Models

III.5.1. Cost Optimization under Reliability Constraint

As early as 1953, Ziermann proposed a model for the inventory control of spare-parts, where the annual cost of inventory holding and ordering is minimized under a reliability constraint. The shortage cost is not explicitly contained in the cost function, it is replaced by the probabilistic shortage constraint: Either the probability, or the expectation, of the shortage must be below a prescribed limit. Here, the demand for the spare-parts is described by a Poisson process. Similar methods have been derived by Brown (1967) for a normally distributed demand. The reliability of the continuous supply, the so-called service level, can be characterized in different ways: besides the probability of continuous supply one can take the expected value of the demand immediately satisfied from stock over the total demand. The latter can be approximated by the ratio: average demand satisfied immediately from stock to the average total demand in a given period. The reorder point belonging to a prescribed service level can often be expressed with simple formulas which have been tabulated for different values of the relevant parameters. The use of such tables has also found application in practice in the case when the demand in a given period is not a normally distributed random amount. In this case, the solution is considered to be an approximation of the optimal decision. Most computer packages for inventory control also use this simple approximation for specifying the reorder level.

For *multi-item* models, the same idea has been extended by Herron (1967). The sum of the joint inventory holding and ordering cost is to be minimized under a common, or several separated, reliability constraint(s) concerning the items. In the model given by Mann (1973), the difference between the amount ordered and that delivered is assumed to have a normal distribution. The expected value of the shortage is constrained in the work of Beesack (1967). The first review of reliability-type constraints appears in the book of Klemm and Mikut (1972). They are denoted by Greek letters, following the notation of Rényi and Ziermann (1960) and are classified in the following groups:

α: the probability that the demand in a given period is satisfied,

β: the expectation of the ratio: the satisfied demand in a given period to the total demand in that period, which is often approximated by the ratio of

$\bar{\beta}$: the average demand satisfied in a given period to the total demand in that period,

γ: the expectation that the demand satisfied will exceed the average demand over the total demand: a reliability measure which was first introduced by Rényi and Ziermann (1960).

Klemm and Mikut (1972) investigated the following problem: which form and value of the shortage cost results in a reliability constraint giving a cost-optimal solution. (This means that a given reliability constraint corresponds to a shortage cost factor for which the ordering strategy calculated on the basis of the reliability constraint ensures the minimal total cost.) The practical importance of this is that the necessary service level (reliability constraint) can usually be more easily estimated in practice than the value of the cost factor. The above investigations were carried out for the order-level system, i.e., the (t_p, S) policy by Klemm and Mikut (1972). They found out that if the shortage cost does not depend on the time and amount of stockout, only a fixed shortage cost appears at each period when shortage occurred, then a reliability constraint of type α has to be applied. The shortage cost proportional to the amount of stockout corresponds to the β service level. The order level which minimizes the inventory holding cost under a reliability constraint is determined by the Lagrange multiplier method. These results have been generalized in the paper of Klemm (1973) for the (s, S) inventory policy and continuous review models.

Under a probability constraint on the shortage and geometrically distributed demand, the reorder level and lot-size which minimizes the expected cost of inventory holding and ordering was determined by Girlich (1973). An iterative procedure is derived in the paper of Das (1975) for an arbitrary demand distribution and random leadtime.

In the model of Cowdery (1976), units are demanded. The expected inventory holding and storage cost is minimized by constraining the expected value of the shortage, and the solution is given in the form of a system of equations. Units are demanded at random times according to a Poisson process in the model described by Magson (1979). This was devised to describe a spare-parts inventory problem, where the leadtime had a gamma distribution. The reorder level is fixed by a prescribed probability of shortage, while the amount of order is determined by cost optimization.

The delivery of an order often occurs not at one time, but, rather, is realized in parts. This practical situation is analyzed in the so-called reliability-type models (described in the next section), where usually no direct cost optimization exists. There are, however, some multi-item versions where cost minimization is carried out under a reliability constraint. The capital invested in initial stocks of different items has been minimized by Prékopa (1973) under the constraint that the probability of shortage may not exceed a prescribed limit for any of the items. The optimal amount of the initial stock for the different items is determined by reducing the original stochastic programming model to a convex programming problem. The model for the delivery of an order is the so-called randomly-scheduled delivery process in which deliveries occur at random times during a given period in random amounts. This will be described in detail in the next section. Studying the same delivery process, Gerencsér (1973) investigated the conditions which allow the decrease of the order level calculated on the basis of an estimation of the shortage cost. The decreased order level has to ensure a prescribed service level.

A multi-item inventory model has been given by Prékopa (1973) in which the capital invested in the initial stock is minimized under reliability constraints. The probability of the shortage occurrence and the conditional expectation of the shortage (under the condition that a shortage occurs) is constrained. The second constraint is especially important in the case when the shortage cost depends on the amount of shortage. An algorithmic solution is described in the paper of Prékopa and Kelle (1976). This is based on a nonlinear programming algorithm where the reliability measure is estimated by a simulation technique. In another form of the model, the cost depending on the expected amount of shortage is also included in the cost function.

III.5.2. Reliability-Type Models

An explicit cost function is not considered in the reliability-type inventory models, where the objective is to determine the minimal stock level which ensures continuous supply with a prescribed probability (service level). The advantage of such models is that the cost factors need not be determined, and that the random character of the delivery process can be more carefully analyzed.

The first reliability-type inventory model was given by Prékopa (1963) and Ziermann (1963). It was developed for the solution of the following practical problem: in an order period T, a commodity is consumed with a known constant demand rate c. The order for this total amount is delivered by the supplier in the order period denoted by (O, T). The delivery occurs not on one occasion, but at random times in the period (O, T) in n parts. In the first version of this random delivery model, the following assumptions were made:

The delivery, scheduled uniformly in time, is disturbed by random factors. The deliveries are assumed to occur at any time during the period (O, T) with the same probability. Thus, the elements of an ordered sample, taken from the uniform probability distribution in (O, T), are considered as a realization of the

40

delivery times. The amounts delivered during the period are assumed to be the same at each time of delivery.

The consumer arranges to have an initial stock M as its safety stock for each period to protect the consumption process against random fluctuations in delivery. The problem can be posed as follows: how to determine the minimal level of the safety stock which ensures a continuous supply during the entire period on a prescribed service level, which is usually a probability denoted by $(1-\varepsilon)$ (ε is a positive number close to zero in practice, e.g. $\varepsilon = 0.01 - 0.2$). If the rate of the shortage and the inventory holding cost factors are known, the probability $(1-\varepsilon)$ can be determined: the corresponding value of the service level results in a cost optimal stock level, as described by Klemm and Mikut (1972). In practice, however, the appropriate choice of the prescribed service level is much easier to decide than to estimate the shortage cost factor. This experience is utilized by the development of reliability-type models.

The optimal level of the initial stock M is the solution of the so-called reliability equation which can be solved by numerical methods for nonlinear equations or using tabulated values. Often an approximate solution is given in a simple explicit form which is sufficiently close to the exact solution for sufficiently large values of n (the number of deliveries).

The first generalized version of the above random delivery process was given by Prékopa (1963), in which the amounts delivered are also random. There is a minimal amount $\delta > 0$ which definitely arrives when a delivery occurs. The remaining amount is randomly subdivided among the lots delivered by $n-1$ random points, which are uniformly distributed on the respective intense representing the randomly delivered part of the ordered amount. This is the so-called randomly-scheduled delivery process for which the asymptotically optimal value of the initial stock level M has been described in a simple explicit form.

The exact optimum for the above model was given by László (1970) in the special case when $\delta = 0$, as the solution of a nonlinear equation. It can be applied when no information is available about the expected minimal lot-size of delivery. The case when the total demand of the order period is different from the total delivery has also been investigated: the general exact solution for these questions was given by Kelle (1980). An analogous multiperiod reliability-type inventory model was investigated by Pintér (1975), where an asymptotically optimal estimate of the initial stock is given for the case when the number of the deliveries and/or demand occurrences follows a joint descrete (possibly unknown) probability distribution.

The delivery of an order often happens in one lot after a random leadtime. If the distribution of the leadtime demand is known, then the safety stock belonging to the prescribed service level can be easily expressed. The formulas are especially in the case of a normal distribution very simple. They are also often applied in practice in the case of other distributions as approximate solutions. Among the first of such results were those of Brown (1967), Gerson and Brown (1970).

The number of deliveries may be so numerous that it is possible to consider this as a continuous process. In the model of Németh (1971), the Wiener process is applied to the approximation of the delivery process. The mean rate of demand and delivery is assumed to be equal. This has been generalized for different mean rates and for random demand or delivery rates by Kelle (1980). The minimal

41

safety stock level belonging to a prescribed probability level of continuous supply is the solution of a nonlinear equation.

Forecasting and stock control methods have been given by Croston (1972) for intermittent demands. The length of the review period is fixed. In most of the periods there is no demand. An order is placed if a demand occurred in a given period. The order is delivered at the beginning of the next period. The order level must ensure a prescribed service level which is measured by the probability of the continuous supply. The specifics of the intermittent demand occurrence are taken into account in the forecasting of the expected demand.

Demand has often trend and seasonal influences. Mann (1973) has proposed, for this, a second-order exponential smoothing technique. The amount of ordering is the sum of the demand predicted for the next period and the safety stock which is determined under reliability criteria based on the estimated error of the forecasting. The demands in the different periods are assumed by Yaspan (1972) to be correlated normally distributed amounts with different parameters. By transforming the reliability criteria of a continuous supply a multi-dimensional correlated normal distribution function is obtained which can be tabulated in the two-dimensional case. In more than two dimensions, a simulation technique must be applied. In the model of Burgin (1975), the amount demanded in unit time has a gamma distribution. Two different types of reliability constraint are considered: an iterative procedure is derived when the probability of the continuous supply is prescribed, while if the expectation of the shortage is constrained then the optimal value of the order level is determined by the gamma function values which are tabulated. In both of these cases, a simple explicit expression is given for the approximation of the optimal order level.

The first model where the delivery may be a random process which is not necessarily homogeneous in time was given by Prékopa (1973). The deliveries in a given order period may be cumulated at any part-period in a random way. Thus the stochastic-type informations available at the time of ordering can be used to construct a realistic estimation of the delivery process. This results in a better planning of the necessary initial stock level which ensures a prescribed service level. Computationally, it means the solution of a nonlinear equation, where the probability of the continuous supply is calculated on the basis of the above model using simulation techniques. The relevant algorithm has been given in detail in the paper of Prékopa and Kelle (1976).

III.6. Multi-Item Inventory Models

The *joint constraints* symbolize as a rule, the connection among different items which makes it sensible to consider a joint inventory policy for a group of items. The total value of the orders placed at the end of a given period is limited in the model of Rényi and Ziermann (1961). Under this capital constraint, the sum of inventory holding and shortage costs is to be minimized for the N different items. The demands of the different items are independent, normally distribute amounts. The vector of the optimal order quantities is constructed by means of the Lagrange multiplier method.

Two different constraints are considered by Gerson and Brown (1963). In one of the models, the total value of the safety stock is limited and the loss of profit is minimized under this constraint. In two other models, the loss of profit or the sum of the ordering and shortage costs is minimized under some limit on the value of the average joint inventories for N items. In all of these cases, the Lagrange multiplier method has been applied. As well as the capital constraint, the rate of delivery is also limited by Buchan and Koenigsberg (1963).

The expected value of the maximal joint inventory level is constrained in the multi-item model of Ryshikow (1969). The objective function is the sum of the expected inventory holding and shortage costs. The total inventory holding cost for a group of items is minimized under a constrained probability of shortage in an other model of Ryshikow (1969). In the papers of Prékopa (1973), Prékopa and Kelle (1976), the reliability constraint may be formulated as the conditional expectation of shortage. These models consider the random delivery process previously described.

The joint inventory capacity is bounded in the model of Page and Paul (1976). The demand is known and the order is placed jointly for all the items. The length of the order period to be determined is that which minimizes the total ordering and inventory holding cost while the maximal joint inventory level is below the capacity limit. A generalization of the above model has been given by Zoller (1977). A constant leadtime and a fixed order amount was considered; thus the length of the order periods may be different. A solution is given only for some special cases.

The *joint ordering policies* for a group of items usually result in a lower total cost, than that resulting from an individual policy for each independently formulated item. The cost reduction may be a consequence of the annual ordering cost decrease due to joint ordering, or a consideration of the demand dependence may result in a saving in inventory holding and/or in shortage costs.

The independent ordering policies of a group of items was first integrated in a joint ordering policy by Balintfy (1964). The (s, c, S) policy, therein introduced, controls the ordering in the following way: if the inventory level of any of the items decreases to its own reorder point s, an order is placed for this item and for all the other items with an inventory level below the so-called can-order point c which is higher than the reorder point s for each item. The order level S is also fixed item by item. For those items with an inventory level above c, no order is placed. The number of resulting orders is generally less which results in an ordering cost decrease but implies higher inventory levels. Optimal joint ordering usually ensures a lower total cost than independent ordering policies. Silver (1965) determined—for a Poisson distributed demand and two items—those conditions which make it possible to reduce the total cost by introducing an (s, c, S) policy. Ignall (1969) showed, however, that the (s, c, S) policy is not always the best joint ordering strategy even for two items. In some cases, the optimal order amount can be specified only by considering the inventory level of the other item. For these cases, the optimal policy can be constructed, for example, on the basis of Markovian renewal theory.

The *dynamic programming* technique was applied to determine the optimal ordering of two items by Iglehart (1965). A recursive relation was derived which can be handled only in special cases and which is numerically very inefficient. This result was widely generalized by Veinott (1965) also using dynamic program-

43

ming. N different items are considered which may be related to each other. The demands for the items in a given period are described by a random vector. The demands for different periods are assumed to be independent and they may have different distributions. Only an ordering cost proportional to the amount of ordering is considered, no fixed ordering cost exists. If some assumptions concerning the cost function are valid, it is sufficient to deal with the cost of a single period. In this case, the optimal ordering policy is characterized by the s and S parameters for each item and a stocking rate for all of the items. If the inventory level of all items is below the reorder point, an order up to level S is placed. If, for one of the items, the inventory level is greater than s then the minimal amounts have to be ordered for each item which ensure an optimal stocking rate. Ignall and Veinott (1969) reduced the multi-period dynamic model to a single period problem under more general conditions taking the inventory capacity constraint also into account and proved the optimality of the above joint ordering strategy. For the case of two items and backorders, Wright (1968) derived an optimal ordering policy with normal and urgent ordering possibilities. The numerical solution of the dynamic programming model is very difficult. Evans (1967) investigated the two items model in the lost sales case. The optimal ordering policy has a complicated structure. A vector of order, two scalar functions and a vector function divide the plane of the inventory level vectors and in each domain a different ordering rule is determined. In another paper of Evans (1969), a multi-item production-inventory model was described. The optimality of a generalized (s, S) policy is proved in the backorders case for production costs with quantity-proportional and fixed factors. In the lost sales case, this statement is valid only for a production cost factor proportional to the amount produced.

The fixed ordering cost which does not depend on the ordered amount, usually causes difficulties in the construction of joint ordering strategies. This is eliminated under quite general assumptions by the so-called (σ, S) policy defined in the papers of Johnson (1967, 1968). This is a multi-dimensional generalization of the (s, S) policy where the order level S is replaced by an N-dimensional vector of the order levels, and the reorder point s is replaced by a subset σ of the N-dimensional space. If the vector of the inventory levels is contained in the set σ, an order has to be placed which increases the inventory levels to the levels prescribed by the vector S. In the opposite case, no order is placed. The inventory model is reduced, in the discrete case, to a Markovian decision process and using the decision iteration method of dynamic programming the optimality of a (σ, S) policy is proved. The numerical solution is not investigated. For the continuous case, the optimality of a (σ, S) policy has been proved by Küenle (1977).

A simplified ordering strategy has been suggested by Ryshikow (1969) for N items. If the inventory level of a single product decreases to a critical level, an order is placed for all of the items. The case of a joint shortage cost, paid according to the greatest shortage, is also investigated. For the optimal parameters of the simplified ordering strategy a simple algorithm is derived: the iterative solution of a system of equations yields a good estimation. In another paper of Ryshikow (1972) an iterative solution is given for the (s, c, S) policy by means of a continuous review approach. The stoachastic interdependence of the demands for two items is considered by Hochstädter (1973) and a nonlinear programming method is proposed for the calculation of the optimal order quantities.

III.7. Multi-Echelon Models

A joint ordering policy of *serially linked echelons* was first formulated by Clark (1958), and later by Clark and Scarf (1960). In the chain of echelons the item gets into the subsequent store by a prescribed rule and the demand is satisfied at the end of the chain. The optimal joint stocking policy was derived only for two subsequent echelons for the case when no fix ordering cost was considered. Under this supposition the cost functions can be separated for the two echelons, introducing a penalty function which replaces the costs concerning the other echelon. For other cases, only an approximate solution method has been suggested.

The *internal storage of a production line* has been modelled by Schussel (1968), where successive production levels are connected to each other by the introduction of internal stores. An iterative procedure is derived for the determination of the optimal lot sizes at each level, resulting in a minimal total cost of setups, production, stock holding and capital investment for all of the internal stores. Here, the joint optimum is approximated step by step starting from the optimum of the individual levels.

A chain of three echelons is described by Ryshikow (1969), where the products of the different production levels are stored. The demand for the finished product in a period is a random amount having a known distribution. If a shortage occurs at any level of the production, then it has a shortage cost consequence, where the cost factor depends on the production level. At a higher level of finishing an increased cost factor appears. For the determination of the optimal stocking policy a complicated algorithm is suggested.

The ordering policy of two serially linked echelons is connected with demand forecasting in the paper of Iglehart and Morey (1971). The optimal stocking policy of the first echelon is of the (s, S)-type, while the second echelon has a complicated ordering strategy characterized by the levels s_n and S_n changing from period to period and the ordering is dependent also on the predicted demand for the next period. The parameters of the optimal ordering policy can be calculated by a recursive procedure.

The joint inventory policy of *parallel stores* has been investigated by Allen (1958). A central depot supplies parallel echelons, where customer demand is satisfied. The connection between parallel stores was also considered by Hadley and Whitin in their paper published in the book edited by Scarf, Gilford and Shelly (1963). A redistribution of inventories is possible between parallel echelons through fast or normal transportation. These redistributions make it possible to reduce the joint inventory level which satisfies the random demands appearing at the stores. The paper of Gross published in the same book considers stock distribution between parallel echelons controlled by a central depot. Here, a single period model is constructed. A multi-period, dynamic stock-distribution model has been given by Bessler and Veinott (1966). This is an adaptation of the multi-item model of Veinott (1965), where the same type of items in different echelons are considered as different items of a single store.

The multi-echelon model of Sherbrooke (1968) has been constructed for the stocking of reparable spare parts. The faulty parts can be repaired with a certain probability in the local depot, otherwise they have to be transported into the central depot for repairing. The $(S-1, S)$ inventory policy provides an optimum for expensive spare parts, i.e., a one-by-one repairing scheme is optimal.

Stochastic connections between the demands of two parallel stores have been taken into account by Iglehart and Lalchandani (1967) in the construction of an optimal inventory policy. The joint inventory level of the two locations is limited. A recursive solution method is given for the calculation of the optimal policy. Only the purchasing cost factor, which is proportional to the quantity, is considered as the ordering cost. The fixed ordering cost has been also included in the model of Hochstädter (1970). The redistribution of stock among the parallel stores is not possible in the above models. The cost function can be separately considered by introducing a penalty function as was done by Clark and Scarf (1960) for serially linked echelons. This procedure yields, in general, only an approximation of the optimal solution.

In the book of Ryshikow (1969) models are also presented for parallel stores. N inventory locations have the same stocking conditions (demand distribution and cost factors). For the minimization of the shortage, distribution and redistribution costs, a complicated algorithm is derived which involves the solution of a transportation problem, too.

The model of a combined multi-echelon system of stores has been given by Chikán and Meszéna (1973). Both parallel and serial stocking locations are involved. The optimization of the system is based on a bottom-up technique. In the lowest level, where the random demands of customers appear, an (s, nq) policy is applied. The central supplier has a fixed ordering cycle and an order-level policy is realized. In connection with the cost factors, the measures of risk and reliability are also investigated.

The joint analysis of *multi-item and multi-echelon* inventory systems has a lot of difficulties, especially in the numerical evaluation of the optimal policy. In the book of Ryshikow (1969), such a complex system is investigated considering stochastic demand. The model consists of two problems: the supply problem based on cost minimization, and the stock distribution depending on the inventory level of the stores.

III.8. Simulation Models

Inventory control was one of the first areas of application of simulation techniques. Inventory management problems are often connected with time-dependent and random processes which can be described by simulation methods in a very natural and straightforward way. The mathematical difficulties arising from the analytical solution of even a simplified model underline the importance of the simulation models which enable us to describe a very complex, sophisticated inventory control system.

The first report on a simulation-type inventory control application was given by Robinson (1957). The interdependence of a central oil depot and some external depots was analysed by a simulation. The characteristics of the central depot have to be specified for which the reliability of the whole system is guaranteed on a certain prescribed level. The simulation program enables the joint of a number of stochastic influences.

A similar multi-echelon system simulation study was presented by Berman (1961). The initial stock of the parallel echelons is known. Until the next central delivery, they can deliver to each other if a shortage occurs somewhere (this is called redistribution). The central order strategy and the redistribution strategy is determined by simulation. A complex multi-item system was solved by Dzielinski and Manne (1961) by using a similar approach.

For research into complex systems, special program languages, the so-called simulation languages, have been developed. The most important of these are DYNAMO (1959), SIMSCRIPT (1962), GPSS (1965) and SIMULA (1967). These program languages provide a suitable and convenient means for the inventory control of complex systems of items and stocks even when considering the stochastic dependence among the different elements of the system. They can be applied to research into an integrated system of production, maintenance, scheduling, marketing and inventory control. A book reviewing simulation methods applied to the above-indicated systems of business and economics has been published by Meier, Newel and Pazer (1969).

Even for relatively simple inventory control models, simulation techniques can be useful. As an example, one can mention investigations concerning the optimal parameters of the well-known (s, S) inventory policy; see, for example, Geisler (1964), Valisalo, Sivazlian and Maillot (1972). This has also been a trend in research during the last 10 years.

In many papers, an ever-increasing number of complex systems are analyzed, applying simulation methods. This is another area of research which has received emphasis in the last few years, examples of which are, amongst others, the papers by Bankaiev, Kostina and Jarovicki (1974).

III.9. Directions and Results of Recent Years

III.9.1. Methodological Directions

The development of inventory control methods—in connection with the general development of the mathematical economic and computer methods—has resulted in many important new results. Here, only a small fraction of these new results is summarized in order to illustrate the most important methodological directions of theoretical results, modelling and solution methods investigated in the last few years.

Methods of mathematical analysis play an important role in the study of multi-item models. A simplified strategy of joint ordering with periods of equal length was optimized by Silver (1976) by means of simple analytical methods. This procedure was simplified by Goyal (1980) and by Goyal and Belton (1980). The fast procedures enable, for practical applications, the determination of the optimal joint policy for many items together. In the case of a demand having a linear trend, optimal ordering was described by Donaldson (1977) using elementary means. This result was generalized by Henery (1979) for an arbitrary increasing demand. The unique solution is given in a simple form, when the time-dependence of the demand can be described by a logarithmically concave function.

Combinatorial methods may increase the speed of the algorithmic solution of known inventory models. A fast algorithm was published by Baker, Dixon, Magazine and Silver (1978) for the dynamic lot-size model under time-dependent capacity constraints. A similar problem was solved in the papers of Richter (1976), (1980) by means of the branch and bound technique of discrete programming, taking into account production and inventory capacity constraints.

Control theory serves a means of describing inventory problems in a very natural way. The methods of solution of inventory control problems based on modern control theory were first summarized in the book of Bensoussan, Hurst, and Naslund (1974). Optimal linear decision strategies for production-inventory systems were described by Pervoswansky (1975). A control theoretical model of a moving-band production with inventory was given by Pervoswansky and Holmach (1978). The solution is based on linear programming. The characteristics of a multi-item inventory model system have been determined by Popplevel and Bonney (1977) using control theoretical results. Schneeweiss (1977) investigated optimal stock control under different (especially quadratic) cost criteria. A multi-item model with stochastic delivery and demand was formulated by Pintér (1977) as a control problem with a stochastic objective function. The solution is determined by simulation. The optimal structure of a deterministic multi-item production-inventory system was analyzed by Rempala (1981) using the control theory methods.

The new results of *stochastic dynamic programming* have been applied to increase the speed of convergence of the value- and decision-iteration method of Bartmann (1975) and Van Nunen (1976). Optimization considering only linear decision strategies was investigated by Inderfurth (1977), and results in a considerable saving in the amount of computation. The optimality of the multi-item inventory strategies was investigated by Kalin (1976). New methods were derived for the determination of the optimal parameter of the (σ, S) policy by Miethe (1978) and for the (s, c, S) policy by Peterson and Silver (1979). A simplified version of the second method was developed by Silver and Massard (1981) for an algorithm executable also on pocket calculators. Theoretical questions concerning the multi-S type generalized multi-item inventory policy were investigated by Bylka (1980) by means of stochastic dynamic programming.

The development of *queuing theory* has also had an effect on research into inventory models. For multi-item models, the stochastic dependence of demands was considered and the state probabilities were derived by Girlich (1977) whilst the Erlang-type demand distribution was investigated by Miethe (1978): both results were based on an analogy with certain queuing models. By means of the theory of stochastic point processes, the validity of the classical economic order quantity model has been extended for a certain type of stochastic demands.

Reliability-type inventory models play an important role from the point of view of applications. Safety stock planning methods based on these models have been considerably developed. Kleijnen (1976) argued that the standard method used in inventory control program packages is not suitable for periodic review. It is based on a continuous-review model and, even with the corrections which are usually applied, is far from the optimal safety stock of periodic review. The exact

48

optimum of the safety stock has been derived by Burgin and Norman (1976) for a gamma distributed demand, and by Schneider (1978) for an arbitrary demand distribution. In another paper of Schneider (1981), simple methods are given which provide a good approximation of the optimum.

Investigations concerning random delivery processes, which cause difficulties in the practical planning of safety stocks, have continued. The joint probability of the continuous supply of n different items is prescribed. Under this reliability constraint, safety stock levels are determined which ensure a minimal of total capital invested. A simplified method for random delivery was derived by Kelle (1977) based on the Lagrange multiplier method. The maximization of the reliability of a continuous supply is completed in another paper of Kelle (1978) under a budget constraint. The case of stochastically interdependent demands is also considered by Kelle (1979). The random delivery process model of Prékopa (1973) is generalized for arbitrary distributed random lots of deliveries: in the paper of Kelle (1980) both demand and delivery processes may be random.

Mathematical statistical methods have been applied to demand forecasting. The adaptive forecast method was improved by Günther (1976) and Girlich (1977). The results of statistical decision theory were applied to derive the type of optimal decision strategy and to approximated this optimal strategy as was given by Waldmann (1976) and Dietsch (1977). Markovian decision theory was used by Küenle (1977) and semi-Markovian decisions were applied by Miethe (1978) to derive optimal decision policies.

Simulation methods are applied for the solution of inventory problems in two main ways. For the numerical solution of stochastic models, the Monte-Carlo technique can be used to calculate multi-dimensional probabilities, see, e.g., Prékopa and Kelle (1967): in most cases, however, the complete inventory system is investigated by simulation. In the theory of stochastic automata, the stochastic search procedures are new theoretical results for the description and optimization of complex production-inventory systems. The papers of Matthes and Müller (1977), Martin (1980) and Müller (1981) may be mentioned as examples. Considerable effort has been made for the control of time-dependent demands e.g., by Ritchie (1981) using simulations technique.

III.9.2. Applications

Applications have developed in two main directions: models developed for specific practical problems, and new general program packages for production and inventory control. Relatively few results have been given for real-life applications. On the basis of these papers, we will present a summary of the most important fields of application.

The inventory control of *deteriorating items* has been investigated since the beginning of the seventies. Among the latest results published, the papers of Nahmias (1975), Fries (1975) and Nahmias (1981) should be mentioned. Their practical importance is that simple, fast procedures are given which provide a good approximation of the optimal policy under more general assumptions than earlier, similar, procedures. As practical uses of these new methods, examples have been described for food, photo material, medicine, etc., stocking. A special field of application is the so-called "blood bank" system detailed, for example, by Jennings (1973). An inventory model applied for hospital pharmacy has been given by Pegels (1981).

The characteristic of *spare-parts stocking* is the time-dependent demand. Increasing, stabilized, and decreasing, demand occurs depending on the age of the machines. An inventory policy was constructed by Ritchie (1981) which considers this fact. A simple fast computer method was developed and applied which provides an approximately optimal policy. A complex country-wide spare-parts supply system control was developed by Sarjusz and Wolski (1980) which operates on the basis of short-range forecasting, normative and economic regulators. A multi-echelon spare-parts inventory system was modelled by Hollier and Vrat (1976) in which stocking and repairing of the parts are to be controlled.

Production-smoothing models determine the amount of stocking for strongly fluctuating demand in such a way that the summed cost of inventory holding and the cost due to the fluctuation of the production level (overtimes, excess of manpower, unutilized capacity, etc.) should be minimal. A description of these models was given by Eilon (1977), an application in building industry was described by Papathanassiou (1981), and an experimental application in six branches of industry has been surveyed by Ghali (1981).

The investigation of *production and inventory systems* has been developed in many directions. The two main fields of research which can be clearly distinguished are the deterministic field of materials requirement planning (MRP) and the stochastic field of multi-stage models. In the production and inventory planning of complex, multi-level, hierarchical production systems, the new idea of MRP has been very successful. This technique was introduced by Orlicky (1975). Reviews about the results of its applications in production control were given by Baker (1977), and Blackburn and Millen (1981), among others. The internal storage model of a production line given by Kelle (1978) considers the stochastic factors and correlations of the demands and production equipments. A reliability-type inventory model applied in commerce has been described by Móritz (1978). A review was given by Kelle (1980) on the application of reliability-type models in industrial, commercial and supply enterprises. The control-theoretical investigations concerning a planning system for internal storage planning in a production line was described in the papers of Tapiero (1977), Stohr (1979), O'Grady and Bonney (1981).

IV. Processing of the Models— Principles and Methods

In the course of the research work on which this book is based, we have sought to achieve results applicable both in theory and in practice, by processing the very ample professional literature as well as by surveying the largest and most characteristic section of existing inventory models. Accordingly, it was necessary to choose a processing method which facilitated this dual goal: theory and practice alike.

In appraising the models, we have started from the basic principles explained in the preceding chapters. Thus, in categorizing the inventory models as decision models, we have taken into consideration the four spheres of these models described in Section I.2.3, whereas, the "properties" of the models have been investigated on the basis of the analysis of the item-level inventory system.

In this book we have considered 336 models. It is our hope that this "sample" appropriately represents the models available in the literature dealing with inventories, which, according to our estimations, comprise in total at least three times this number. The English, German, Russian and Hungarian literature has been examined, including the many notable books and most of the international journals in this field. Forty percent of the models originate from 11 books, 50% from 12 journals and the remaining 10% from other publications. As for the chronological origin of the models: 7 models were published before 1960, 66 between 1961 and 1965, 113 between 1966 and 1970, 99 between 1971 and 1975, and 51 after 1976. The bibliography of the models considered are given in Appendix I.

We adopted two approaches for processing and analyzing the models. On the one hand, we prepared a 1—3 page concise description of each model—the slightly modified (mostly abbreviated) versions of these descriptions constitute the second part of this book. On the other hand, we prepared a coded version of each model based on a previously elaborated uniform code system consisting of 45 elements (this is practically the coded model of the original model), where the single code values represented the main features of the model. In the following, we review these two processing approaches. It is necessary to detail the code system and its application, even if this may seem to be a bit dry for some readers, since this formed the basis for analysis and evaluations as well as for the classification of the models, which we consider to be one of the main results of the book. Moreover, this may form the basis of practical applications.

IV.1. The Model Descriptions

The description of the models considered has been prepared partly as a reminder and as a reference. It was possible to record a great deal of information in this way, which could not be captured by the coded version due to the character of the models. Our other, similarly important goal has been to represent models in a form which is easy to understand for managers, who might want to implement a given model and wants to know under what circumstances a certain model can be used. After reading the brief description, the user should be able to decide whether the model fits precisely enough to the actual inventory system he wishes to model.

It is not necessary that the original model be completely reconstructed from the description, since it can be traced back to the bibliography. We do not touch upon the derivation of the mathematical models, neither will verification of the propositions be found in these descriptions. Nevertheless, we must refer to the route of solution and to the existence of the exact optimum or approximate solution, for the way in which the decision parameters are determined is a matter of importance.

Each description contains—in a consistent way for each model—the denomination, the assumptions system, the objective function, the solution algorithm and bibliographic reference (summarized in alphabetical order in this book).

The *denomination* of the model stresses the most important traits of the model, only. The description of the model's *assumptions system* makes up the major part of the summary. The features of the model are detailed according to the characteristics of item(s), store(s), replenishment (input), demand (output) and constraints.

Knowing the assumptions of the mathematical model makes possible to decide whether to accept or reject a given model for a particular practical application.

According to the approach for analyzing the item-level inventory system given in Chapter II, we describe the components of the system, the connections among the components, the environment and the resources of the system by means of the assumption system.

The presentation of the *objective function* is also essential in a user-oriented description since the objective function contains those variables which decisively determine the performance of the inventory system and can thus be directly influenced by the user (decision maker).

Should the objective function be either to determine the minimum cost or the maximum profit, or to meet a reliability requirement, in all cases the desired behaviour of the inventory system is expressed in it, irrespective of the mathematical approach involved.

On the basis of our system approach, the objective function measures the efficiency of the system.

For the expert, the solution of a given inventory problem means determining the best values of the decision variables (which he can influence), under the existing circumstances, for achieving the set objectives as closely as possible. In other words, the solution algorithm is focused on the determination of the optimal values of the decision parameters which satisfy the given constraints. In numerous cases, either there is no unambiguous solution or the optimum values of the decision variables cannot be expressed by a closed, analytically tractable mathematical formula. Nevertheless, this information is useful in practice also, since it often turns out that from among the set of the possible values of the decision variables some are definitely better than the others, or they reasonably approximate the optimal solution. Determining the values of the decision variables (by the solution algorithm) the behaviour of the system may be affected so that it will tend towards the desired condition.

The bibliographic *Reference* gives the original source of the model. In this way, full particulars of the deductions, verifications, etc., not provided in our description can be easily checked. The model descriptions are sufficient to enable an idea of the models and to select the likely suitable models for application. A practical application requires, in any case, a comprehensive knowledge of the original model.

The second part of this book has been prepared on the basis of the original model descriptions. (The description of some important models have been given more completely, whereas others have been significantly abbreviated.) The grouping, according to which the model descriptions are given, as well as our basic knowledge concerning the models, is supported by the computer outputs of the coded model versions which are based on the code system described in the following section.

IV.2. The Code System

IV.2.1. Aspects for Developing the Code System

In devising the code system we had to keep in mind the fact that the unified code system was to be suitable to identify—"describe"—several hundred models. This meant that we had to proceed very carefully in elaborating the system so that the coding system was flexible enough for evaluating models with very different approaches and containing very different assumptions, on the basis of consistent principles. For this reason, we studied a great number of inventory models systematically prior to compiling our code system in order to obtain an impression of the possibilities, circumstances and traits. We also carried out numerous trials.

As the developing of the code system we had to take into account which different aspects could be discussed in one model. As an example to illustrate this: an important question concerning every inventory management model is how many items can it handle. It is quite obvious that two sharply distinguishable cases are: one, or several, items making up the inventory. In compiling our code system we considered this question to be the most important, but we also insured that the specific, or the parametrically set, number of items should also be included in the code system. In other cases, however, the logical separation of "possible cases" was not so evident at all.

Thus, the possible versions of the model's characterizing criteria (assumptions, conditions), constituted one group of factors which required attention. The other group were the characterizing criteria themselves. As for a given criterion, certain versions of the criterion were also to be taken into account. In the same way, inserting some criteria in the code system was also obviously necessary. One soon realizes that, for example, the deterministic or random character of the demand and replenishment process, the number of items and stores, the objective of the model, the cost factors, etc., are all indispensable criteria characterizing a model, and, as such, are of primary importance (with a few exceptions only).

Nevertheless, the situation is different in many cases. There are a lot of criteria which exist or become important only with regard to a small part of the models. As far as developing the code system was concerned, the objective was to find a proper proportion: on the one hand, for the majority of the models, criteria of minor importance about which, therefore, only rather loose information, or just nothing, could be put into the code system, were neglected.

We also had to take care not to miss important criteria, maybe not in every model, but in a considerable fraction of them. In such cases, the coding system would not be able to "accept" a non-negligible part of the models.

IV.2.2. The Structure of the Code System

It became more and more apparent in the course of our research that the importance of the information to be transferred from the original model into the coded version is rather varying. Based on theoretical considerations and practical experiences we have distinguished ten criteria which describe the most essential features of any inventory management model. These are the following:

— number of the items stored in the system,
— number of stores in the system,
— character of the replenishment (the inflow or input) process,
— character of the demand (the outflow or output) process,
— mode of treating the temporal changes (dynamics) in the system,
— objectives of the system,
— operational mechanism of the system (depending mainly on the ordering rule)
— mode of inventory reviewing,
— mode of treating the shortages of items on stock,
— handling of leadtime.

The codes representing these criteria are brought together in the so-called "main codes" and have been treated separately from all the other codes both as to their form and contents. It has been our intention, because of the primary importance of the criteria involved, to accord the role of primary model identification, model characterization and model grouping to the main codes. At the same time, the main codes and the other codes are closely related in that the criteria succinctly expressed by the main codes are described in more detail in the other codes.

IV.3. Detailed Description of the Code System

IV.3.1. Formal Characteristics of the Code Card

The coded version of the inventory management model is the completed code card. All the information required to ensure that the coded version conforms unambiguously to the original model has to be present on the code card. A typical code card is shown in Fig. 6.

In the upper left-hand corner of the code card appears an identity number (serial number). Each serial number identified one model. To the right of the identity number are 10 squares for the main codes.

The major part of the code card is reserved for the ten horizontal code rows. The initial letters in each row indicate the contents which provides—as mentioned before—the particulars of information delivered by the ten main codes (MC1—MC10).

The meaning of the letters given in the left-hand column is:

I   = item
S   = store
R   = replenishment (input)
D   = demand (output)
DY = dynamics
OF = objective function
DV = decision variables
SM = system and mathematics
SO = shortages and orders
C   = costs

At the bottom of the code card there is room for bibliographical references. The code card has three "side-fields" for

— remarks,
— mathematical predictions, and
— the optimal solution.

These fields are separated not only formally but their contents differ from most of the positions in the code card, as well, since—as with the information in the rows "bibliography" and "objective function" (OF)—, the information here cannot be transferred to a computer, or, at least, only after considerable processing. We will describe the side-fields later.

IV.3.2. General Code Values

Prior to giving a comprehensive description of the individual codes, we have to mention three code values of general meaning.

It has been necessary in developing the code system to distinguish three cases (though, they sometimes merge into one another in the practice). One such case is when a model gives a "response" (criterion value) to a certain "inquiry" (criterion) which cannot be found among the set possible responses. This is the so-

Order-level system with backlogging

	1	2	3	4	5	6	7	8	9	10
	1	1	0	0	0	1	1	-1	1	0
	1	2	3	4	5	6	7	4	1	5

Code							
I	1 1	2 -1	3 0	4	5 7	0	
S	1 1	2 -1	3 0	-1		0	
R	1 0	2 3	3	4 1	5 -1	6	7
			-1				
D	1 0	2 0	3	4	5		
			-1	-1			
DY	1 1	2 0					
OF	1 1						
DV	1 s	2+ 0	3+ 1	4 0	5 1		
SM	1 -1	2 0	3 0	4 0	5 1		
SO	1 1	2 1	3 1				
C	1 2	2 0	3 0	4 0			

DV3 t_p is determined by q_p

DV2 in the discrete version of the model S is integer, so DV2=1

$$C(S) = c_1 \frac{S^2}{2q_p} + c_2 \frac{(q_p - S)^2}{2q_p} \qquad (0 \leqq S \leqq q_p)$$

Mathematical assumptions: 0

Optimal solution:
Continuous case:
$$S_0 = q_p \frac{c_2}{c_1 + c_2}$$
Discrete case:
$$S_0 - \frac{u}{2} \leqq q_p \frac{c_2}{c_1 + c_2} \leqq S_0 + \frac{u}{2}$$

Reference: Naddor, E.: *Inventory systems.* J. Wiley, N. Y. 1966. p. 63—68.

Fig. 6. Sample code card

called "other" case. To denote this we have always used a code values 9, reserved for this purpose. In each case of this sort, a comment has been given on the essentials of the "other" case.

The situation is somewhat different with the "not treated" case to which we have allocated code value 7. In these cases, the author of the model does not treat a given aspect in his model, neither according to our method, nor in another way. Nevertheless, in numerous cases, even if the author does mention certain features, these can be concluded from other characteristics of the model. In these cases, we used common sense to answer the question in the code square.

The third case is when a certain criterion does not make sense in a given model. For example, if the model has one store only, then the question: "what type of connection exists between the stores?" has no sense. Code value −1 has been given to these cases.

After these preliminary comments we may now consider the detailed description of the code system.

IV.3.3. The Main Codes

Number of items stored in the system (MC1)

From the point of view of the structure of the model one of the basic questions is whether one or several items are treated. Only this distinction has been made in the main code, and we did not ask the following question here: how much is the "several"? (There is more specified information concerning this in the code row of items (I).)

The codes:
  1 : one item stored
  8 : several items stored

Number of stores in the system (MC2)

The idea is very similar to that of the previous one.

The codes:
  1 : one store in the system
  8 : several stores in the system

Character of the inflow (input) process (MC3)

By the character of the inflow (replenishment) process we identify whether there are probabilistic (random) elements in it or whether every feature of the inflow is known (deterministic).

(The features or elements of the inflow process are: time of inflow, intensity of inflow, its discrete or continuous character, duration, quantity of the item coming in, etc.) If all these criteria of the input process are known with certainty, we speak of a deterministic inflow; if any uncertainty exists concerning any of the essential features (e.g., if we know only the respective probability distribution), it is considered to be a stochastic inflow.

The codes:
  0 : input is deterministic
  1 : input is stochastic

Character of the outflow (output) process (MC4)

With appropriate modification according to the sense, all the statements of the previous paragraph are valid for the outflow (output) process, too. Since we suppose—as is implicitly accepted in inventory theory—that the decision maker's specific intention is to satisfy all demands if there is sufficient inventory on hand, the deterministic or stochastic character of demand determines the character of the outflow—which makes it possible to consider the outflow instead of demand in that respect. This is favourable for the symmetrical treatment of input and output.

The codes:

    0 : output is deterministic
    1 : output is stochastic

Way of treating time in the system (MC5)

This criterion characterizes whether a model is static or dynamic. In studying the models and evaluating them by using the code system, it was, in many cases, this character of the models which was the most difficult to judge. The main reason for this was that in the literature there are no generally accepted relevant definitions and, thus, the authors of the various models have described their models static or dynamic according to different criteria.

We have considered as being dynamic those models for which the objective function (valid for the whole planning period) depends, in its general form, at any given time, on the values of the same objective function taken at some other time. Thus the actual form of the objective function may be determined on the basis of the interrelations between the various substitute values of the objective function. This criterion has proved to be acceptable in practice, although, several marginal cases have occurred.

The codes:

    0 : the model is static
    1 : the model is dynamic

The objective of the system (MC6)

The general goal of operating an inventory management system is always the satisfaction of demand for a stored item. However, the operation of the system can be developed in different ways, and it depends on the concrete objective to be achieved by operating the system as to which way will be chosen from among the existing possibilities. Most of the models have the objective that the inventory management system is to be operated in the most economical way possible under the existing conditions, i.e., the operation cost shall be minimal, or the positive income balance (incomes minus expenditures) deriving from the operation of the system (covering the demands for the stored item) shall be maximal. These types of models are called optimization models.

Another group of models does not want to, or is unable to, take into account the operation costs. Usually, in these models the objectives are directly related to satisfying demands. The typical goal of these models is that the system shall be able to satisfy the demands for a stored item at any time at a given probability (reliability) level. These models are the so-called reliability models.

58

Some models can be found, for which no other objective for specifying the mode of operation exists beyond the general goal of satisfying demands. These are called descriptive models. The descriptive models often represent inventory systems by instruments of control theory, describing the reaction, and its mechanism, of the system to the impulses of the "external world".

These three types of models have been distinguished in our code system.

The codes:

1 : optimization model
2 : reliability model
3 : descriptive model

Operation mechanism of the system (ordering rule) (MC7)

The operation mechanism of the system is decisively influenced by the decisions concerning the inflow of the inventory items into the store; that is, by the ordering rules regulating the inflow (input) process. These are the parameters that the decision maker (the operator of the system) can utilize—completing the environmental effects (conditions) which exist independently from his intentions—to influence the operation of the system.

As indicated in the previous chapters, the four basic types of operation mechanism of inventory systems are—with the usual symbols: (t, q), (t, S), (s, q) and (s, S).

In numerous models, versions of these basic types are used where any of the parameters (decision variables) have been previously fixed, i.e., they are not the subject of optimization.

The prior fixing of parameters takes place either in order to achieve a desired objective or is justified by some external constraints. The fixed parameter is indicated by the index p. Our code system distinguishes between the above basic mechanisms (ordering rules) and the versions of these which come into existence by fixing certain parameters. The mechanisms occurring very rarely—e.g., the mechanisms (s, q_p), (t, q), (t, q_p) and (s, S_p)—have been attached to the appropriate "basic mechanisms" so as not to diverge from the usual quantitative barriers of code values.

(The original mechanisms arise from the model descriptions in these cases, too.)

The codes:

0 : (t, S) mechanism
1 : (t_p, S) mechanism
2 : (s, q) or (s, q_p) mechanisms
3 : (s_p, q) mechanism
4 : (t, q), (t_p, q) or (t, q_p) mechanisms
5 : (s, S) or (s, S_p) mechanisms
6 : (s_p, S) mechanism
8 : (t, S_p) mechanism

Mode of reviewing inventory (MC8)

In order to follow the operation of the system, and to ensure a proper basis for ordering decisions with information, the recording, checking, "reviewing", of the inventories is of course indispensable in most of the cases. Inventory models distinguish two basic modes of inventory reviewing: the continuous and the periodic one.

In the case of certain special models there is no inventory reviewing. For those models describing processes of a deterministic character only, we qualify inventory reviewing as "meaningless". If we suppose we know with full certainty the states of the system in advance, then reviewing the inventory has no point.

The codes:

0 : no inventory reviewing
1 : periodic review
2 : continuous review

Mode of treating shortages (MC9)

In general, the models treat the shortage of a stored item in three essential ways. One possibility is that shortage is not allowed. Several models assume that though shortage may occur, the demand for the item in short supply will be covered later when the item is available. This situation is called "backlogged demand".

In numerous cases, however, there is no point in considering the size of shortages, since a demand for an out-of-stock item is "turned away" from the system, and is "lost" to the system. In this case, items of only such quantity are to flow in—if other conditions remain unchanged—as the quantity needing to be procured without a shortage, when the inventory level falls to zero.

The codes:

0 : shortage not allowed
1 : shortage allowed, demand backlogged
2 : shortage allowed, demand for the missing quantity is lost (lost sales)

Character of leadtime (MC10)

In practice, there is always a certain interval of time between the placing of an order and receiving the ordered quantity—this is called replenishment leadtime.

Models disregarding this leadtime may do so for two reasons. The first is that, if this time is very short compared with the other time intervals considered (e.g., to the time interval between two subsequent orders), then its effect is negligible. In the second case, it can be shown that for a considerable number of cases the replenishment leadtime, known in advance, does not change the structure of the model or the decision parameters.

The definition of the leadtime is not unambiguous. Another concept is also realistic—and many authors insist on this—namely, leadtime is defined as the interval between placing an order and initiating delivery. Despite this, we interpret the leadtime to equal the time between the placing of an order and the total delivery of the ordered quantity. These two intervals do not always coincide because supply often takes place either by several partial deliveries, or continuously. (A good example of continuous delivery is supply through a pipeline.) This means that, according to our code system, the leadtime is zero only if the total ordered quantity arrives immediately in one lot. Supplies occurring in more than one lot, or continuously, have been separated and denoted a separate code; in this way, all the cases in the following refer to one lot supply.

Besides a zero leadtime, the most simple case is the constant leadtime. In this case, a known period of time passes between the placing of an order and receiving the ordered lot, and this will not change in the course of operating the system.

For some models, the length of the leadtime is known in advance, but will change in the course of the operation of the system. This case is called deterministic variable leadtime.

Much more important is when the leadtime is a random variable. In this case, we only know some probabilities concerning the arrival of the ordered quantity. The codes:

0 : no leadtime
1 : constant leadtime
2 : deterministic variable leadtime
3 : probability variable leadtime
4 : fulfilment in more than one lot, or continuously

This completes the description of the main codes. In the following, we review the other codes. These extend the information presented in the main codes, and are listed according to code card rows.

IV.3.4. The Items Code Row

Number of items (I1)

Whereas, in the main codes, the question which is addressed is whether one or several items are stored in the system, the information here concerns the number of sorts of items. Some multi-item models define concretely (numerically) how many items are stored (a good example is the two-item inventory model), however, in most cases, the number of items is indicated in general terms (e.g., the number of items stored is N).
The code:

an autocode (i.e., the specific number of items or the number of items given in general terms)

Connection between the items (I2)

Of course, this aspect is meaningful only in the case of multi-item models. We have identified the following cases in our code system. The first is that though there may be several items in the model they bear no essential relation to each other. The next possibility is that the items have a substitute character, i.e., one sort of item may substitute partly or entirely another one. If the items complement each other, this means that one item is suitable (able to cover demands) together with another item only (complementary items). A further form of connection between items is when a given model applies some joint condition, restriction, or constraint relating to several items (e.g., the quantity of a given item, which may be ordered depends on the ordered volume of other items, since the quantity to be procured simultaneously from all items is limited).
The codes:

0 : no connection between the items
1 : substitute items
2 : complementary items
3 : joint constraint for various items

Type of inventory (I3)

By "type of inventory" we refer to the role of the inventory in the production process. The following types of inventories have been distinguished: raw materials, work-in-progress (or semifinished goods), finished goods, spare parts.

As a matter of course, an appropriate code is used when the model does not specify the type of inventory.

The codes:
- 0 : type of inventory is not specified
- 1 : stock of raw materials
- 2 : work-in-progress inventory of semifinished goods
- 3 : finished products inventory
- 4 : stock of spare parts

Changes in the value of the item stocked (I4)

By depreciation, we mean, here, decrease in the use value, i.e., the fact that an item will be less suitable to satisfy demands, or maybe entirely unsuitable, after a certain period of stocking.

In inventory theory, models treating items which can be used to satisfy demands only within a given limitation are usually called models with perishable items. An increase in value (appreciation) may occur, if, for example, the item can be sold at a higher price after a certain time due to market conditions, or if the physical properties of the items, as well as its ability to meet demands, improve in the course of storing.

The codes:
- 0 : no change of value
- 1 : depreciation
- 2 : appreciation

Measurability of items (I5)

By measurability of an item we mean the ways it can be measured or counted, depending on its physical character. A discrete measurability exists if the units of the item can be counted. The measuring of the items is called continuous if the quantity of the item expressed by a given unit can take any real number value, within a given interval.

It must be stressed that measurability is explained here exclusively according to the physical properties of the item in the model, and not on the basis of the set of possible values of the model's decision variables (see DV2 below).

The codes:
- 1 : the item is continuously measurable
- 2 : the item can be measured in discrete units

IV.3.5. The Code Row of Stores

Number of stores in the models (S1)

Similarly as for items, the actual number of stores in the models, which has been qualified in the main code as one or several stores, is considered by the detailed codes only. The number of stores may be indicated by a specific value, or in general form (e.g. "*N*").

The code:

   autocode (i.e., the specific or general number of stores)

Connection between the stores (S2)

This aspect is appropriate only in the case of multi-store models. Three possible cases are distinguished in our code system. In the first case, the stores have no connection *in merito* with each other. In the second case, the stores are connected linearly. This means that there is no hierarchical relation among the stores in the sense of having a "central" store. The characteristic flow of items is when the items proceed from one store to another. The third case is when there are several parallel stores linked to a central store. The central store is the warehouse that allocates the inventory items ordered and stored by itself; according to some principles, among the parallel (e.g., regional) stores in compliance with their orders. Each of these regional stores may have an independent inventory management.

The codes:

   0 : no connection between the stores
   1 : linear connection between the stores
   2 : parallel stores connected to a central store (warehouse)

Storing capacity (S3)

In practice, the space at disposal for storing items is always limited. Nevertheless, this constraint is often hardly of impotance because—due to other conditions and circumstances—the quantity of items actually stored is far less than the capacity of the storage facilities. This is reflected by the fact that there is no storing capacity constraint in a considerable fraction of the inventory models.

On the other hand, with numerous models—just as in many practical cases—the size of the store may become a "bottleneck". In such cases, store capacity limitation is an integral part of the model.

For the sake of completeness, in our code system we have taken into account also the case—though it has less practical importance—when a lower (minimum) limit is set for the items to be stored (this may be an aspect for the economical utilization of the storing capacity).

The codes:

   0 : unlimited storing capacity
   1 : minimum limit
   2 : maximum limit

IV.3.6. The Row of Input Codes

The input and output processes are the actual realizations of replenishment and satisfying demand respectively—two processes which have a basic influence on the operation of an inventory system.

In addition to their primary importance, input and output processes may be the most difficult to consider from the point of view of the code system. There

63

exist many types of processes with significant differences between them. Input and output have a great number of characteristic parameters; from among these only the most important ones can be coded.

The input and output code rows have quite the same structure, since—as mentioned in the discussion of the item-level system—the separation of these "two" processes is a matter of viewpoint only: a process denoted "output" from the point of view of one of the item-level systems is an "input" from the aspect of another system; this situation always exists in practice. This is why only the row of inputs will be discussed in full. The codes refer to the date(s) of inflow and outflow, to their temporal distribution, to the quantity following in or out simultaneously, as well as to the temporal distribution of the quantity of stocks flowing in and out.

Character of inflow (input) replenishment process (R1)

The code is identical with the third main code.

The codes:
  0 : input is deterministic
  1 : input is stochastic

Reception (arrival) dates of the ordered quantities (delivery dates) (R2)

Concerning the reception date(s) of the amounts of items inflowing due to a given order, we have to distinguish first of all whether delivery takes place in one or several (at most: countably infinite) time moments or whether the supply is continuous. In the case of continuous supply, our code system distinguishes between supplies with uniform and non-uniform rates.

Should delivery take place in one single lot (i.e. the ordered quantity arrives in one delivery), our code system distinguishes between the two most frequent cases: the complete ordered quantity arrives at the beginning of the reorder period, or it arrives at the end of the reorder period. The case—of no great practical importance—when the whole quantity arrives as a single consignment on a known date between the beginning and the end of the reorder period—i.e., after a deterministic leadtime has passed—has been ascribed as a delivery at the beginning of the reorder period.

If the shipment of the lot ordered by a single order takes place on several (countable) dates definitely known in advance, this case has been coded as "delivery (reception) in definite time intervals". Two codes have been reserved for such cases, if reception takes place on one or several dates but we do not know these date(s) exactly in advance.

The codes:
  0 : delivery is continuous and uniform
  1 : delivery is continuous, but not uniform
  2 : delivery takes place in definite intervals (on several dates known in advance)
  3 : the whole lot arrives at the beginning of the reorder period or delivery takes place in one lot after a deterministic leadtime
  4 : the whole lot arrives at the end of the reorder period
  5 : delivery occurs in one lot at a random time moment
  6 : delivery occurs in several lots at random time moments

64

Probability distribution of delivery dates (R31, R32)

As a matter of course, this code makes sense only if the dates of inflow are uncertain (i.e., if R2=5 or 6). Our code system forks with the third input code into two sub-codes. The first of these sub-codes makes clear whether the probability distribution of the arrival date(s) are specified by the model or not. The second sub-code shows—assuming a specification—to which, from among the practically important types of distributions, the model applies. Thus, this R32 code can be explained only (i.e., its value is not -1), if R31=1.

The codes:

Specification of the probability distribution of the shipment dates (R31)
1 : probability distribution is fixed
2 : probability distribution is optional

Type of the specified probability distribution (R32)
1 : uniform distribution
2 : normal distribution
3 : exponential distribution
4 : gamma distribution
5 : Poisson distribution
6 : binomial distribution

Quantity of items arriving at one time (R4)

In the case of continuous delivery, the definition of quantities arriving at one time has no meaning, because the expression "at one time" presumes one or more discrete date(s) of delivery. One of the most important practical cases is if shipment takes place in one single lot.

The greater fraction of the models in which delivery takes place at several random times postulate that on these random dates one unit of the item is received.

Some of the models do not directly determine the quantity arriving at one time, but they establish the volume of the items received into the store up to a certain date (series of dates). Our code system labels this case: "treating of cumulated quantity received". Finally, we have distinguished the case when the quantity arriving at one time is neither a unit of the item (at a random time) nor the entire lot.

The codes:
1: total ordered quantity arrives in one lot
2: the model treats the cumulated quantity received
3: a unit of the item arrives at random moments
4: a discrete given quantity but not the complete lot arrives at one time

Probability distribution of the quantities delivered during the reorder period (R51, R52)

In our code system, we consider the total quantity of items being delivered to fill one order with optional scheduling as a basis. (This, in general, means delivery within one reorder period.) This code is reasonable if this quantity sum has a random character. (If the ordered lot arrives on several dates in different random quantities, but its total sum equals the ordered quantity within a reorder period, then we consider this case—in a slightly simplified way—as being deterministic.)

65

Coding is identical with what has been demonstrated with the probability distribution of the reception dates.

The codes:
Specifications of probability distributions of the quantities received during reorder period (R51)
  1: probability distribution is fixed
  2: probability distribution is optional
Type of probability distribution specified (R52)
  1: uniform distribution
  2: normal distribution
  3: exponential distribution
  4: gamma distribution
  5: Poisson distribution
  6: binomial distribution

IV.3.7. The Row of Demand (Output) Codes

Most of the considerations described in connection with the row of input codes are valid for the row of output codes with some obvious modifications (e.g., instead of inflow, outflow is to be considered, and instead of delivery, we consider demand, etc.). Therefore it is not necessary to repeat the aforesaid, however the codes themselves are given for the sake of completeness.

Character of outflow (output) process (D1)

The codes:
  0: output is deterministic
  1: output is stochastic

Dates of Demand Occurrence (D2)

The codes:
  0: demand is continuous and uniform
  1: demand is continuous but not uniform
  2: demand arises in deterministic intervals (several times during the reorder period with known dates)
  3: total demand occurs at the beginning of the reorder period
  4: total demand occurs at the end of the reorder period
  5: demand emerges within the reorder period in one lot in a random moment
  6: demand emerges within the reorder period in more batches in several random moments

Probability distribution of the dates when demands arise (D31, D32)

The codes:
  Specification of the probability distribution of dates when demands emerge (D31)
  1: probability distribution is fixed
  2: probability distribution is optional

66

Type of specified probability distribution (D32)

1: uniform distribution
2: normal distribution
3: exponential distribution
4: gamma distribution
5: Poisson distribution
6: binomial distribution

Volume of demand arising at one time (D4)

The codes:

1: demand emerges in one lot within the reorder period
2: the model treats the cumulated quantity of demands
3: unit demand arises at random times
4: a given, discrete volume of demand, but not the total demand of the reorder period

Probability distribution of the volume of total demand arising during the reorder period (D51, D52)

The codes:

Specification of probability distribution of the volume of demands emerging during reorder period (D51)

1: probability distribution is fixed
2: probability distribution is optional

Type of the specified probability distribution (D52)

1: uniform distribution
2: normal distribution
3: exponential distribution
4: gamma distribution
5: Poisson distribution
6: binomial distribution

IV.3.8. The Code Row of Dynamics

Periodicity of decisions made in the model (DY1)

This code reveals essentially whether the modelled system operates during only one period or several periods related to each other. There are one-period models when only one decision is made for ordering at the beginning of the period, and at the end of the period the functioning of the model finishes.

Those models, in which the subsequent reorder periods functionally depend on each other, are called multi-period models.

In the multi-period models, in general, it is not a single decision which is made, but a sequence of decisions. Normally, there exists an (optimal) ordering rule valid for all periods but the concrete decision concerning the actual ordering—in accordance with this rule—generally has to be made for each reorder period separately.

The codes:
- 0 : single decision
- 1 : sequential decisions

Mode of treating temporal changes in the system (DY2)

The code is identical with main code 5.

The codes:
- 0 : the model is static
- 1 : the model is dynamic

IV.3.9. The Code Row of the Objective Function

The objective of the system (OF1)

The code is identical with main code 6.

The codes:
- 1 : optimization model
- 2 : reliability model
- 3 : descriptive model

After the code character the objective function itself (or the state equation) has been given on the code card.

IV.3.10. The Code Row of Decision Variables

List of decision variables (DV1)

The decision variables are the parameters, the values of which will be determined by the decision maker in order to control the operation of the system. In fact, the decision variables usually represent the parameters (t, s, q, S) by which the various ordering rules can be defined—see main code 7.

Nevertheless, the actual decision variables of a model (of a certain policy)—due to various reasons—do not always correspond exactly to the theoretical decision variables identifying this ordering rule. Therefore, it is necessary to specify the decision variables actually occurring in the model in addition to the theoretical decision variables described by the main code. This code serves this very purpose. (Where necessary, we have taken into account the theoretical and the actual decision variables in the remarks.)

The code:
- autocode (list of symbols of the decision variables actually present in the models)

Character of the set of values of decision variables (DV2)

By the character of the set of values of the decision variables we mean whether the variables are—in a mathematical sense—discrete or continuous. We emphasize again the difference between this code and item code 5. In the latter case, the description discrete or continuous character refers to the physical properties of the item, while here it refers to the mathematical properties of the decision variables. (The two aspects, of course, are frequently not independent of each other.)

We have identified three possible cases in our code system. According to this, the codes are the following:

0 : all decision variables are continuous
1 : all decision variables are discrete
2 : both continuous and discrete decision variables are present

Decision variables with a prefixed value (DV3)

As indicated in the discussion of the main code 7, some decision variables may exist, with certain types of policies, which are fixed at a definite level in advance (prior to optimization).

The code here indicates if any fixed (prescribed) decision variable exists, taking into account the type of policy of the given model.

The codes:

0 : no prescribed decision variable
1 : there is (are) prescribed decision variable(s)

A remark has been made following the code character stating which is (are) the prescribed decision variable(s).

IV.3.11. Systems and Mathematics Code Row

In this code row rather different issues have been handled; namely, some general characteristics of the inventory system modelled and some computational and mathematical aspects.

Reviewing inventory records (SM1)

The code is identical with main code 8.

The codes:

0 : no inventory review
1 : periodic review
2 : continuous review

Relation of the inventory system to systems of higher level (SM2)

In reality, any inventory system is always a sub-system of some hierarchically higher management system. This is the reason for introducing this code.

The codes:

0 : the system is not explicitly connected to any higher level system
1 : the system is connected to some company management system
2 : the system is connected to some sectorial management system
3 : the system is connected to the national economic management system

We used a character other than 0 only if the connection was handled explicitly in the model.

Special constraints of the model (SM3)

In some of the models there are special preconditions: constraints of an economic character restricting the potential "scope of operation" of the model.

The special constraints of an economic character have been separated from the so-called mathematical preconditions. These will be treated later on in Section IV.3.14.

The codes:
   0 : no special economic constraint
   1 : volume of demand is constrained (maximized)
   2 : replenishment (continuous shipment) rate is limited
   3 : the possible smallest ordering lot-size is prescribed
   4 : the possible largest ordering lot-size is prescribed
   5 : the average inventory level is limited
   6 : there is a budget constraint on the maximum inventory
   7 : the permissible maximal volume of shortage is prescribed
   8 : the maximal length of shortage is constrained
More than one code can be used if necessary.

Computerized version of the model (SM4)

This code shows whether, in the bibliographical source of the model computer programs, there appear detailed flow charts, solution algorithm or other calculation techniques. If we merely refer to such techniques, no numerical realisation was given, i.e., we have taken into account the numerical realizations only if they can be utilized directly on the base of the source.

The codes:
   0 : no calculation technique is given
   1 : computational technique is specifical

The mathematical apparatus predominantly used in the model (SM5)

In the greatest number of the models studied, various methods related to different branches of mathematics are simultaneously present; however, one predominant technique can almost always be selected from among these methods. The characteristic mathematical apparatus is the most complex and sophisticated one comprising simpler methods applied in the model.

The codes:
   1 : calculus
   2 : probability theory
   3 : game theory
   4 : heuristic methods
   5 : linear programming
   6 : dynamic programming
   7 : stochastic programming
   8 : computer simulation

IV.3.12. Code Row of Shortages and Orders

Mode of treating shortages (SO1)

This code is identical with the main code 9.

The codes:
   0 : shortage is not allowed

1 : shortage is allowed, demand backlogged

2 : shortage is allowed, demand for the missing quantity is lost (lost sales)

Prescription of the reorder period (SO2)

This code value addresses the question whether the length of period between two successive orderings is prescribed in the model, or whether its determination is one of the tasks of the model.

The codes:

0 : length of reorder period is not prescribed

1 : length of reorder period is prescribed

Length of reorder periods (SO3)

This code indicates whether the reorder periods are necessarily equal or not, as given in the particulars of the policy used in the model.

The codes:

0 : length of reorder periods is not necessarily equal

1 : length of reorder periods is equal

IV.3.13. The Code Row of Costs

List of unit cost factors (C1)

As is well known, the usual cost factors of inventory models are: the inventory carrying (holding) costs, the shortage costs and the ordering costs. Their notation is: c_1, c_2, c_3, respectively.

As a matter of course, a given model may, in addition, consider many other costs; and it often happens, too, that from among the three basic cost factors one or more factors are omitted. In the course of coding, the most important cases have been given separate code values, the remaining ones have been given the category "other".

The present code gives the list of the cost factors (unit costs) actually occurring in the model. Besides the usual cost factors, c_o denotes the "other" types of costs.

The codes (the costs handled in the model):

1 : c_1, c_2, c_3

2 : c_1, c_2

3 : c_1, c_3

4 : c_2, c_3

5 : c_1, c_2, c_3, c_o

6 : c_1, c_3, c_o

Dimensions of unit cost factors (C2)

All three basic unit cost factors have a typical, "usual" dimension postulated by most of the models. These typical cost dimensions are the following:

— holding cost: \$/unit time/unit quantity

— shortage cost: \$/unit time/unit quantity (this means that the backlogged demand has been considered as the "usual" case)

— the ordering cost (c_3) refers to one ordering.

In many models, of course, the actual cost factors corresponding to the costs c_1, c_2, c_3 have dimensions different from the "usual" dimensions. This code shows whether the cost dimensions are typical or not. If not, it is classified as "other case".

The codes:

0 : the dimensions of the costs are "usual"
9 : other dimensions occur

Character of unit costs (C3)

In some of the models, not only constant unit costs are postulated but the actual cost factors are (not alway linear) functions of inventory level or time, etc. In such cases, we consider the unit costs to be varying.

The codes (the name of varying costs):

0 : none
1 : c_1
2 : c_2
3 : c_3
4 : c_1, c_2
5 : c_2, c_3
6 : c_1, c_3
8 : c_1, c_2, c_3

Changing of the purchase price of items on stock (C4)

Some models investigate the functioning of the inventory system under the condition that the purchase price of the item depends on the system's operation (e.g., on the quantity ordered or the time of ordering).

These models are grouped according to whether the changing purchasing price plays a role in the operation of the system, or not.

The codes:

0 : purchase price is independent of the decision
1 : purchase price depends on the decision

IV.3.14. The Side Fields of the Code Card

The side fields of the code card provide additional information on the models. Nevertheless, only one part of this information can be processed by a computer since the content and character can only be coded to a limited extent, or, in fact, not at all. However, we have considered it desirable that all the essential features of the models should appear on the code card, the more so since our objective from the very beginning—besides the computerized processing of the code cards —has been their direct, "personal" utilization.

Mathematical preconditions (MP)

In contradiction to the third code of the "System and mathematics" (SM) code row which indicates economic constraints and conditions, the mathematical preconditions provide the distinct possibility to categorize a model as well as to solve it by means of mathematical procedures.

All important mathematical preconditions are relevant, but, here, we record only whether there are mathematical preconditions or not.

The codes:

0 : no mathematical preconditions
1 : mathematical preconditions exist

Optimal solution (OPT)

The efficiency of the operation of an inventory system can be influenced by the actual value of the parameters, or decision variables, which the decision maker must determine. Thus, in general, the objective of the system may be fixed depending on the determination of the optimal values of the parameters. These values may be considered as the (optimal) solution of the model.

In this code field, the optimal solution or the mode of solution have to be described if this exists (or can be given), provided this is indicated by the model. The form of the optimal solution may differ significantly depending on the model.

The optimal solutions or the mode of solution actually indicated are coded according to their main types.

The codes:

0 : no optimal solution is given
1 : the optimal solution is given by a closed (analytical) formula
2 : a relation is given providing a basis for an iterative approach
3 : a simulation algorithm is given for the optimal solution
4 : an approximate solution is given
5 : any other calculation process is given for the optimal solution

Remarks

The remarks given are intended to improve the accuracy, ensuring complexity, explanation and identification of information of code values in the code card. A great variety of types of remarks may exist. Four examples are listed below:

a) Should a given model possess a general decisive property of importance for evaluating the model, but which is not involved in any of the codes, this must be covered in a remark.

b) If the code value, chosen for describing a certain characteristic, is essentially exact but requires supplementary information, then a supplementary remark is needed.

c) The identification of symbols used to describe the objective function, or the decision parameters, etc., must be given in the remarks.

d) If a particular code has the value 9 (i.e., if it is considered "other"), this must always be explained in a remark.

There is a two-way formal relation between the code squares and the remarks. Every code square to which some remark is attached has to be marked in its upper right-hand corner, and, vice versa, each remark has to be identified by a latter and figure indicating to which code it belongs. (With the exception of the remarks which do not directly belong to any of the codes: these have to be numbered.)

The information given in the remarks is not coded.

IV.4. Evaluation of the Code System, Experiences Obtained in Applying It

In general, the code system has fulfilled our expectations. It has proven to be suitable for a concise, and relatively precise description of inventory models and for the display of their most important features, thus providing a good basis for the study of the various models, for revealing similarities and differences, for demonstrating relations between the characteristics of models, and for enabling systematic considerations.

The basic construction of the code system (enabling a connection between the main codes and those providing further details) has proven to be good. The main codes have verified our assumption that the most fundamental features of the models could have been revealed by them. The main codes have also proven to be very suitable for the purpose of computer processing: their conciseness and characterizing power have considerably simplified the detection of relations and classification.

After these general and rather positive remarks we shall discuss the drawbacks and weak points, rather than the qualities and strong points, of the code system. This is a reasonable thing to do as far as future research work is concerned, but, here it is important to draw the attention of the reader to the constraints of application. According to our ex post judgement, some of the drawbacks could have been avoided by an even more comprehensive, more careful, prior analysis, or by a more prolonged trial processing, but many of the drawbacks are unavoidable.

These can be classified in six groups:

1. Criteria, or criterion, values carrying important information are missing from the code system (e.g., with the code relating to the purchase price it would have been reasonable to separate the time-dependent price changes from the changes depending on the order quantity).
2. Criteria, or criterion, values conveying unnecessary information, have been taken up in the code system (e.g., the code relating to the type of inventory).
3. The definition of the possible versions of a criterion (code) is not exact, correct or sufficiently complete (e.g., with the code of the optimal solution the definitions "iteration approach", "approximate solution" and "any calculation process" have not been distinguished unambiguously).
4. The desired correlation does not exist between some criteria of the code system (e.g., in the case of, for example, "measurability of items" and "set of values of decision variables").
5. Certain definitions, and categories, have not been properly identified (e.g., in the case of "dynamics").
6. Due to the reasons given above and also independently of them, contradictions emerged during the application of the code system (e.g., as a result of not applying the general code values $(9, 7, -1)$ precisely enough).

Only the most general weaknesses having the most significant consequences will be discussed in detail.

One weakness of our code system of general validity is that we have not managed to separate unambiguously the different operation periods in the models, i.e., the planning period, reorder period and reviewing period. An explanation for this is that the different models discuss the operation periods in different ways; moreover, their treatment is not always clear within a given model.

An exact definition and distinction between the periods with regard to the operation of the system is particularly important because the various elements of the model may refer to different periods. It is not at all indifferent, for example, what unit time the parameters describing the demand refer to: is planning, ordering or reviewing period the very period of time at the beginning of which the demand arises, or—when demand is continuous—in which period does the rate of demand remain unchanged; from among the three periods which does its probability distribution refer to? According to our experience, the separation of the reorder and the reviewing periods may be especially difficult. Sometimes models do not provide the required information, and, in other cases, the explanations given are unsatisfactory.

A problem related to that above is that the code system does not clarify how long the modelled inventory system is operated during an infinite or any finite period.

The most unfavourable aspect of the code system is that we have not been able to code the input and output processes as precisely as desired—or we have only been able to do this by means of additional remarks. Some of the problems have already been pointed out: namely, the difficulty of referring to the time of input and output processes. Numerous other—more or less severe—inaccuracies or unfortunate choices could be mentioned in addition, insofar as the coding of the input and output processes provide examples for almost all of the six types of errors listed above.

We do not find sufficient the relation between the codes describing the time and quantity of replenishment (delivery) and the occurrence of demand (i.e., code pairs R2 and R4 as well as D2 and D4). Sometimes the two codes together are redundant, while in some cases they are not able to elucidate certain problems even together.

Due to the imprecise definition of the reference period, the code system does not clarify unambiguously in which reorder period(s) the items come in upon a given order. More simply: the connection between orders and deliveries is not exactly defined. This is due, to a great extent, to the fact that the explanation and coding of the lead-time is not accurate, detailed and precise enough. The problems are more serious since the lead-time—being among the main codes—has completely been separated from the input process. The coding of the input row would be much more precise if a more detailed characterization of the leadtime would be organically linked to it.

While we consider it to be a failure of coding the input process that the deliveries during a given reorder period have not been clearly connected to the order(s) generating them; in the output process we have not taken into consideration that the demand cannot be linked to some single act, as the placing of an order in the case of replenishment. Therefore, it is always relative whether demand emerges as one batch if the system operates for a long time—during several reorder periods or over an infinite period of time. Accordingly, judgement is questionable with some models as to whether demand emerges in one or more batches. (In practical terms, the question is: which successive demand occurrences can be qualified as part of the same demand emerging in several batches.)

Finally, we have to point out again that—in spite of the drawbacks indicated—our code system operates satisfactorily; its acceptable accuracy, comprehensiveness and applicability have been verified by the successful computerized processing which has been carried out on the basis of the coded model versions.

IV.5. The Coded Form of Inventory Models, Applications of the Coded Models

Based on the code classifications described, all 336 models have been coded. In this way, every model is available in two forms after processing: as a complete code card and as a model description. Cross-referencing from one to the other has been solved so that both have the identification number of the model, and, in addition, the ten main codes are indicated in the model description, and the denomination of the model can be read in the code card. Both forms exactly refer to the source of the model, too. The preparation of the coded form of a model necessarily entails a loss of information, even with the remarks inscribed on the code cards. The description of an inventory model and its coded version contain different information. The qualitative information of the original model has first been condensed, then transformed into quantitative information; and because so many kinds of inventory models may exist, accordingly, their description may differ very much, too. On the other hand, since there is a finite number of possible code values in our code system, information loss is inevitable in both echelons. The question emerges, whether we may gain anything if so how much with this way of handling the information. By processing the code cards we have the opportunity to obtain a series of models summarizing the lengthy original model descriptions and emphasizing the elements necessary to the adaptation. The coded versions are easy to treat for both further theoretical investigations and use in practice; moreover, neither of them can be realized without the coded model forms.

Let us review what we can gain by using the coded models. The two major factors which encourage coding have been the need to systemize the models and give the possibility of selection for practical utilization. In any case, it is easier to select a set of models, suitable for a given purpose, from among the models summarized in the code cards, than to make a selecting from among the original models, since by looking at the code card it is immediately obvious whether the model corresponds to some given condition or not (e.g., identifying models having a stochastic input and output). Nevertheless, in the case of more complicated conditions of selection (e.g., selection of multi-location, demand restricted models treating backlogged demand in the case of shortage) surveying all 336 code cards as well as their selection is by no means so simple, and neither is the simultaneous handling of all codes of all models solved. This has justified feeding the code card data into a computer as well as the investigation of the data bank by computer programs for the purpose of inquiry and analysis.

A computer data bank has been established having sequential construction and consisting of 336 records; each record referring to one model. The records contain the identification number of the model and the series of the code values of the model by characters as well as a code referring to the relevant literature. (The bibliographical reference code is a series of characters made up of the initial letter of the author's name, an identification number of his works, the initial letter of the original language of the source of the model, a code referring to a book or periodical, as well as the last two figures of the year of issue.) The identification number in the data basis is identical with that indicated in the code card and the description. The records consist of 60 characters. From these 60 characters, 57 contain valuable information for the time being (4 characters serve for identification, 45 are code values, 8 are references to the literature), the remaining characters have been left over for extension and serve for certain technical purposes to the inquiry program.

The computer data bank carries even less information than the set of code cards: neither the form of the objective function nor the qualitative information are present in the records. This data bank could not have been established without filling in the code cards, and this form of demonstration and preservation of the models provides an extremely broad opportunity both to deepen our knowledge of the models and to promote their adaptation.

The recognition of models and inventory modelling have been advanced by this form of model in such a way that we can become acquainted with the features of the models by the computer programs for inquiry and analysis. The analytical programs determine the frequency, the relative frequency and the conditional relative frequency of a code value of the models according to their characteristic features. The relation between two codes can be analyzed by means of contingency tables. The data basis enables analysis by multi-variable statistical methods, the grouping of models and the characterizing of individual groups, too.

The adaptation of the models is supported by the computer data bank so that the computer selects a model, defined by the user, from the data bank, or seeks those models "similar" to a given model. The question of practical use will be discussed in detail in Section IV.5.2.

IV.5.1. Utilization of the Inventory Model.
Data Bank to the Analysis of Models

As a first step, a computer inquiry program has been established for identifying models having given features. This program is essentially identical to a model selecting system. The program is capable of selecting model(s) from the data bank by a given code value or set of code values. The input parameters of the program contain the code values of one or several features of the searched model(s). The program examines which models possess the given code combination and writes their identification numbers or, upon special command, the coded models themselves. The inquiry program makes it more efficient, quick and accurate to treat simultaneously several criterion (code) values as well as to select the corresponding models than could be achieved by handling the code cards. For the time being, 10 model features/conditions can be treated at the same time (obviously, the program can be extended and made suitable for building in more conditions). For example, by calling the multi-item, single location models of stochastic input and output we search the models corresponding to the condition

$$MC1=8 \wedge MC2=1 \wedge MC3=1 \wedge MC4=1.$$

The program examines in the records for which models is the character combination 8111 present at characters 5—8, then the identification numbers and the codes of these models can be displayed. In this way, anyone who knows the code values to which the characteristics of a model are transformed—i.e., knows the coding command—may call up the models of interest. By displaying the identification numbers and code values of the models of a given character we will see how many such models we have and what are the features of these types of models.

For the investigation of the complete set of models, statistical evaluation tables can be prepared from the data bank by computer programs which enable a comprehensive analysis of the model system.

The element $[a_{ij}]$ of the frequency matrix shows how many times (from among the 336 models) the ith code takes the jth possible value of the code. The relative frequency matrix shows the relative frequencies of code values (j) by codes (i). The conditional frequency and the conditional relative frequency matrices show the sub-set of the 336 models according to some given aspects (specified by the condition), the frequencies and the distribution of code values by codes, respectively.

By using the contingency tables, the relation between two features of models can be analyzed. The rows of the table will be determined by the possible values of a given code, and their columns by the possible values of another code; the elements of the table indicate how many models fall to the given pair of code values. Also the contingency tables may be worked out either for all 336 models or for a subset of models created according to certain conditions. Based on the contingency tables, by calculating the χ^2 values the closeness of the relation between two codes can also be measured.

In analyzing the models by means of the frequency and the contingency tables, we have sought the characteristic features of economic and mathematical modelling of inventories. Our statements have been based on the sample of 336 models. In our opinion, this sample is sufficient to consider the major part of our statements—at least regarding tendency—to be valid for a wider circle of inventory models as well.

. These issues will be discussed in detail in Chapter V.

IV.5.2. Adaptation of the Inventory Model Data Bank

An important constraint to the practical use of the mathematical models is that managers and mathematicians do not understand one another to a desirable extent. To overcome this difficulty—as an intermediate solution—the coded and computerized form of the models can be applied in the following way.

Naturally, the practical manager is able to explain verbally the specific features of his inventory system. He knows the characteristics of the items, the storing method and, in the case of more than one items and stores, their interrelations. He may characterize the demand and procurement traits of the items. He knows the costing methods in the company environment. He sees clearly the limiting factors such as storage and financial constraints. He knows which factors can be influenced by him and what effects exert influence on the stocking of the item in question (e.g., how the date and volume of orders affect the satisfaction of demands). He also knows what the goal of stockholding is in compliance with the company's objective, and how to appraise inventory management. In so far as the inventory manager is capable of drawing up these factors, this may be understood as an economic inventory system, and we seek the inventory model which provides a solution to this system.

The inventory parameters given by the inventory manager (having been summarized, for example, in a properly prepared questionaire) will be transformed into our code values and compared with our set of models. Then we select that model from our computer data bank which corresponds most to the problem. For the selection it is not necessary to determine all the codes we use, only the ones occurring in the real system. The inquiry program described above may be used for selection. Thus we choose a model, every code of which corresponds to the series of codes pertaining to the concrete inventory problem.

Should such a model not be at our disposal, there may be one not too far removed from our problem and this one may help to find the required solution. We may examine, model by model, "how remote they are" from the requested model. The model sought is characterized by k code values, $k \leq 45$ (k is a function of the completeness of the description that is of the real problem). The distance between the stored models and the requested model can be defined in two ways: (1) how many code values of the stored models are identical with, or different from, those of the requested model; (2) by treating the models as points of a k-dimension linear space we examine which point (model) is the nearest to the model type sought. Method (2) requires that the calculation of distance be performed from the stock of code values arranged according to complexity; the proper transformation of the code system and the computer data bank have been accomplished in the course of applying multivariable statistical methods. Selecting those models from the computer data bank which are near enough to the models sought, we will know the identification numbers of the models, and possibly an index, in respect of the similarity of the models sought and the selected ones.

By tracing back the original model from the bibliographical reference, it is possible to examine whether the selected models are indeed suitable to provide a solution to the given problem; how much the condition system of the models deviates from that of the model sought; whether the model can be modified so that it might be applied to the actual problem.

It is not sure that from amongst those selected one model will really be used which is "the nearest" to the model sought because it may happen that a model less "close" is more suitable for the solution of our problem.

The selection of the models requires the cooperation of managers and the operations researcher. The manager does not need to have a profound knowledge of the terminology of mathematical models but he has to know his own inventory system and to be able to describe it. On the basis of the series of codes attached to the description and the computer data bank, models suitable for application can be selected. Neither does the operations researcher need to create a new model from the elements of the inventory models he knows in order to solve the actual problem; he can obtain a more comprehensive knowledge of the models selected from the set of models. The utilization of the stock of models ensures conditions for the quick and efficient application for both managers and the operations researcher.

One of the essential preconditions for the implementation of the models is to find the very model(s) that actually help to solve the given problem.

Another important factor of implementation concerns how the quantification of the model elements takes place. Also in this phase, efficient application is imaginable only with the active participation of management. A decisive question is whether a computerized data processing system is available which facilitates the quantification of the model parameters, or whether the data are to be collected from manual records.

The possibility of combining the stock of models with the inventory data processing system is a question leading too far away from our original subject. But it should be emphasized that on the basis of our knowledge of the recent trends in the development of computerized inventory management systems, we believe that an appropriate subset of the existing inventory models can be efficiently used in these systems.

V. Statistical Analysis of the Model Sample

V.1. Distribution of the Models with Respect to Their Studied Features

The simplest, most obvious, way of statistically analysing the features of the 336 collected models—as a first step—is the investigation of the distribution of models according to their individual properties. A survey of the various codes establishes how the models are distributed among the possible criterion values. In this way, a very simple yet meaningful characterization of the set of models may be achieved. In the following, we give a detailed review of the distribution of models according to various criterion values.

In the course of the discussion, the individual criteria are grouped according to their logical connections (mostly connected to one of the main codes).

V.1.1. Characteristics of the Items on Stock

More than four fifths (82%) of the models are single item models. In studying the connections between the sorts of items for the remaining 18% (i.e., the multi-item models) we have found that both complementary and supplementary relations are very rare between the items: their share is 1% each. A relatively frequent condition connecting the items is some joint constraints—this occurs in 7% of cases. For another 7% of the models there is no real connection between the items, in spite of the fact that several items are handled in the model.

The type of stocked items is specified in 10% of the models only (which means that 90% of the models can be, in principle, adapted to any type of inventories). We have not found any models specifying raw material inventories. 2% of the models treat in-process inventories, 4% treat finished goods, and 3% of the models assume spare parts inventories.

In the major part of the models (94%) the value of the stocked item does not change during storage. We have not found any model dealing with an increase in the item's value; every model which assumes a change in the value of items has treated a value decrease. (These are mainly the models of the so-called perishable goods.)

V.1.2. Characteristics of the Locations

93% of the models are single-location models. In the remaining 7% of multistore models, locations are linked—parallel with each other—to a central warehouse. There are only very few examples of a linear connection between stores.

In 94% of the models, the storing capacity is not limited. For the remainder, the storing capacity is maximized.

We point out here that the models—contrary to our expectations and with 1 or 2 exceptions—do not investigate the connections of the inventory system to higher system levels.

V.1.3. Characteristics of Input and Output Processes

According to the input and output character, it is conspicuous that the distribution of the models is inversely polarized: the great majority of the models treat a deterministic input (79%) and stochastic output (71%).

It is typical in the case of an input process—characteristic of two-thirds of the models—that the full ordered quantity arrives at the beginning of the ordering period or after a deterministic leadtime. Apart from this case, only the one- or several-lot shipments on one or more random dates contribute a remarkable percentage (9% each) in the set of models. The stochastic input models contribute up to 50% each of specified or non-specified probability distribution of reception dates. (Among the specified distributions, uniform and exponential distributions are the most typical.)

Contrary to the input process, the continuous arising of demands is characteristic of the output processes and, at the same time, such models are more diversified as to the types of processes. In the greater part of the models (52%) demand is continuous: from among these models 37% postulate a uniform rate, and 15% a non-uniform rate, of demand.

Much less significant is the demand arising in one lot at the beginning of the period (with 16% of the models) compared with the similar types of deliveries with the input process. With the output processes, only those models postulating multi-item demands arising on several random dates contribute a considerable fraction (18%), that of models assuming demands on one random date is not important.

With 40% of the models, the time distribution of output is stochastic as well. With two-thirds of these models, the probability distribution of dates when demands arise is not specified. (For specified probability distributions, Poisson and exponential distribution are the most typical.)

It is very characteristic to the input process that the complete ordered quantity is delivered in one lot (this is the situation with 84% of the models). Continuous shipment occurs in 6% of the models and a cumulative treatment of input in only 5%. With just 7% of the models, input is stochastic as far as quantity, too.

The type of the output process relating to the volume of simultanous demands is also more diversified than that of the input process.

With almost one half of the models (47%) demand arises continuously. Roughly every fourth model shows demand in one lot (the share is 27%). Unit demand on random dates occurs in 11%, and cumulative treatment of demands in 9%, of the models.

In two-thirds of the models, the output is stochastic also with respect to quan-

tity. With the great majority of these models (about three-quarters) the probability distribution of the volume of demands is optional, and non-specified. (In
the case of specified distributions, a normal distribution occurs slightly more
frequently than some other types of distributions.)

V.1.4. Time Treatment in the Models

More than three-quarters of the models (77%) are static, i.e., the share of
dynamic models is 23%. In 18% of the models there is one decision, in the majority
(82%) the system is controlled by a sequence of decisions.

V.1.5. Objectives of the System

In the majority of the models (87%) the objective is optimizing. The share of
the reliability models (7%) slightly exceeds that of the descriptive ones (5%).

V.1.6. Operation Policy (Ordering Rule) of the System

From all the possibilities, four types of policies have proved to be frequent in
inventory models: one of these can be found in nearly four-fifths of the models.
The relatively most frequent ordering mechanism (s, q) occurs in more than
a quarter of the models. (We refer here to the fact that also (s, q_p) has been coded
along with (s, q) i.e., both are included—although the percentage of the former
is rather low.) With a share of more than 20%, second is the (t_p, S) mechanism
followed by the three (t, q) mechanisms (dominant is (t_p, q)) as well as by the
mechanism (s, S) $((s, S_p)$ occurs rarely). The rest of the policies are represented
in our sample by a smaller percentage only; from among these the share of (t, S)
and that of (s_p, q) with 5% each are worth mentioning.
Whereas the above data refer to the "theoretical" decision variables determining
the policies of the models, by investigating the actual decision variables of the
models it can be established that there is no decision variable having a prescribed
value (with index p) in about 60% of the models, while in 40% there is. The major
part of the models (60%) presume continuous decision variables; the share of
models treating only discrete variables is slightly higher than 30%, and relatively
few models (about 6%) have both continuous and discrete variables.
There is a close connection between the ordering rule and the prescribed or
non-prescribed, equal or non-equal, character of the ordering periods. Namely,
it is obvious that—apart from some incidental, very special models—the length
of the ordering period is treated at least in the models postulating t_p variable
(it may be the case with other models too), and that the length of ordering periods
is necessarily equal at least in the models treating t and t_p variables. As verified
by the data, in 33% of the models the length of the ordering period is considered
(the share of (t_p, S) policies is 21% within this). The length of the ordering periosd
is necessarily uniform in more than half (51%) of the models (from these the
percentage of (t_p, S) is 21%, that of (t, q) and (t_p, q) is 16%, and that of (t, S)
is 5%).

V.1.7. Review of Inventory

The review of inventory is in compliance with the character of the input-output processes as well as with the ordering rule of the model. According to the numerical data, in roughly one-third of the models (34%) there is no inventory review. From these, 26% are deterministic models (both by input and output), where reviewing inventory serves no purpose; in the remaining 8% there is no inventory review, in spite of the stochastic character of the model.

Substantially more frequent is the periodical, rather than the continuous, inventory review: the former occurs in 44% and the latter in 21% of the models. The explanation for the higher percentage of periodical inventory control is that continuous review of the inventory makes no sense where input and output both occur at fixed intervals and/or orders are placed at fixed (t) intervals according to the ordering rule of the model.

V.1.8. Mode of Treating Shortages

In nearly a quarter (24%) of the models, shortage must not occur. The most general method of treating shortages—applied in 55% of the models—is the assumption of backlogged demand. Relatively rare is the case of lost sales: it occurs in 14% of the models. (Relatively high is the percentage of models which do not treat the problem of shortages: their share is 7%.)

V.1.9. Types and Characteristics of Costs

From among the six specified combinations of costs the most frequent—with a share of 25%—is the combination comprising (c_1, c_2, c_3) costs alike (i.e., inventory holding, shortage and ordering costs). A further 7% is the ratio of those models where other cost factors are present besides the above three. In 14% of the models the combination of costs is (c_1, c_3) and in 7% it is (c_1, c_3, c_e); i.e., these are models without shortage costs. (Obviously, a considerable part of these models are identical with those where shortage must not occur.) For 9% of the models the cost combination (c_1, c_2) is characteristic (no ordering cost). In a quarter of the models, the set of costs differs from the six specified cost combinations. About one tenth of the models do not treat costs (these models are mainly reliability models).

As for the dimensions of the inventory unit costs, the figures show that the minority of models treat exclusively the dimensions we judged typical. (Inventory carrying and shortage costs are quantity- and time-related, while the ordering cost is a fixed amount for each order.) 35% of the models apply these cost dimensions. 55% treat, in addition, costs different from this. One tenth of the models do not treat the problem of cost dimensions (obviously, they are the same which do not treat the costs themselves, either). Specific costs are, for the great majority (70%) of the models, constant. In 10% of the models the problem of a constant or varying character of specific costs cannot be discussed in the absence of cost treatment. In the rest of the models (20%) some of the unit costs are not constant.

In the majority of cases (87%) no connection dealt with in the model existed between decision and purchasing price; in only 3% of the models was assumed a decision dependent on purchasing price. In 10% of the models the question was not relevant.

V.1.10. Economic Constraints in the Models

In nearly four-fifths of the models there is no special economic precondition or constraint. The remainder of the models is distributed according to the concrete types of constraints specified by us. (10% of the models employ economic constraints differing from those not specified by us.) From among the relatively frequent types of constraints the restriction on maximal inventories from a capital aspect has to be picked out (4%) along with the prescription of the largest permissible shortage (3%). The other types of constraints are represented in 1—2% of the models.

V.1.11. Mathematical Characteristics of the Models

Probability theory dominates in most models (51%). The second mathematical method most frequently applied is calculus with 26% (this is characteristic mainly of the deterministic models). Besides these, only dynamic programming has a considerable share: 11%. Game theory, linear programming, stochastic programming and other methods do not amount to more than 1—3%. In the majority of the models there is no special condition prescribed for the sake of easier mathematical treatment.

In 30% of the models some calculation method is given for the optimal solution. In a further 28%, the optimal solution can be written as a concrete, closed formula. In 8% of the models, the solution may be approached by iteration, and in 4% other approximate solutions can be given. For 3% of the models there is no optimal solution whereas in 2% of the cases simulation is the proposed method of investigation. In about one-fifth of the cases an optimal solution may be defined (reached or accessible), though not in the frame of our specification.

In only a negligible minority of the models (about 3%) one can find—in the bibliographical source treated—a detailed flow diagram or computer program translating the solution algorithm or any other important calculation method into a computer code.

V.2. Characterization of the Model System Based on Contingency Tables and Relative Frequency Matrices

This analysis is very important from the aspect of recognizing the structure and characteristics of our model system. The contingency tables show the connection between two criteria (codes) in the models. Analyzing them shows how one of the concrete values of a criterion affects the values of another investigated criteria, i.e., which properties of the models attract or repel one another, which of them occur always together and which preclude the other.

For example, we show a simple (two dimensional) but important contingency table for the connections between the criteria MC3 and MC4 which demonstrate the deterministic or stochastic character of input and output processes. (See Table 1)

Table 1

Contingency matrix of the input and output processes

Output (MC4) \ Input (MC3)	Deterministic 0	Stochastic 1	Not treated 7	Total
Deterministic 0	84	13	—	97
Stochastic 1	182	54	2	238
Not treated 7	—	—	1	1
Total:	266	67	3	336

As shown in the table, occurring most frequently is the deterministic input-output connection, and stochastic input "attracts" stochastic output (in only 68% of the models operating with a deterministic input is the output stochastic, whereas in the case of a stochastic input the percentage of stochastic output is 81%), etc.

On the basis of a contingency table, the connection between two criteria can always be studied. Should we intend to reveal the complete system of connections by this method, then contingency tables should be available for every possible pair of criteria. Based on previous considerations we pointed out 18 criteria (the 10 principal codes as well as codes I2, I4, O2, O4, SM3, SM5, SO2 and C1) and prepared contingency tables for each pair of them. In this way we were able to analyze $\binom{18}{2}$, i.e., 153 contingency tables.

Similar information, as presented in the contingency tables, is contained in the relative frequency matrices. These matrices show under any given condition the distribution of the models according to the given condition (thus, for example, the characteristics of models operating with a deterministic input can be evaluated by the prescription of condition MC3=0). By comparing these conditional criteria with each other as well as with the criteria distributions characteristic to the whole of the model system, further valuable information can be gained.

The conditional relative frequency matrix has been elaborated for the same criteria as the contingency tables, based on the same 18 variables.

In this way, we managed to characterize comprehensively the connections inside the criterion system and, at the same time, the set of models. It turned out in the course of systemizing the information that our statements could be summarized in the most characteristic way if we applied the following grouping:

— statements related to the complexity of processes represented in the inventory models (in our interpretation, the complexity of processes is qualified mainly by the static or dynamic treatment of time, by the number of items, and that of the locations);

— statements related to the decision maker's actions expressed in the inventory

models (this is expressed mainly in the objectives and ordering rule of the system);
— statements related to the input and output processes represented in the inventory models (including reviewing inventory treatment of shortages as well as the statements in connection with the leadtime).

V.2.1. Complexity of Processes Represented in the Models

It is a general tendency that the models which describe complex processes—postulating several items, several locations and being dynamic—belong insofar as many other aspects are concerned to the relatively simple models. Should we choose complex, sophisticated conditions, only more simple versions can be elaborated for the remaining criteria.

The following general properties — indicating relatively simple modes of treatment—are characteristic to the multi-item and/or multi-location and/or dynamic models alike:

— either they do not treat the input process at all (like some multi-item models) or they assume the simplest, i.e., one-lot delivery (taking place usually at the beginning of the ordering period);
— in conformity with the aforesaid there is no leadtime in them;
— the interval between two subsequent orders (ordering period) is determined for the system externally (i.e., the optimalization of the order period is precluded from the system).

It is characteristic to all three types of complex models mentioned above, that they treat other costs, too, than the three "typical" cost factors. With the multi-item multi-location models these other costs are connected with the ordering cost in general, and they obviously arise because special expenditures may arise in the case of more than one items and locations as well. With the dynamic models—besides the relatively frequently occurring other costs—it is also conspicuous that non-varying specific costs exist very rarely compared with the usual cases. It can be assumed that in the majority of the dynamic models specific costs are time-dependent.

The above general characteristics are valid for the multi-item, multi-location and the dynamic models alike. Two from these three types—multi-item and multi-location models—possess especially similar properties. In addition to what has been listed above, they have the following common characteristics:

— as a tendency, the models are deterministic (both input and output processes are completely known in advance);
— there is no review of inventory, in consequence of the deterministic character (inventory reviewing is unnecessary in deterministic models, since the volume of inventory is known—or can be calculated—at any time in advance); the simpler version, namely: periodic review occurs in some cases;
— relatively frequent is the "prohibition" of shortages due also to the deterministic character: namely, in most cases, shortages can only be prevented when both input and output processes are known in advance;

— the models in question apply usually the simpler operation policies like the (t_p, S), (t_p, q) and (t, q) ordering rules.

Among the above three types the multi-item models are the simplest. In addition to all features mentioned previously, their static character is very strong. They are the more simple, since they have more restrictions, or constraints, than the average. (Among these restrictions very typical is the capital constraint on maximal inventory.)

All statements listed above are valid for the multi-location models as well, nevertheless, they are somewhat more "complicated" which is manifested in the fact that—compared with the multi-item models—they are mostly of dynamic character and there are associated less constraints than the average. (Among the multi-location models, multi-item models occur more often than the average.)

As already said, many multi-item models do not treat the input (replenishment) process. It is an interesting contrast that the multi-location models often do not discuss the output process.

The situation with the dynamic models is different. In many aspects also these are simple—as shown in the first list, they discuss simple input processes, there is no leadtime and the ordering period is fixed—but, on the other hand, they are more sophisticated than the average regarding other aspects. The obvious reason for this is that the dynamic construction and approach of the models have sense and contents only if the modelled system, and the processes, are complex to a certain extent. This is the explanation for the phenomena—contrary to the features of the multi-item multi-location models—that in the dynamic models:

— the output is more frequently stochastic than on the average;
— due to the stochastic character, an inventory review is employed and usually in the simpler version: the periodical review;
— due also to the stochastic character, shortages are in general allowed;
— the more complex operation policies, (namely the ordering rule (s, S)) are characteristic;
— they treat both input and output processes, and one-lot input and output at the beinning (or end) of the period are typical.

Further characteristic features of the dynamic models are that stochastic input can hardly be found among them, and that reliability and descriptive models are almost entirely missing. (It is worth mentioning that with the multi-item multi-location models, also reliability and descriptive models occur with an average frequency.)

Although our statements do not "have a very firm basis"—due to the rather small number of models in question—it still has to be noted that deterministic leadtime of varying length occurs only in dynamic models. Discrete quantities are characteristic also in the dynamic models. The assumed reason for this is that dynamic programming gives preference to discrete variables. The length of the ordering period is fixed, the inventory review is periodic, input and output are performed in one lot (i.e., both time and quantity are discrete), and usually the decision variables themselves are discrete, too.

Dynamic programming as a mathematical method is very characteristic of the dynamic models: mathematical preconditions and concrete formulae for solution are less frequent here, than on the average.

V.2.2. The Decision Maker's Attitude

Under the title "decision maker's attitude" we summarize the knowledge gained about the objectives and operation policy of the inventory system. As for the objective of the inventory models, we distinguish optimization, reliability and descriptive models. Their distribution in our sample of models is very disproportionate as can be expected: the greater number of models are optimization models, fewer are reliability models, and only a very few descriptive models can be found among them.

We give first a brief review on the easily identifiable, small group: the descriptive models. These models—according to our experience—are more sophisticated than the average as far as some of their features are concerned but their remaining properties are simpler. It is characteristic that they often postulate several locations and several items, and they employ the (s, q) and (s, S) ordering rules. At the same time, descriptive models are usually deterministic: both input and output processes are known in advance. An interesting characteristic of them is that, more often than average, they apply mathematical simulation as methodical tools. The great majority of these models do not treat cost factors.

Reliability models have rather characteristic features. A considerable part of them (about 60%) employ the (t_p, S) ordering rule and, in this way, they transfer their characteristic features partly to the model family (t_p, S), as well. The traits of the reliability models may be summarized as follows:

— due to the essentials of the model, they do not treat in general the behavior of the system in the case of shortages but give only reliability conditions to limit shortages;
— originating from the essentials of the model, too, they usually do not deal with cost factors but their proportions are implicitly expressed in the probability level of the reliability condition;
— the models are static;
— they discuss stochastic input processes;
— replenishment does not take place in one lot in most cases; the cumulative handling of the delivered quantities along with the deliveries in several lots on random dates are very characteristic, the time distribution of deliveries is determined more often than on the average: a mainly uniform distribution is assumed;
— characteristic are the continuous and uniform output as well as the random demands arising in several lots. They often handle cumulative demand, the probability distribution of demand is frequently specified—it is usually a normal distribution;
— decision variables are in general continuous;
— because of the dominant (t_p, S) mechanism these models often have fixed variables;
— periodical inventory review is characteristic.

The characteristic features of the total of the optimization models—just because they make up the great majority of the models—do not deviate significantly from the traits of the complete 336 element set of models. Therefore, the optimization models will be divided into groups according to the operation policy (ordering rule) of the models and the properties of these subgroups will be compared.

There is a fairly conspicuous correlation between the ordering rule and the grade of complexity of the models. There are types of ordering rules which can be found in models of explicitly "simpler" condition systems, and there are ordering rules used mainly in essentially "more complex" models. The policies pertaining to the simpler models are (t, q), (t_p, q) and (s_p, q), while the ordering rules (s, q), (s, S), (s, S_p) and (s_p, S) are applied in complex models.

The important features—as tendencies—of the "simpler" and "more complex" models are compared in Table 2.

Table 2

Main characteristics of "simple" and "complex" models

Simpler models (t, q) (t_p, q) (s_p, q)

1. Both input and ouptut are deterministic.
2. Due to the deterministic character, inventory review has no sense.
3. Usually there is no shortage allowed or if there is a shortage with models (t, q), (t_p, q), then lost sales are more frequent than on the average.
4. Order periods—except (s_p, q) models—are uniform.
5. It is also characteristic of the (s_p, q) models that there is no leadtime, demand is continuous; decision variables are continuous, too.

More complex models (s, q) (s, S) (s, S_p) (s_p, S)

1. Stochastic output is typical.
2. Usually there is inventory review and it is continuous with (s, q) and (s_p, S) models, otherwise it is periodical.
3. Shortages may occur: general is the backlogged demand (it is true however, that the prescriptions for s_p are often aimed at excluding shortages, therefore, no shortages occur with these models).
4. Order periods are not prescribed and not uniform.
5. Leadtime is characteristic to the (s, q) models and in many cases it is random (stochastic input).
   Demands at random are usual.
   One-lot delivery is characteristic to the (s, S) mechanism.
   Continuous demand is rare; the number of dynamic models is relatively high.

In characterizing the reliability models we already indicated that these models influence to a great extent the characteristics of the whole set of (t_p, S) type models—the majority of the reliability models contain the (t_p, S) ordering rule—though a considerable percentage of the (t_p, S) models do not belong to the reliability models. For this reason, in the traits of models with (t_p, S) policy a certain duality can be observed. It may be explained also as an effect of the reliability models that it is characteristic of the (t_p, S) models that

— deliveries in several lots and cumulated treatment of the delivered quantities dominate;
— they apply various constraints.

At the same time, further features of the (t, q) and (t_p, S) models will separate themselves from the traits of reliability models (maybe they are even in contradiction to the latter group). An example for this is that in these models one-lot delivery often occurs. Also the continuity of demand is characteristic, but it does not contradict the properties of the reliability models.

Decision variables of mixed value set characterize the (t, S) models, while (t_p, S) models are characterized by continuous decision variables. Finally, it can be considered as a simple consequence of the ordering rule employed that inventory review is periodic—continuous inventory review would make no sense if orders are to be placed at intervals with length t anyway—and that ordering periods are uniform.

V.2.3. Input and Output Processes

In analyzing the two processes which decisively influence the features of the inventory system, first we investigate the role of uncertainty, afterwards we will analyze the time and volume dependent characteristics of the in- and output.

As for the treatment of uncertainty, the great majority of models in the model sample have—as expected—a stochastic output but the models of stochastic input are in a minority. Though there are models postulating stochastic input and deterministic output simultaneously, the tendency is that models discussing stochastic input treat stochastic output more frequently than the average. (The most frequent combination is deterministic input and stochastic output (54%) followed by the models with deterministic input and output (25%), then that of stochastic input and output (16%) and finally by the combination stochastic input and deterministic output (4%).)

It is rather a simplification and not quite exact to say that it is one step forward from the models of deterministic input-output towards the more sophisticated case of stochastic output, and it is another step if we postulate, along with stochastic output, stochastic input as well.

Apart from the few models of stochastic input and deterministic output it can be stated that what makes the model stochastic is the stochastic output. In this way, it is not surprising that the stochastic or deterministic character of the output divides very explicitly the features of the models in many respects. Similarly, as was done with the ordering rule of the inventory system, the essential features of models of a deterministic output and those of a stochastic output are compared in Table 3. It may be seen that the properties listed with the simpler ordering rules and deterministic output, and those listed under complicated ordering rules and stochastic output are often the same, indicating that the classification according to the complexity of the operation policy and the character of output leads to the separation of similar groups of models.

The greater part of the models having a deterministic output have a deterministic input too, and deterministic models are, in many respects, simpler than stochastic ones. This is reflected in the upper part of Table 3. For the deterministic models as stated previously, inventory review has no sense; shortages can be prevented, thus usually no shortage cost (c_2) is to be treated. In the simple (deterministic) input, the process leadtime usually has no role. The output process is also very simple: continuous and even. Onto all these simple conditions simple ordering rules can be built up.

Table 3

Main characteristics of models with deterministic and stochastic output

Deterministic output

1. Ordering rules (t, q) (t_p, q) and (s_p, q) are characteristic.
2. Due to the deterministic character, there is no inventory review.
3. Usually shortage is not allowed.
4. The length of order period is not prescribed in general but often uniform.
5. Usually there is no leadtime but replenishment in several lots or continuous input are characteristic.
6. Replenishments are more often continuous than the average, cumulated treatment often occurs.
7. Demand is in general continuous and even.
8. The typical cost combinations (c_1, c_3), (c_1, c_2, c_o).
   At (t, q)-type models c_o cost occurs in general.
9. Mathematical preconditions are rarely given, and optimal solutions are frequently given by closed formulae.

Stochastic output

1. Ordering rules (s, q) (s, q_p) (s, S) and (t_p, S) are typical.
2. Inventory review is usually periodical.
3. Shortages may occur in general.
4. Length of order periods is often prescribed, if not, they are not uniform.
5. Leadtime exists more often than the average.
6. In general the complete ordered quantity arrives in one lot.
7. Demand arises on one or more random dates; or in one lot at the beginning of the period (typical is the cumulated treatment of demands).
8. Characteristic cost combinations are: (c_1, c_2, c_3) and (c_1, c_2, c_3, c_o);
   (To the (t_p, S) models also (c_1, c_2) is very characteristic).
9. Mathematical preconditions are often set, concrete formulae for solution are rarely given.

The situation is usually just the reverse with the models having a stochastic output. Due to the uncertainty of the output process (maybe that of input too), inventory reviewing is necessary, and the occurrence of shortages cannot be excluded in general, therefore, among the operation costs also shortage cost has to be taken into account. The more sophisticated, stochastic output process is characterized in addition by the fact that—in contradiction to the simple continuous and uniform character of the deterministic output—demands arising on random dates or at the beginning of the period with random quantities become frequent here. We are not able to verify this but assume that the generally prescribed length of ordering periods as well as the assumption of delivery in one-lot are necessary simplifications to the otherwise sophisticated conditions. More complicated ordering rules pertain to the above complicated condition system, too.

Besides what has been discussed above, worth mentioning is the tendency that the share of dynamic models grows to a certain extent with stochastic output. At the same time, the output character does not affect the distribution of models according to their objectives and neither are the number of constraints influenced.

We obtain features similar to those listed in connection with the models having a deterministic output, if we investigate the groups of models with no inventory review, those excluding shortages as well as the models without leadtime. Should we deal with any of the listed types of models, we will find properties similar to those of models having deterministic output. Of course this is not surprising since we have found among the special properties of the deterministic models the non-existence of inventory review, the exclusion of shortages and the lack of leadtime. All these criterion values are closely connected with each other: they all refer to the groups of simple, deterministic models.

The closest to the traits of stochastic output models (with deterministic input in general) are the properties of the models of periodic inventory review. A smaller part of the models of periodic inventory review are reliability models (but the reliability models treat periodic inventory review as a rule). It originates probably from the features of reliability models, too, that among the models treating periodic inventory review the number of models treating no shortage and cumulated inputs is relatively large. Since the majority of the models of periodic inventory review are not reliability models, their further characteristics deviate from the properties of the latter, and several of their features are similar to those of the stochastic output models. The following features of the models treating periodic inventory review are worth mentioning:

— dynamic models are relatively frequent;
— (t_p, S) and (s, S) operation policies are relatively frequent;
— prescribed and uniform ordering periods are relatively frequent;
— delivery in one lot is relatively rare;
— demand arising in one lot is relatively frequent;
— characteristic cost combination is (c_1, c_2, c_3).

As seen, the features of the models of periodic inventory review—apart from the third one in the above list—coincide with the properties of the stochastic output models. As for the treatment of shortages, the models postulating controlled shortages coincide the most with the stochastic output models (disregarding the reliability models and those without inventory review).

There are several among the stochastic output models—particularly the types (t, q), (t_p, q) and (t_p, S)—in which there is no inventory review. (Since we speak of models of stochastic output, inventory review could be explained, however, the model does not accomplish that.) In these models shortages are allowed, the demand for a missing item often becomes lost to the system. (This latter fact is a consequence of the lack of inventory review.) In addition, one-lot delivery is typical along with the missing, or not treated, leadtime, the set ordering period, single decision and the lack of cost factor c_3. Presumably because of the lost sales, costs of deviating dimensions often emerge. It has to be remarked that the—relatively small—group of models postulating lost demands may be attached most reasonably to the "stochastic models without the inventory review category described above.

If we compare the characteristics of stochastic output models with the different groups based on various types of leadtime—the model family of constant leadtime may be the most similar. Their common traits are:

— deterministic input,
— stochastic output,
— (t, s), (s, q), (s, S) operation policies,

— shortages allowed,
— multiple-lot demand.

Proceeding to the analysis of the input processes, we state first of all that the characteristic features of the deterministic input models do not deviate very much from the features of the complete set of models, this being a simple consequence of the fact that the dominating percentage of models have a deterministic input. The models are made stochastic in most of the cases by the uncertainty of the delivery times; much less is the case of random quantities.

By a rough classification, the stochastic input models can be divided into two large groups: (t_p, S) type reliability models and (s, q), (s, q_p) type optimization models (the share of the former is 27%, that of the latter is 48%). Although the criteria of the reliability models have already been listed, for the sake of comparison we consider now the features of the stochastic input reliability and optimization models with those of the whole set of stochastic input models. (See Table 4 where the row is empty, the respective group of models does not have that characteristic feature.)

It turns out immediately from the table that the stochastic input models—especially as conclusions of the process descriptions of complicated input and output—are more complex and intricate than the typical, "usual" models. It can easily be traced back how the characteristics of the complete set of stochastic input models take shape as a result of the—partly similar and partly quite different—properties of the two main groups: reliability and optimization models. The main differences between the traits of these two main groups may be summarized as follows:

— Reliability models—resulting from their character—do not treat shortages and cost factors. According to the sense, this situation is the reverse with the optimization models.
— Reliability models postulate the simplest (t_p, S) policy; a straight consequence of this is that ordering periods are fixed and equal, and that only periodic inventory review is required. Optimization models often employ the more intricate (s, q) policy with non-prescribed ordering periods of different length; continuous review of the inventory may become necessary.
— Reliability models treat cumulative input and output and have continuous decision variables. On the other hand, in optimization models only discrete decision variables are typical.

On the basis of the aforegoing, we may take the risk of stating that, within the set of the stochastic input models, reliability models are relatively simpler while optimization models are relatively more sophisticated. This is supported by the fact that among the reliability models the dynamic treatment of the time-dependent processes is entirely negligible.

On the other hand, the group of the optimization models of stochastic input can well be approached with the selection of models having continuous inventory review and random leadtime. These three properties are in relatively close connection, and identify a rather complicated group of models.

In the aforesaid, we have dealt only with the deterministic or stochastic character of input and output processes. Additional knowledge completing and confirming

94

Table 4

Stochastic input models

1. (t_p, S) and (s, q) policy.
2. Usually static models.
3. Shortage may occur but also non-treated shortage is frequent.
4. Constraint referring to shortage volume is frequent (economic constraint is also in general more frequent).
5. Inventory reviewing is in general continuous.
6. Random leadtime is typical.
7. Delivery at one or more random times (the time distribution of deliveries is typically uniform or exponential).
8. More complex than average output process (more frequent random time demands; continuous or non-uniform demands).
9. The cumulated treatment of input and output is typical.
10. —
11. —
12. Costs are frequently not treated.

Stochastic input reliability models

1. (t_p, S) policy.
2. Static models.
3. Non-treated shortage.
4. Reliability conditions to limit the occurrence of shortages.
5. Inventory survey is usually periodic.
6. —
7. Delivery does not occur in one lot in general (but in several random time moments); usually uniform distribution.
8. Multiple-lot random demand; continuous demand with uniform intensity.
9. The cumulated treatment of input and output is typical.
10. Order periods are prescribed and uniform.
11. Usually continuous decision variables.
12. No cost factors.

Stochastic output optimization models

1. (s, q) policy.
2. —
3. Backlogged demand.
4. Shortage maximization; frequent mathematical preconditions; rare concrete formulae for solution.
5. Inventory observation is, in general, continuous.
6. Random leadtime.
7. Random time deliveries (rather one than more random time moments; usually exponential time distribution).
8. Demand occurrence in several random moments.
9. —
10. Order periods are not prefixed and not uniform.
11. Usually discrete decision variables.
12. "Other" costs and non-treated costs are typical.

our statements can be obtained by analyzing the models for input and output delivery dates as well as for the quantities arriving and leaving the store. In the final section of our assessment we will do that.

Considering the input process it is typical—and valid for two-thirds of the models—that delivery takes place in one lot at the beginning of the ordering period (including shipments arriving after a deterministic leadtime). It is evident that the features of the members of this extended model family do not significantly deviate from the traits of the complete set of models. (Maybe, only so much has to be added that continuous inventory review occurs somewhat more frequently than on the average.)

Based on a small group consisting of about 20 models, interesting conclusions can be drawn for the continuous input models. It is true, as a tendency, that demand arises continuously in models of continuous input. Moreover, if the rate of inputs is even, so is demand, too; if not, nor is demand. Surprisingly, the even or uneven rate of input (and output, respectively) closely relates to the deterministic or stochastic character of input and output. An even rate can be found in models of deterministic input and output, whereas an uneven rate occurs in the models of stochastic input and output. As a consequence, the ordering period is not prescribed in general with the former types (deterministic models); but it is with the latter (stochastic models). (We have already referred to this feature when discussing output characters.)

Those models postulating one shipment (reception) on a random date correspond very well to the—usually (s, q) type—optimization models of stochastic input. The characteristic features of these models could be listed point by point, too, (stochastic output, controlled shortage, continuous inventory review, etc.). We point out here only the tiny difference that whereas the (s, q) mechanism can unambiguously be referred to as a characteristic feature of stochastic input/optimization models; with the random delivery models also (s, S) can be picked out, besides (s, q). Apart from this, both model categories can be characterized by the same properties.

Among the models assuming deliveries on several random dates, the—usually (t_p, S) type—stochastic input reliability models are also available (their share is about 40%). It is obvious that a considerable percentage of this group originates from among the stochastic input optimization models.

The following properties derive from this dual character:
— the models are of (t_p, S) and (s, q) type (50% each);
— the models are usually static;
— inventory review is always accomplished;
— demand arises at several random time moments;
— the volumes of demand are often cumulative;
— closed formulae are rarely used, much more often computing procedures are given.

As is seen, if models discussing one random time delivery are in conformity with the stochastic input optimization models, then the models treating several random shipments reflect the properties of all the stochastic input models, simply because they unite the—partly very different—features of the reliability and optimization models of stochastic input in a similar way.

It is also interesting that the characteristics of models treating accumulated delivered quantities entirely coincide with the characteristics of the reliability models as for their essentials. (Nevertheless, the cumulative treatment of input is not the exclusive trait of reliability models.)

The most typical realization of output is when demands arise continuously and uniformly (more than one-third of the models belong to this category). These types of models do not form a characteristic group in the total set of models. We list some of their features below without presuming a clear-cut separation of these models. Characteristic are

— the relatively frequent presence of several items;
— static character;
— continuous decision variables;
— equal ordering periods and
— the existence of mathematical preconditions.

A group separating itself a little more clearly—and having about 50 members— is the model family postulating continuous, unevenly distributed demands. Characteristic here are the (s, q) and (s, q_p) policies occurring more frequently, while (s, S) and (s, S_p) mechanisms are hardly encountered compared with the average. A further typical feature is that — in accordance with what we have said about input processes—the input process is usually also continuous and have uniform intensity. As described in the case of the input process, when input and output do not have uniform intensity, input and output will generally be a stochastic process.

Recalling the similar properties indicated for the stochastic input models— namely, that demand is continuous and does not have uniform intensity—and, furthermore, the fact that the (s, q) mechanism is characteristic of the group in question, although it cannot be stated quite definitely, we may assume that a considerable part of the models with a continuous demand that does not have uniform intensity come from among the (s, q) type, stochastic input optimization models. Neither do the models postulating a one-lot input at the beginning of the ordering period, (counting about 50 models too), constitute a characteristic category. Only the relatively frequent occurrence of the (s, S) policy, the relative scarce (s, q) policy as well as the higher than average share of the fixed leadtime may be cited as characteristics. Apart from this, the characteristic relations of the complete set of models are reflected in this model family as well.

A few (16) models only postulate demands arising in one lot at the end of the ordering period. This group—it seems—is in close connection with the category of the dynamic models, namely, their characteristic features are in almost all details the same as the traits listed with the dynamic models. Accordingly, the share of models of

— (s, S) policy,
— lack of leadtime,
— single decision and
— treating additional costs

is high.

The interrelation between the two groups of models can, of course, be found out more directly from the fact that the share of dynamic models is high among the models discussing one-lot demands at the end of the ordering period.

According to evidence, the set of models which treat demands arising at one random date is a group of mixed composition and hard to identify (a small group consisting of less than twenty models). They have no really characteristic properties, only so much can be mentioned that the share of those models is relatively high in which:

— (t, q) and (s, S) mechanisms,
— the lack of inventory review,
— no leadtime,
— single decision and
— the solution by concrete formulae

are characteristic.

Nearly 60 models belong to the group postulating demands at several random times. A considerable number of these come from the (s, q) type stochastic input (and output) optimization models, but in addition also a certain percentage of the stochastic input reliability models can be found among them. The characteristic features of the group refer also to this construction:

— stochastic input,
— (s, q), ordering rule,
— continuous inventory review,
— non-treated shortages and lost sales,
— random leadtime.

V.3. Analysis of the Model System with Multi-Variable Statistical Methods

V.3.1. Summary of the Methods Applied

To undertake the comprehensive investigation of the structure of our model system, multi-variable statistical methods have been applied. These methods enable taking into account simultaneously the variables characteristic to the models (codes, and code values, respectively) and to analyze their combined effects. The primary objective of the analysis has been to contribute to the foundation of a model classification. The issues of these investigations are summarized in this chapter, assuming that the reader is familiar with the applied methods, the results to be achieved by using them as well as their constraints.

For the successful adaptation of multi-variable statistical methods, an approach based on the parallel or subsequent application of versatile complex methods is indispensable. We have been led by the following concept.

As a first step we repeatedly and carefully examined the contents and meaning of our variables. For the purpose of multi-variate investigations, 37 parameters have finally been selected. The code values of each category of these parameters have originally been equivalent to measurements on a nominal scale. However, as the internal nature of the inventory models as well as their complex or simple features have been known, the originally nominal categories are suitable for classifying them according to their complexity. By transforming the code values according to this sequence an ordinally arranged (quasi-ordinal) scale has been gained. These transformations have been carried out for every characteristic feature, and in this way our data system has become considerably easier to treat and suitable to draw more stable conclusions.

We have divided the obtained ordinal scales into two parts (dichotomized) on the basis of practical and professional considerations. Thus, the information carried has become much more concise and, at the same time, the parallel performance of trials has become possible, supported by this divalent variable system, too.

Let us take an example: the main code MC10 in Table 5—the categories in the table are ordered according to their complexity. For our codes of different types, also including dichotomization, we have prepared frequency distributions in order to check the traits of both our model system and our decisions applied to dichotomization.

Table 5

An example of the code transformation
Leadtime code MC10

Categories	Original codes	Transformed code values	Dichotomization
Leadtime has no sense in the model	−1	1	
Non-treated	7	2	} 1
No leadtime	0	3	
Constant leadtime	1	4	
Leadtime is deterministically known but altering	2	5	
Leadtime is a random variable	3	6	
Delivery in several lots or replenishment rate is uniform	4	7	} 2
Other	9	8	
		Ordinal complexity scale	

Within the framework of multi-variate studies we applied factor analysis and cluster analysis. In order to avoid misunderstandings it should be noted that from factor analysis the factor weights have been observed only and no metric conclusions have been drawn. On the basis of the factor analysis exclusively, further experimental hypotheses have been put forward, to be checked by examinations of a different nature, and to the selection of the variable combinations serving as a base for systematization. In this way, the problems emerging around the preconditions of this methodology did not severely jeopardize our work. The investigations carried out are summarized in Scheme 1.

The multi-variate statistical study has been concluded by specifiyng the grouping criteria and the systematization of the set of models. The "state space" spanned by the model criteria, the distribution of models inside this, the evaluation of the empty sections of space as well as the analysis of the properties of cluster-centroids (typical models) belong to future research tasks.

99

Scheme 1
Multi-variable analysis of inventory models

I. Ordinally arranged
codes for each
variable

II. Dichotomized
variables

In the following Block I is detailed (every test has been repeated with Block II)

I
336 models
37 variables

Factor analysis

Cluster analysis
Step 1
336 models
37 variables
↓

Cluster analysis runs to gain experience and
to establish the number of clusters with
non-hierarchic techniques

Cluster analysis *Step 2* with various variable combinations

a. 10 variables
(10 main codes)

b. 12 variables
(12 factors in
correspondence
with 1 variable
each)

c. 23 variables
(selected by
subjective
evaluation)

d. 24 variables
(without asymp-
totic frequency
distributions)

e. 28 variables
(using variables)
of primary im-
portacne in the
factor)

After summing up the experience gained

f. *7 variables*
(using the determinant variables
of the previous investigations)

g. *11 variables*
(extended by several features
of the 7 variable combination)

The *a, b, c, d, e, f, g* blocks run with 5 various non-hierarchic cluster techniques, by creating
8 clusters on the basis of the trials. In other words, the number of analyzed runs in this phase:
(7 variable combinations) x (5 techniques) x (2 code systems) = 70 (ordinal, dichotomous)
Complementary control tests in the versions *f* and *g*:

✳

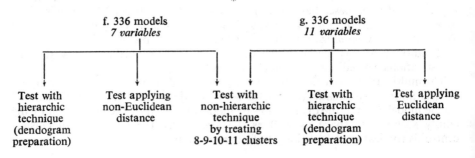

f. 336 models
7 variables

g. 336 models
11 variables

Test with
hierarchic
technique
(dendogram
preparation)

Test applying
non-Euclidean
distance

Test with
non-hierarchic
technique
by treating
8-9-10-11 clusters

Test with
hierarchic
technique
(dendogram
preparation)

Test applying
Euclidean
distance

100

Table 6

Evaluation of factor analyses

Variables / Factors	F_1	F_2	F_3	F_4	F_5	F_6	F_7	F_8	F_9	F_{10}	F_{11}	F_{12}
MC3	■											
R2	■											
R31	■											
MC10	■											
MC4		■										
D51		■										
MC8	≡	■										
MC9		≡										
C2			■									
C1			■									
MC6			■	≡								
C3			■							≡		
C4			≡								≡	≡
SM31				■								
SM32				■								
MC1					■							
I2					■							
SO2						■						
SO3						■						
MC7						■					≡	≡
MC2							■					
S2							■					
D2								■				
D4								■			≡	
D31	≡	≡							≡			
MACO									■			
DY1									■			
MC5										■		
SM5										■		
DV3											■	≡
R4	≡								≡		≡	
DV2												■
S3					≡			≡				
R51	≡		≡	≡								
I4				≡		≡						
I5	≡									≡		
SM31				≡	≡							≡

■ Variable of major importance in the factor

≡ Variable of minor importance

☐ Negligible share in this case

Table 7

Relation between factors and main codes

Factors	Corresponding main codes	Percentage of information represented by the factor
F_1	MC3 (or MC10)	14
F_2	MC4 (or MC8)	10
F_3	MC6	8
F_4	SM31 (or SM32)	6
F_5	MC1	6
F_6	MC7	5
F_7	MC2	5
F_8	D2 (or D4)	4
F_9	MACO	4
F_{10}	MC5	3
F_{11}	DV3	3
F_{12}	DV2	3
		71

V.3.2. Concluding Remarks on the Factor Analysis Study

The results displayed in most cases only very weak connections between our variables. There were only a few exemptions where strong interdependences had direct reasons.

The result of comparing the runs carried out in the two basic coding versions has been the following:

	With ordinal codes	With dichotomization
Number of factors (based on eigenvalues)	12	12
Information contents (%) represented by the factors	71	69

From among the 37 variables, 28 can be found in the 12 factors—as seen in Table 6—essentially without overlappings. The remaining 9 variables are present in several factors, but with less emphasis. The relationship between the main codes and the factors is interesting to observe. From the 12 factors, 7 correspond to one main code each (from the first 7 factors 6). Thus the information contents represented by the main codes is 51% (in the case of ordinal coding).

The computations with dichotomized variables comprised 36 variables (one variable showing extreme asymmetry has been left out of the trial). The essentials of the calculations supported the previous investigation.

To demonstrate the relations between the factors and the main codes based on the runs with ordinally arranged codes, Table 7 has been complied.

102

V.3.3. The Results of Cluster Analysis

Scheme 1 shows the echelon of cluster analytical investigations:

1. Trial calculations for preparing the non-hierarchic program runs, in order to set a limit to the number of clusters.
2. Cluster analysis of the complete material.
3. Trial examination of the primary variable combinations.
4. Applying the new variable combinations formed on the basis of partial results for grouping.
5. Examination of the variable combinations fixed as resultants in their hierarchy, by applying a non-Euclidean distance concept.

The steps described have comprised nearly 100 cluster analytical investigations based on professional considerations and we have striven to determine our conclusions methodically step by step.
(We indicated at the beginning of this chapter that parallel calculations have been carried out applying 5 techniques and two coding versions.)

Table 8 shows clearly the comparison of the essential variables combinations in the series of calculations.

The goal of our previous considerations has been the selection of the most promising combination of variables and—based on this—to create a basis for classification. Our sequence of trials has verified the most expressively that column of Table 8 which contains 7 variables. The number of variables could not be reduced further, since the properties in the combinations have behaved characteristically in several cases and displayed a decisive character in the formation of groups.

The increase of the number of variables disturbed the state space, because a considerable part of the variables did not possess fundamental decisive endowments from the viewpoint of grouping or classification and, therefore, taking them into account would result only in confusion among the groups. It was precisely the elimination of these confusing effects which yielded the most work in the course of our endeavours.

The results obtained by the various versions of cluster analysis constituted the starting point for the classification which is the basis of our model descriptions in the second part of this book. As a matter of course, the automatic classification obtained has in many cases been supervised. In this way, the final classification has been built up on the common base of intuitive knowledge concerning the models, and other previously described analyses as well as the cluster analysis.

Table 8

Comparison of cluster calculation results

No.	Original variables	7 variables resultant combination	10 variables main codes	11 variables extended resultant combination	12 variables corresponding to the 12 factors	23 variables based on subjective evaluation	24 variables without asymmetric distributions	28 variables selected on the basis of factor analysis
1.	MC1		×		×	×	×	×
2.	MC2		×		×	×		×
3.	MC3	×	×	×	×	×	×	×
4.	MC4	×	×	×	×	×	×	×
5.	MC5		×		×	×	×	×
6.	MC6		×		×	×	×	×
7.	MC7	×	×	×	×	×	×	×
8.	MC8		×	×		×	×	×
9.	MC9		×			×	×	
10.	MC10		×	×		×	×	×
11.	I2	×		×			×	×
12.	I4							
13.	I5						×	
14.	S2							×
15.	S3							
16.	R2					×	×	×
17.	R31						×	×
18.	R4					×		
19.	R51							
20.	D2	×		×	×	×		×
21.	D31						×	
22.	D4	×		×		×		×
23.	D51			×			×	×
24.	DY1					×		×
25.	DV2				×		×	×
26.	DV3				×		×	×
27.	SM31					×		
28.	SM32					×	×	×
29.	SM33				×	×		×
30.	SM5							×
31.	SO2					×	×	×
32.	SO3					×	×	×
33.	C1	×		×		×	×	×
34.	C2			×		×	×	×
35.	C3						×	×
36.	C4						×	
37.	MACO				×	×		×

Part Two

CLASSIFICATION AND DESCRIPTION
OF THE MODELS

Introduction

To create a well established classification system of the inventory models was one of the most important objectives of our research. In the second part of the book we first give our classification system then compare it with other classifications in the literature—and, after this introductory chapter we give a summary of all 336 models, in the order of our classification system.

Classification of the Models

The basis for creating this classification system has been provided by the analytical examination of the individual models and statistical analyses (first of all the multivariable statistical approaches). Thus our classification system has been derived from the analysis of our 336 element model sample—but we certainly hope that it can be used more generally (this expectation is supported by the comparison of our system with others, as we shall show.)

The first step in creating our classification was the use of results of the cluster analysis. As a result of the many clusteranalysis computations we found that the following seven characteristics of the models play the most important role in forming the various groups of models:

— the number of products,
— the number of locations,
— the character of input,
— the character of output,
— dynamics (the handling of time),
— the ordering rule (mechanism),
— the objective of the model.

We have found that these characteristics (which in fact, are described by the first seven main codes) can be used for a hierarchical classification, with a careful selection of the order of applying these codes. This way we have created the seven-level hierarchical classification scheme shown in Figure 7(a). At the bottom of this system there can be found the twelve "main groups" which seem to have different basic characteristics. It can be seen that we did not apply all possible combinations of the seven codes mentioned above, since some of these combinations were either irrelevant or meaningless. This way this classification is not strictly hierarchical—but, on the other hand we get a meaningful group at the end of each branch.

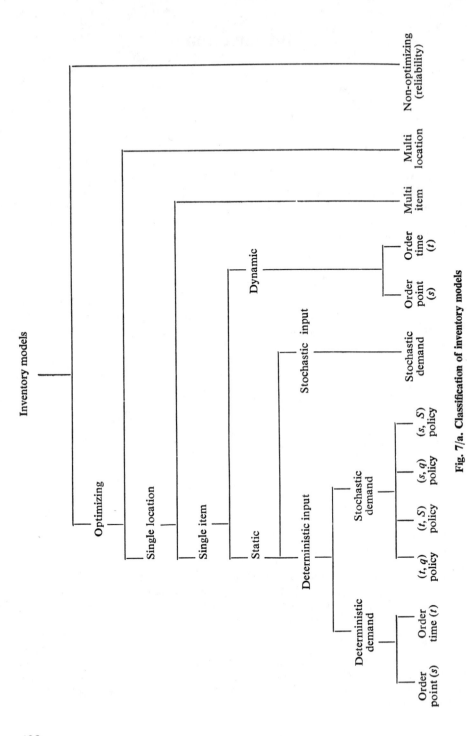

Fig. 7/a. Classification of inventory models

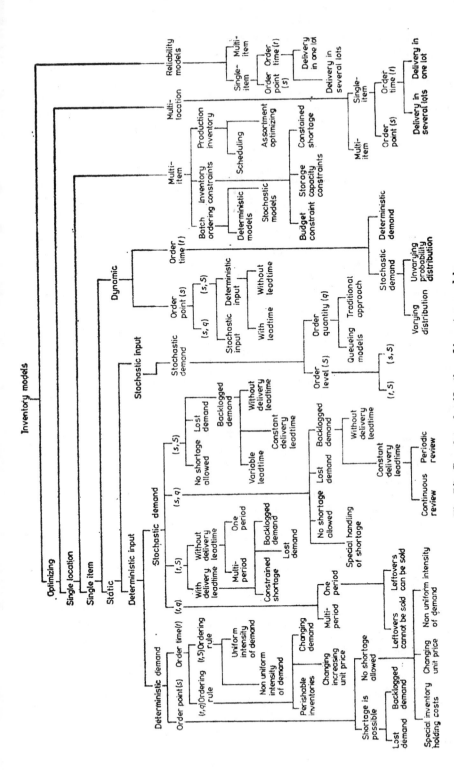

Fig. 7/b. Complete classification of inventory models

In the subsequent analysis, we have created subgroups of these main groups. The further decomposition has not been made by a unified approach to all main groups but rather by considering the special characteristics of the individual main groups. These subgroups do not have such a sound basis as the main groups —they have been created by intuition as well as by using the results of the previous analyses. We show the complete classification scheme on Figure 7(b), and give the subgroups also at the introduction of the description of all the twelve main groups.

Comparison with Other Classifications

It was an obvious idea to compare our classification system with others from the literature.

Our classification system worked quite well when applied to our sample of models—with two exceptions we could group all the 336 models in an appropriate class. The distribution of models in the twelve main classes is almost uniform. Considering that our sample is a rather random one, this shows that the system is workable. It was, however, obviously a good test of not only the classification system, but our whole approach to handling the models, to make a comparison with other approaches.

Let us emphasize at the beginning that we did not aim to evaluate the different classification systems on a basis of saying "this one is better or worse than the other one". This would be an oversimplified approach; the problem is much more complicated than to be handled in such a one-dimensional way as "ranking" the various classification systems. Any of them can be useful and important from some point of view which might not even be considered in another system.

We have considered several classification systems in the literature, and tried to apply our code system to the aspects used by the authors. In some cases (like Handley—Whitin 1963, Veinott 1968, Ryshikow 1969, Tersine 1976, Hollier—Vrat 1978) we had to reject the idea of comparison for various reasons, but we found six systems which could be included in the analysis: Naddor 1966, Hochstädter 1969, Klemm—Mikut 1972, Aggarwal 1974, Nahmias 1978 and Silver 1981. (As a seventh system, we had ours: for comparison we used the seventh hierarchical level shown in Figure 7—where we have the twelve main groups.) The classification systems of the other authors are given in Figures 8—13.

Of course, the logic behind these classifications is quite different, but one of the most interesting aspects of the analysis is just that: if we use the same characterizing (code) system and give a different emphasis to the various characteristics, how will that affect the classification? Does that lead to a very different classification or are there stronger ties between the "similar" models and how they stay together? Questions like these confronted us when we started the analysis.

It was relatively easy to apply our code system to describe the various classification systems since the main classifying aspects are rather similar; the difference is greater in the combination than in the definition of the basic characteristics. Nevertheless, there were cases when we had to use intuition to identify some aspects. In other cases we had to interpret the intention of the authors. We certainly hope that we have not failed any of the classification systems. Anyway,

110

we must emphasize that the interpretation given here does not necessarily reflect that of the authors, for which we apologize and take the responsibility.

We have used our code-system to describe the classification of the authors, and applied all classifications to the 336 member sample of models. The analysis given below is based on a series of computer runs from elementary statistical evaluation to discriminant-analysis. When evaluating the results of our analysis it must be considered that the comparison has been made on the basis of our system of aspects and our sample of models. Therefore, we must be very cautious in judging the various classification systems—though it is promising, from the point of view of the value of the results, that it can be shown rather well that the system of aspects considered by the various authors corresponds well to our system of aspects. Of course, there can be differences in the interpretation of the various aspects, i.e., it is quite possible that we mean something else under a particular aspect than another author—this fact leads to the "same code—different consideration" type of error.

The various classification systems are very different, a priori, from at least three points of view:

— Do they attempt to give a systematic classification (Aggarwal, Naddor, Hochstädter, Klemm—Mikut), or to establish the most important groups based on a priori knowledge (Nahmias, Silver)?

Naddor (1966)

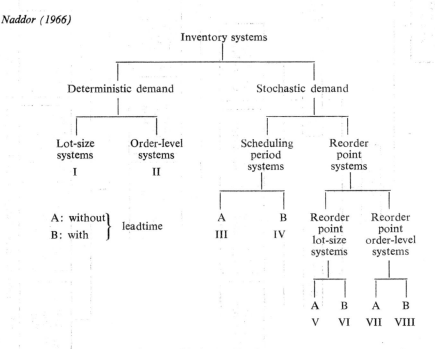

Fig. 8. Naddor's classification system

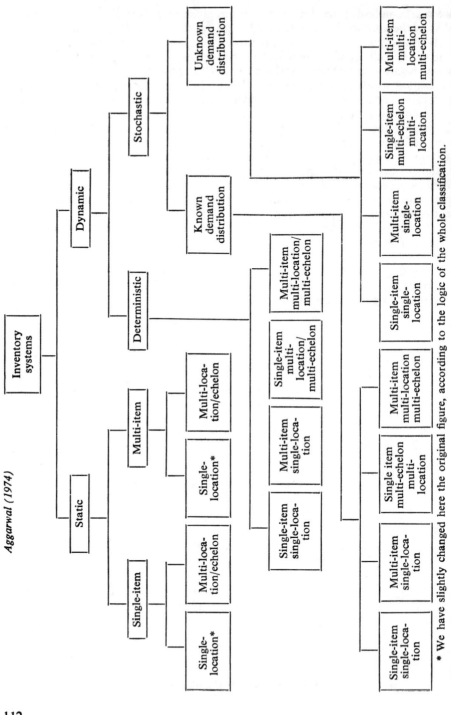

Aggarwal (1974)

* We have slightly changed here the original figure, according to the logic of the whole classification.

Klemm—Mikut (1972)

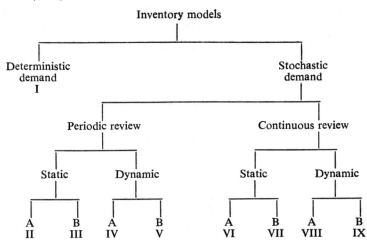

A: deterministic leadtime
B: stochastic   leadtime

Fig. 10. Klemm—Mikut's classification system

Hochstädter (1969)

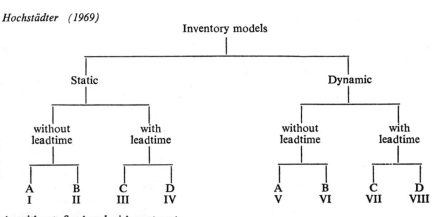

A: without fixed replenishment cost
B: with   fixed replenishment cost
C: backorder case
D: lost-sales case

Fig. 11. Hochstädter's classification system

113

Nahmias (1978)
  I. The EOQ model and extensions
 II. Continuous review models with random demand
III. The dynamic lot-size periodic review model with concave/convex costs
 IV. Single period models with stochastic demand
  V. Multiperiod stochastic models with/without set up cost and zero/non zero leadtime
 VI. Multiple products
VII. Multiple echelons
VIII. Obsolescence, decay and perishability

Fig. 12. Nahmias's classification system

Silver (1981)

 1. Single item with deterministic, stationary conditions
 2. Multi-item deterministic, stationary situation under budget, space, replenishment or work load constraints
 3. Single item with deterministic but time-varying parameters
 4. Single item with stationary, probabilistic demand with known distribution
 5. Multi-item probabilistic stationary situation under a budget or space constraint
 6. Single item, single period situation
 7. Multi-item, single period situation with a budget constraint
 8. Coordinated control of items under deterministic, stationary demand
 9. Coordinated control of items under probabilistic stationary demand and discount opportunities
10. Single perishable item with stationary probabilistic demand
11. Multi-echelon, stationary situations

Fig. 13. Silver's classification system

— How many aspects are considered in the classification? Of course, a systematic classification can only consider a smaller number of aspects, (since a great number of aspects would lead to an exponentially growing number of groups). As it can be seen in Table 9, Nahmias and Silver handle many more aspects than any other groupings.
— What aspects are considered? It can also be seen in Table 9 that in the classifications a great variety of aspects appear (19 out of the 45 we had originally considered), subsets of which are handled by the various authors. Here we have to add that some of the authors considered only a subset of the models: e.g. Naddor deals only with single item, single location, static models; Hochstädter considers only deterministic input–stochastic output models. We found it interesting that our analysis explicitly shows that if we apply the classifications with or without these "filters" the results are much the same.

As for the number of models within the different groups of the various classifications (see Table 10) one can see that the more "mechanical" classifications (Aggarwal, Hochstädter, Klemm—Mikut and Naddor) excluded fewer models of the 336 element sample. This can be explained by the fact that these classifications are based on a much smaller number of aspects than the system of Nahmias and Silver (See Table 9.) The models considered are distributed among the groups most uniformly in Naddor and Nahmias, and there is an extreme case in Aggarwal's classification. There are a fairly large number of empty, or almost empty groups. If our 336 model sample represents the whole set of inventory

Table 9

Aspects of classification considered by the various authors

Aspects	Aggarwal	Barancsi et al.	Hochstädter	Klemm—Mikut	Naddor	Nahmias	Silver
Single vs several items	×	×	0	0	0	×	×
Single vs several locations	×	×	0	0	0	×	×
Deterministic vs stochastic input	—	×	0	×	×	×	×
Deterministic vs stochastic output	×	×	0	×	×	×	×
Static vs dynamic	×	×	×	×	0	×	×
Cost optimizing vs reliability	—	×	—	—	0	—	—
Ordering rule	—	×	—	—	×	×	—
Periodic vs continuous review	—	—	—	×	—	×	—
Handling of shortage	—	—	×	—	—	×	—
Handling of leadtime	—	—	×	×	×	×	—
One vs several periods	—	—	—	—	—	×	×
Time pattern of demand	—	—	—	—	×	—	—
Probability distribution of demand	×	—	×	—	—	—	×
Budget, capacity or other constraint	—	—	—	—	—	—	×
Perishability	—	—	—	—	—	×	—
Arrangement of locations	—	—	—	—	—	—	×
Mathematical tools used	—	—	—	—	—	×	—
Types of costs considered	—	—	—	—	×	—	—
Cost dimensions	—	—	—	—	—	—	×
Number of aspects explicitly considered	5	7	4	5	6	12	10

0 implicitly fixed
× used in classification
— not considered

models then these empty groups can be considered blank spots, i.e., inventory situations which are not handled by models.

If we take a look at Table 11, interesting results can be discovered from the point of view of the stability of various groups of inventory models. The Table shows that from all models considered in the various classifications how many have been put into our 12 classes. It can be seen that groups I, II and XI are very stable, while with some exceptions groups IV, V, VI and X can be also considered stable. Groups III, VIII and IX show a rather mixed picture, while groups VII and XII are unstable (the latter has practically dissolved).

Table 10

Number of models in the groups of the various classifications

	Aggarwal	Barancsi et al.	Hochstädter	Klemm—Mikut	Naddor	Nahmias	Silver
I	190	26	80	84	59	34	46
II	12	16	55	78	16	10	4
III	51	14	39	3	31	0	9
IV	4	29	8	47	20	25	0
V	5	24	21	2	12	17	7
VI	1	28	26	26	24	56	4
VII	1	38	11	33	36	22	2
VIII	0	39	7	3	17	12	9
IX	21	25		2			0
X	1	51					6
XI	0	21					16
XII	0	23					
XIII	15						
XIV	2						
XV	1						
XVI	1						
	305	334	247	278	215	176	103

Note: There is no connection between the characteristics of the groups having the same number in the different classifications. (There is a slight tendency to proceed to more complicated models moving off from group I in all classifications.)

The results show that there is a rather unanimous approach to the classification of the simple deterministic models and to the multi-location models. Authors are largely in agreement in classifying the deterministic input—stochastic output models (except those with a (t, q) ordering rule—which is quite understandable since this ordering rule can be applied in a stochastic system under very special circumstances only). There are big differences in handling the models with stochastic input.

We have analysed the two-dimensional cross-relations tables of all two element combination of classifications. We have used several methods to analyze the connections and found the results summarized in Table 12.

The centroids of the various groups of models have been represented in the space of discriminant functions; for an example, see Figures 14 and 15. In most cases both the discriminant functions and the map of the centroids can be explained rather well, which is an indication of the meaningfulness of a given particular classification system. Naturally, the groups should have been represented in a multi-dimensional space: the figures are not for proof, but for illustration only.

On the basis of the discriminant functions we considered the automatic classifications. We could interpret a "predicted group membership", which means

Table 11

Number of models considered by different classifications and included in our groups

	Aggarwal	Barancsi et al.	Hochstädter	Klemm—Mikut	Naddor	Nahmias	Silver
I	26	26	26	26	24	22	26
II	16	16	16	16	16	14	16
III	14	14	8	3	—	10	5
IV	29	29	28	26	29	4	1
V	24	24	22	24	24	15	3
VI	28	28	27	25	25	3	—
VII	37	38	2	34	—	4	—
VIII	26	39	32	36	25	17	3
IX	15	25	22	24	16	9	6
X	49	51	42	36	39	51	18
XI	19	21	19	19	14	21	16
XII	22	23	3	9	3	6	9
	305	334	247	278	215	176	103

Table 12

Cross connections between the various classifications

	Aggarwal	Barancsi et al.	Hochstädter	Klemm—Mikut	Naddor	Nahmias	Silver
Aggarwal	×	0	—	—	—	0	0
Barancsi et al.	2	×	+	+	—	+	—
Hochstädter	3	3	×	0	0	—	0
Klemm—Mikut	4	2	4	×	0	0	0
Naddor	5	2	4	2	×	—	—
Nahmias	2	1	4	3	4	×	0
Silver	1	1	3	3	3	1	×

As another aspect of the analysis, in the upper triangle we have put a 0 sign at the places where the columm classification in terms of its strength of explanation value in forming the common groups dominates the row classification. The sign + shows the opposite case, while in case of a — no significant connection of that type could be discovered.

Strength of connection (lower triangle of the table):

1 — very close; 2 — close; 3 — medium close; 4 — loose; 5 — very loose.

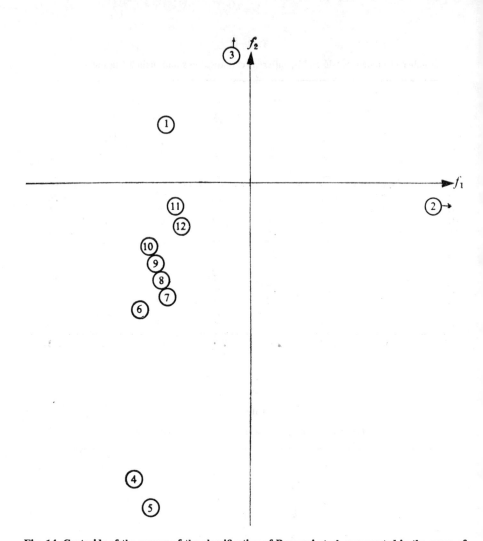

Fig. 14. Centroids of the groups of the classification of Barancsi et al. represented in the space of the first two discriminant functions
f_1: dominated by the number of items, f_2: dominated by the number of locations.

the member of models classified automatically to the various groups. The "probability of hitting" (i.e., the ratio of the models put to the same group by the automatic classification such as by the original classification (see Table 10) was very high (above 85%) in the case of our system and that of Nahmias and Silver, medium (between 30 and 50%) in the case of Hochstädter, Klemm—Mikut and Naddor, and low in case of Aggarwal). This result shows two things:

118

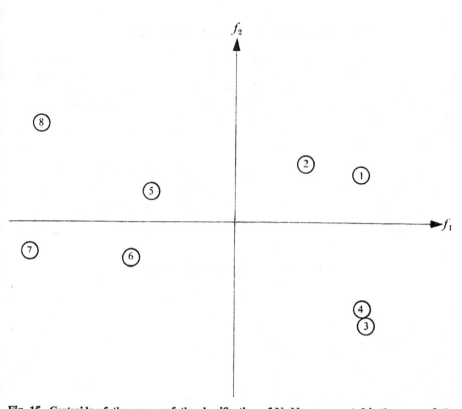

Fig. 15. Centroids of the groups of the classification of Naddor represented in the space of the first two discriminant functions

f_1: dominated by the ordering rule, f_2: no dominating aspect, cost factors with major weights.

— the consistency between our grouping and the others,
— the stability of the system of the various authors,

The results are not surprising. In the case of Nahmias and Silver, many models have been excluded when we fitted their system to the others, so the remaining models form rather stable groups, while as a consequence of the character of Aggarwal's system far too large a number of models have been put into a few groups (mainly to the first one) so they are necessarily regrouped when other aspects have been considered as well. The system of Naddor, Hochstädter and Klemm—Mikut could handle a fairly large number of models, but the classifications of the remaining models could not correspond to the automatic regrouping dominated by our aspects.

These results show the significance of the difference between the various aspects used.

Conclusions of the Comparison

The different classification systems show fairly different characteristics when applied to our sample of models. The various features of classifications and differences among them can be explained and some interesting conclusions can be drawn. We have seen that the logically strict hierarchical classifications create too many statistically unstable groups with few elements in several groups, while those classifications which consider only the a priori known "relevant" groups, though this creates stable classes, leave out a great many existing models. So one can arrive at the conclusion that the two approaches need to be combined.

Basically all authors agree on the most important aspects—the first few hierarchical levels can be created considering these aspects at the different groups. Then a variety of sets of aspects can be used to form the subgroups, considering the specific characteristics of each particular main groups. As we mentioned before, we have applied this procedure in creating our classification system.

On the Description of Models

In the following we give the description of the 336 models analysed by us. In publishing these descriptions we wanted to give an idea to the reader: what are the basic characteristics of the individual models, what are the assumptions, objective functions and methods used. For accomplishing this goal we necessarily had to give descriptions very different in exhaustiveness. We have advanced one or two "basic" models from each main group, which we think represent the main characteristics of the whole group, and give a detailed description of these models, while the further models in the same groups will be described only to show the essence of the model.

We warn the reader against using these descriptions either for theoretical analysis or for practical purposes. Our objective here was only to give basic information through publishing these abstracts—which can be the basis for further study of the original publications, the exact reference to which can be found in Appendices I and II.

Before turning to the descriptions we still have to make three practical remarks:

— It would have been nice to use a unified system of notions but we found it absolutely impossible. In such an extensive model system the various authors give such different meanings to otherwise "similar" parameters that "unifying" would have been more misleading than useful. So we had to give the meaning of all notations in the individual descriptions.
— There are some special models which cannot be classified unanimously into the groups or subgroups used. We did not put them into a general "other" groups, but—referring to their specialities—gave their description in the "closest" group.
— We used the following general notations when giving the dimensions of the unit costs:
  [$]: value in money term,
  [Q]: quantity
  [T]: time.

120

I. Deterministic Input and Demand, (s, q) Policy

This is the classical family of inventory control models, based on one of the most simple system of assumptions therefore it is probably the most elaborated group of models. Two subgroups can be distinguished. No shortage is allowed in one of the subgroups, here the parameter s is fixed on the level $s=0$ (i.e. replenishments are made whenever the inventory has run out). In the other subgroup shortage is allowed, thus the parameter s is a decision variable of the model. Further classification of the first subgroup can be given on the basis of the specific assumptions of the models, which may differ from those of the basic model concerning the inventory carrying cost, the purchasing cost and the type of demand. A special model with uniform replenishing rate belongs also to this group of models. The second subgroup can be divided into two parts on the basis of the shortage reaction, whether there is a backorder or a lost sales case. Three special models, two descriptive type and one based on control theory appear in the group of models with backordering.

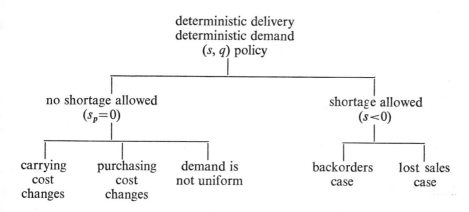

1 and 2. *The classical economic order quantity model*

Main codes:

$$1\ 1\ 0\ 0\ 0\ 1\ 3-1\ 0\ 0$$

Assumptions:

There is a deterministic, continuous and uniform demand with a constant rate r. The whole amount ordered arrives at one occasion with zero lead-time. The order rule is as follows: replenishments are made whenever the inventory reaches the prescribed zero reorder level ($s_p=0$). The replenishment size is constant, thus it is the lot size q which is the decision parameter of the system. If the lead-time is a known constant L, the reorder level must be chosen for $s_p=Lr$.

The above order rule has the consequences that:

— no shortage is allowed, thus only the carrying cost factor c_1 and the replenishment cost factor c_3 has to be considered,
— the order periods have the same length $t=q/r$.

The model considers one item and one inventory location, it has a static character.

The assumptions of both models are the same, the difference is only in the derivation of the objective function. Here we follow the logic of model 2.

Objective:

The number of the replenishment orders is $1/t=r/q$ in a time unit. The inventory level fluctuates in the interval $(0, q)$. The mean level of the inventory is $q/2$, because of the constant demand rate. The objective function of the system is the total of the carrying and replenishing costs in a time unit:

$$C(q) = \frac{c_1 q}{2} + \frac{c_3 r}{q}$$

which has to be minimalized with respect to q.

Solution:

For the optimal value of q, $\dfrac{dC(q)}{dq}=0$ provides a necessary condition. The equation

$$\frac{dC(q)}{dq} = \frac{c_1}{2} - \frac{c_3 r}{q^2} = 0$$

has the unique solution

$$q_0 = \sqrt{\frac{2rc_3}{c_1}}$$

which, in fact, gives the minimum of the objective function, since the second derivative $\dfrac{dC^2(Q)}{dq^2}=2c_3 r/q^3$ has a positive value for $q=q_0$ (which gives a sufficient condition for optimality).

I.1. Lot-Size Models with No Shortage

I.1.1. Lot-Size Models under Specific Inventory Holding Costs

3. *Lot-size system for perishable items*

Main codes:

$$1\,1\,0\,0\,0\,1\,3 \quad -1\,0\,0$$

122

Assumptions:

Perishable goods are considered, thus the total inventory carrying cost is not a linear function but is supposed to be proportional to the nth power of the time of storage. The total carrying cost can be expressed in the form $K = a \cdot c_1 \cdot q \cdot t^n$, where n is a constant $(n > 1)$ and $a = 1/n + 1$. The cost factors (c_1 and c_3) have the following dimensions:

$$[c_1] = \frac{[\$]}{[Q][T]^n}; \quad [c_3] = [\$] \quad (n > 1).$$

Objective:

The cost function is:

$$C(q) = \frac{c_1 q^n}{r^{n-1}(n+1)} + \frac{c_3 r}{q} \quad \text{(for given } n > 1\text{)},$$

which has to be minimized according to the lot size q.

Solution:

The necessary condition $\dfrac{dC(q)}{dq} = 0$ yields

$$q_0 = \sqrt[n+1]{\frac{n+1}{n} \cdot \frac{c_3 r^n}{c_1}},$$

which in fact gives the minimum value of the cost function.

4. The "expensive storage" system

Main codes:

$$1\,1\,0\,0\,0\,1\,3 \;-1\,0\,0$$

Assumptions:

In this version of the classical lot-size system (1) the total carrying cost is not a linear function but is supposed to be proportional to the mth power of the quantity stored: $K = a c_1 q^m t$ where m is constant $(m > 1)$ and $a = 1/(m+1)$. The dimensions of the cost factors are

$$[c_1] = \frac{[\$]}{[Q]^m[T]}; \quad [c_3] = [\$] \quad (m > 1).$$

Objective:

The cost function is

$$C(q) = \frac{c_1 q^m}{m+1} + \frac{c_3 r}{q}.$$

Solution:

Applying differentiation as before, the optimal lot size is

$$q_0 = \sqrt[m+1]{\frac{m+1}{m} \cdot \frac{c_3 r}{c_1}}, \quad (m > 1).$$

123

5. *Carrying cost is a non-linear function of both the inventory and the stocking time*

Main codes:
$$1\ 1\ 0\ 0\ 0\ 1\ 3\ \ -1\ 0\ 0$$

Assumptions:

In this version of the lot-size system, the carrying cost is not linear but is proportional to a constant power both of the amount of inventory and the time of storing. The total carrying cost is expressed in the form $K=ac_1q^m t^n$, where m, n and a are constants, $(m+n-1)\geq 0$. The special case $n=1$ is the model for perishable goods (Model 3_1), while the case $m=1$ is the model for expensive storage (Model 4_1), which have been described previously. The dimensions of the cost factors are

$$[c_1] = \frac{[\$]}{[Q]^m[T]^n}; \quad [c_3] = [\$], \quad (m\geq 1, n\geq 1)$$

Objective:

The cost function can be expressed in the form

$$C(q) = \frac{ac_1\,q^{m+n-1}}{r^{n-1}} + \frac{c_3 r}{q}.$$

Solution:

The optimal value of the lot size results by differentiation:

$$q_0 = \sqrt[m+n]{\frac{c_3 r^n}{ac_1(m+n-1)}} \quad \text{for} \quad m+n-1\geq 0$$

6. *Optimal lot size in the case of discounted costs*

Main codes:
$$1\ 1\ 0\ 0\ 0\ 1\ 3\ \ -1\ 0\ 0$$

Assumptions:

The lot size q (which is a production order) arrives from the production sector at a constant rate in the time interval $(0, t_1)$. It is stored during the interval (t_1, t_2) and it is consumed in the interval (t_2, t_3).

Objective:

The cost of ordering is given by the sum of the set-up cost of a production lot (which is effective in the time interval $(0, t_1)$) and the production cost. The latter is a monotonously decreasing function of the lot size. The inventory carrying cost consists of two parts: the actual inventory holding and the capital cost which appear during the time interval $[0, t_3]$ with different intensities. The costs are discounted according to their effect as a function of time with a discount rate $1/(1+r)^t$ at the time t ($r>0$ is a given policy parameter). The sum of the above four discounted costs gives the objective function, as a function of the lot size q. The discounting is performed in an exact and in an approximate way. Different, rather complicated cost functions can be formulated.

Solution:

This cannot be achieved in an explicit analytical form. Therefore numerical methods are suggested for the minimization of the objective functions, but they are not detailed by the author.

I.1.2. Lot-Size Models with Cost/Price Changes

7. *Lot-size model with purchase price changes*

Main codes:
$$1\ 1\ 0\ 0\ 0\ 1\ 3\ \ -1\ 0\ 0$$

Assumptions:

There is a constant demand rate r. A purchase order is given when the stock level decreases to the reorder level $s_p = 0$. The unit cost of an item depends on the quantity q purchased. For lot-sizes q in the range Y_i to Y_{i+1}, the price is b_i for all units. The prices $b_1, b_2, ..., b_n$ will, in general, be decreasing. Thus there is a quantity discount applied to every unit purchased. Without loss of generality, $Y_1 = 0$ and $Y_{n+1} = \infty$ can be assumed.

Objective:

The cost function is defined in a piecewise way,

$$E_i(q_i) = \frac{f \cdot b_i q_i}{2} + \frac{c_3 r}{q_i} + b_i r, \quad \text{if} \quad Y_i \leqq q_i < Y_{i+1}$$

for $i = 1, 2, ..., n$. Here f denotes the carrying charge (in percentage) and c_3 the fixed purchase order cost which is independent of the lot size.

Solution:

The algorithm is based on the following scheme:

1. Calculate $q_n = \sqrt{\dfrac{2rc_3}{f \cdot b_i}}$. If $q_n \geqq Y_{n-1}$, then q_n is the optimal value of q.
2. If $q_n < Y_{n-1}$, then calculate q_{n-1}.
   If $q_{n-1} \geqq Y_{n-2}$, then compare the value of $E_{n-1}(q_{n-1})$ with $E_{n-1}(Y_{n-1})$ and the point belonging to the smaller value yields the optimal value of q.
3. If $q_{n-1} < Y_{n-2}$, then calculate q_{n-2}. If $q_{n-2} \geqq Y_{n-3}$, then the minimum of $E_{n-2}(q_{n-2})$, $E_{n-2}(Y_{n-2})$ and $E_{n-1}(Y_{n-1})$ yields the optimal value of q.
4. In the case of $q_{n-2} < Y_{n-3}$ the value of q_{n-3} will be calculated. If $q_{n-3} \geqq Y_{n-3}$, then the q, for which the minimal value among $E_{n-3}(q_{n-3})$, $E_{n-3}(Y_{n-3})$, $E_{n-2}(Y_{n-2})$ and $E_{n-1}(Y_{n-1})$ is attained, yields the optimum.

$\vdots$

The calculation proceeds the same way until the minimum is reached. Obviously, the outlined procedure consists of at most n steps.

8. *Lot-size system with quantity discounts*

Main codes:
$$1\ 1\ 0\ 0\ 0\ 1\ 3\ \ -1\ 0\ 0$$

125

Assumptions:

The optimal lot size is influenced by the fact that the purchasing price $b(q)$ depends on the lot size of the reorders. The holding cost per unit is a certain fraction of the purchase cost $c_1 = fb(q)$. The unit cost of replenishment consists of a fixed cost and of the purchase price: $c_3 = e_3 + qb(q)$. There is a constant demand rate r and no shortage is permitted.

Objective:

The total cost of the system is

$$C(q) = fb(q)\frac{q}{2} + \frac{e_3 + qb(q)}{q/r} = fb(q)\frac{q}{2} + \frac{e_3 r}{q} + rb(q).$$

Solution:

The method for minimizing $C(q)$ depends on the explicit form of the function $b(q)$. The author considers the example $b(q) = b_0 - b_1 q$, where $b_0 \gg b_1$. In this case, the optimal lot size q_0 can be found by solving the equation

$$q_0 = \sqrt{\frac{2re_3 + 2fb_1 q_0^3}{fb_0 - 2rb_1}}$$

from which q_0 can be obtained by successive iterations. In the case of discrete quantity discounts we have the previous model (Model 7).

9. Lot-size system with price increase

Main codes:

$$1\ 1\ 0\ 0\ 0\ 1\ 3\ \ -1\ 0\ 0$$

Assumptions:

A single price increase situation is considered. Purchases before time-moment T_0 will cost d per unit quantity, while purchases after T_0 will cost $d + k$. A single ordering decision just before T_0 will be optimized. There is a constant demand rate r and no leadtime. Shortage is not permitted. The unit holding cost is a fraction of the purchase price: $c_1 = cd$. The ordering cost is c_3.

Objective:

Let K' designate the total cost during the period T_0 to T_1 when an amount $q' > 0$ is purchased at the cost d, then: $K' = dq' + \frac{q'}{2} dp \frac{q'}{r} + c_3$. In the case of buying at the new price, the cost function is:

$$K = (d+k)q' + \frac{q_0}{2}(d+k)p \frac{q'}{r} + \frac{q'}{q} c_3.$$

Compare the cost of not taking advantage of the price change with the purchase price, when an amount q' is ordered just before the price increase. The cost difference is

$$G = K - K' = q'\left[k + \sqrt{2c_3(d+k)\frac{p}{r}}\right] - \frac{dp}{2r}q'^2 - c_3.$$

The amount q_0' which maximizes this difference provides the solution.

126

Solution:

Can be given (applying differentiation) in explicit form,

$$q_0' = \sqrt{\frac{2c_3 r}{(d+k)p}} \cdot \frac{d+k}{d} + \frac{kr}{pd} \quad (k > 0).$$

which can be expressed by using the optimal lot size q_0' belonging to price $(d+k)$ (See Model 1) in the following form:

$$q_0' = q_0 + \frac{k}{d}\left(q_0 + \frac{r}{p}\right).$$

10. Lot-size system with variable price change

Main codes:

$$1\ 1\ 0\ 0\ 0\ 1\ 3\ -1\ 0\ 0$$

Assumptions:

Are similar to the previous model (Model 9) except that the amount of the price change is random, with expected value $\bar{k}$ and probability density function $f(k)$.

Objective:

The expected gain, for a purchase of quantity $q' > 0$, just before the time of price change equals

$$C(q') = \int_{-d}^{\infty} q'\left[k + \sqrt{2c_3(d+k)\frac{p}{r}}\right]f(k)\,dk - \frac{dp}{2r}q'^2 - c_3.$$

If the ratio $\frac{k}{d}$ is sufficiently small, then $\sqrt{d+k}$ can be approximated by $\sqrt{d}\cdot\left(1+\frac{k}{2d}\right)$. In this case

$$C(q') = q'\left[k + \sqrt{2c_3\,dp/r}\cdot\left(1+\frac{k}{2d}\right)\right] - \frac{dp}{2r}q'^2 - c_3.$$

Solution:

Is given in explicit form in the case of the above approximation:

$$q_0' = \bar{q}_0 + \frac{\bar{k}}{d}\left(\bar{q}_0 + \frac{r}{p}\right),$$

where $\bar{q}_0$ is the optimal lot size based on an expected price change of $\bar{k} > 0$, i.e.

$$\bar{q}_0 = \sqrt{\frac{2rc_3}{(d+\bar{k})p}}.$$

11. Lot-size with uniform replenishment rate

Main codes:

$$1\ 1\ 0\ 0\ 0\ 1\ 3\ -1\ 0\ 0$$

127

Assumptions:

There is a demand with constant rate r. The replenishment order is given, when the stock level decreases to the reorder level $s_p=0$. The replenishment starts immediately after having the order placed with a uniform rate p, where $p>r$.

Objective:

For a lot size q, the total cost of the system to be minimized is

$$C(q) = \frac{c_1 q \left(1 - \frac{r}{p}\right)}{2} + \frac{c_3 r}{q},$$

where c_1 is the inventory carrying cost factor and c_3 is the fixed order cost.

Solution:

Can be given explicitly by

$$q_0 = \frac{\sqrt{2rc_3/c_1}}{\sqrt{1-r/p}}.$$

I.1.3. Lot-Size Models with Non-Uniform Demand Rate

12. *Reorder-point lot-size system with increasing demand*

Main codes:

$$1\ 1\ 0\ 0\ 0\ 1\ 3\ \ -1\ 0\ 0$$

Assumptions:

The system operates only during a prescribed period, which consists of H time units. At this time, there exists a total demand of D units. The rate of demand r changes linearly with time T, i.e., $r=aT$. The optimal reorder point is given by $s_0=0$ so when the stock level decreases to 0, an order is given for amount q. There is no leadtime. In consequence of increasing demand and uniform lot size the order period decreases in time.

Objective:

The total cost of the system

$$C(m) = \frac{c_1 Dh(m)}{m} + \frac{c_3 m}{H},$$

where $m=D/q$ is the number of orders during period H, (m is a positive integer) and

$$h(m) = \frac{2m}{3} - \frac{\sqrt{1}+\sqrt{2}+\dots+\sqrt{m-1}}{m}.$$

128

Solution:

For the optimal (integer) value m_0 of m, the following constraints are valid

$$F(m_0-1) \leqq \frac{c_1 DH}{c_3} \leqq F(m_0),$$

where

$$F(m) = \frac{m(m+1)}{(m+1)h(m) - mh(m+1)}.$$

The optimal lot size is $q_0 = D/m_0$.

13. Lot-size model with power demand pattern

Main codes:

$$1\ 1\ 0\ 0\ 0\ 1\ 3 \quad -1\ 0\ 0$$

Assumptions:

There is a constant lot size and the orders are given, when the stock reaches the reorder level $s_p = 0$. There is no leadtime and there are constant order periods. Demand is known to have a power pattern which is defined by

$$Q(T) = S - x \sqrt[n]{\frac{T}{t}},$$

where

$Q(T)$ is the inventory level a time T,
S is the initial stock (at time $T=0$),
x is the demand of the period with length t,
n is the power index characterizing the time dependence of the demand fluctuations.

In the system there is a basic prescribed period V during which there is a power demand pattern with index n. Let W be the known demand during V. The order amount is $q = mW$ where m is assumed to be an positive integer.

Objective:

The total cost

$$C(q) = \frac{c_1 q}{2} + \frac{c_1 \dfrac{W}{2}(1-n)}{1+n} + \frac{c_3 r}{q}$$

is to be minimized, where $q = W, 2W, 3W, \ldots$

Solution:

For the optimal q_0 (by the convexity of $C(q)$) a sufficient condition is $C(q_0) \leqq$ $\leqq \min \{C(q_0 - W), C(q_0 + W)\}$, which yields

$$q_0(q_0 - W) \leqq \frac{2rc_3}{c_1} \leqq q_0(q_0 + W)$$

where $r = W/V$.

14. Deterministic lot-size model—a general root law

Main codes:

$$1\ 1\ 0\ 0\ 0\ 1\ 3\ 2\ 0\ 0$$

Assumptions:

There is a deterministic demand which is a power function of time in the form $b(t)=kt^r$, where $k>0$ and $r>-2$. The full amount ordered is delivered immediately. No shortage is allowed, thus only the inventory holding cost c_1 and the cost of ordering c_3 has to be considered. An order is given when the inventory state is $s_p=0$. A known finite time horizon T is considered.

Objective:

The total cost equals

$$C(m, t) = c_1 Y(m, T) + c_3 m$$

where $Y(m, T)$ is the total inventory during the time horizont T and m is the number of orders in $[0, T]$.

Solution:

For the optimal m_0 we have the inequalities

$$\Delta L_{m_0} \leqq \frac{(r+2)c_3}{kc_1 T^{r+2}} \leqq \Delta L_{m_0-1},$$

where $\Delta L_m = L_m - L_{m-1}$,

$$L_i = [(r+2)-(r+1)L_{i-1}]^{-1/(r+1)} \quad \text{for} \quad r \neq -1$$

and

$$L_i = \exp\{-(1-L_{i-1})\} \quad \text{for} \quad r = -1 \quad (i = 1, 2, \ldots),$$

and

$$L_0 = 0.$$

The time moment of the ith order t_i may be calculated on the basis of the inequalities

$$R_i(m_0)\left[\frac{(r+2)c_3}{k\cdot c_1}\right]^{1/(r+2)} \leqq t_i^0(m_0) \leqq S_i(m_0)\left[\frac{(r+2)c_3}{k\cdot c_1}\right]^{1/(r+2)}$$

where

$$S_i(m_0) = \frac{\prod\limits_{j=i}^{m_0-1} L_j}{[\Delta L_{m_0}]^{1/(r+2)}}$$

and

$$R_i(m_0) = \frac{\prod\limits_{j=1}^{m_0-1} L_j}{[\Delta L_{m_0-1}]^{1/(r+2)}}.$$

The lot-size is equal to the demand occurring between two consecutive time insstants of orders.

130

15. *Optimal lot-release size*

Main codes:

$$1\ 1\ 0\ 0\ 0\ 1\ 3\ \ -1\ 0\ 0$$

Assumptions:

There is a uniform, deterministic input (production) in batch size q. After producing a batch, it will be stored, thereafter it will be consumed with a constant intensity r.

Objective:

The inventory holding cost consists of two parts: the storage cost and the cost of capital tied up in inventories. The setup cost corresponds to the ordering cost. There is still the production cost which is a monotonously decreasing function of the batch size (corresponding to the exponential "learning" curve). The sum of these four costs—as a function of the batch size—is the objective to be minimized.

Solution:

The derivative of the objective function results in a sufficient condition for the optimum in the form of an equation which can be solved by numerical methods (e.g. by Newton—Raphson iteration).

I.2. Lot-Size Models When Shortage is Allowed

I.2.1. Backorder Case

16. *Deterministic lot-size model with possible shortage, continuous case*

Main codes:

$$1\ 1\ 0\ 0\ 0\ 1\ 2\ \ -1\ 1\ 0$$

Assumptions:

There is a continuous demand with known constant rate r and an immediate delivery. The time of ordering may be specified as a periodic order system with the length of a period $t=\dfrac{q}{r}$, or, as an order-level system with the reorder level $s=S-q$. The order period and the lot size q are both constants. Demand is waiting when there is a shortage, then it will be first satisfied when an order arrives. The cost factors of inventory holding and shortage depend on time and quantity, and the cost of ordering is a constant value at each ordering. The variables S and q are continuous.

Objective:

The total cost per time unit

$$C(S, q) = c_1 \frac{S^2}{2q} + c_2 \frac{(q-S)^2}{2q} + c_3 \frac{r}{q} \quad (0 \le S \le q)$$

is to be minimized.

Solution:

By differentiation of the objective, using the necessary condition of optimality we obtain the explicit solution

$$q_0 = \sqrt{2c_3 r}\,\sqrt{\frac{c_1+c_2}{c_1 c_2}} \quad \text{and} \quad S_0 = q_0 \frac{c_2}{c_1+c_2},$$

17. Deterministic lot-size model with shortage, discrete case

Main codes:

$$1\ 1\ 0\ 0\ 0\ 1\ 2\ \ -1\ 1\ 0$$

Assumptions:

They are basically the same as for the previous model (Model 16); the only difference is that the values of the decision variables (the order level S and the lot size q) arediscrete, they may be only integer multiples of the unit u (for S), and positive integer multiples of the unit v (for q).

Objective:

The total cost per time unit

$$C(S, q) = c_1 \frac{S^2}{2q} + c_2 \frac{(q-S)^2}{2q} + c_3 \frac{r}{q} \quad (0 \leq S \leq q)$$

has to be minimized for $S=..., -2u, u, 0, 2u, ...$ and $q=v, 2v, ...$

Solution:

The necessary condition for optimality is

$$C(S_0, q_0) \leq C(S_0 \pm u, q_0 \pm v).$$

The selection of the (not necessarily unique) solution pair (S_0, q_0) can be based on a straightforward enumeration of all relevant pairs (S, q) (i.e. by substituting them sequentially into the function $C(S, q)$).

18. A static inventory with perturbed demand

Main codes:

$$1\ 1\ 0\ 0\ 0\ 1\ 2\ 2\ 9\ 0$$

Assumptions:

There is a continuous demand with a constant rate r_0. The model has a back-orders and a lost-sales version. In the case of a shortage the loss of goodwill is expressed not in the form of a cost, but by the decrease of the demand in the

form $r = \dfrac{r_0}{(1+Hi)}$, where H is the ratio of the shortage relative to the demand

i is a constant characterizing the level of goodwill. The ordered amount is q, when the inventory exceeds the reorder level s. There is an immediate delivery and the minimal amount which can be ordered is L_0. The income is proportional (with rate d) to the amount sold. The inventory carrying cost is linear (with cost factor c_1).

132

Objective:

The profit per time unit, as a function of the reorder, equals
a) in backorder case

$$p(s) = \frac{r_0 d}{1 + \frac{si}{q}} - \frac{H}{2} \frac{(q-s)^2}{2q},$$

or

b) in lost sales case

$$p(s) = \frac{r_0 c(q-s)}{M+si} - \frac{H}{2} \frac{(q-s)^2}{q};$$

thus, the respective functional form has to be maximized.

Solution:

The profit function is analyzed under each of the following conditions for both the backorders and lost sales cases:
1. $q = L_0$;
2. $q - s = k$, where k is the initial stock;
3. $T_0 - \frac{q+si}{r} = 0$, where $T_0 = \frac{q}{r}$ is the prescribed order interval.

The optimum may be given as a solution of the equation $dp(s)/ds$ on the interval $0 \leq s \leq q$, that means the solution of a cubic equation in all of the above cases.

19. Application of servomechanism theory for solving a production control problem

Main codes:

$$1\ 1\ 0\ 0\ 1\ 3\ 2\ 2\ 1\ 0$$

Assumptions:

The production of an item is continuously controlled, depending on its inventory level. Demand is met from stock, the shortage is backlogged. There is no delay in production, the amount ordered is immediately produced, and delivered to stock. The costs are only implicitly considered, in connection with the changes of the inventory level. The model may be described by the following three equations:

$$\Theta_0(t) = K_1[\mu(t) - \Theta_L(t)],$$

$$\mu(t) = K_2[\varepsilon(t)],$$

$$\varepsilon(t) = \Theta_1(t) - \Theta_0(t),$$

where $\Theta_0(t)$ is the optimal, $\Theta_1(t)$ is the actual level of inventory, $\Theta_L(t)$ denotes the demand, and $\mu(t)$ the actual level of production, as a function of time t. K_1 and K_2 are operators of the state description and of the decision.

Objective:

Description of the system and of its stability.

10 Chikán

Solution:

Based on the Laplace-transformation, the behaviour and stability of the operators for state description and for decision are analyzed in the case of different demand patterns.

20. *Servomechanism theory model for production with delay*

Main codes:

$$1\ 1\ 0\ 0\ 1\ 3\ 2\ 2\ 1\ 1$$

Assumptions:

The production of an item is continuously controlled, depending on its inventory level and on the orders not fulfilled yet. The other difference in comparison with the previous model (Model 19) is that here a production delay is assumed which is given by a constant time interval τ. Thus the equations for the model description change in the following manner:

$$\Theta_0(t) = K_1[\mu(t) - \theta_L(t)],$$

$$\mu(t) = \eta(t - \tau),$$

$$\eta(t) = K_2\varepsilon(t) + K_3\theta_L(t),$$

$$\varepsilon(t) = \theta_1(t) - \theta_0(t),$$

where the decision is determined by two operators K_2 and K_3 (corresponding to the decision rule). The other notations are similar to the previous model.

Objective:

Description of the system and its stability.

Solution:

Similarly to the previous model, the consequences of different demand patterns are analyzed (in terms of system behaviour and stability).

21 and 22. *Order-level, lot-size system for uniform replenishing rate*

Main codes:

$$1\ 1\ 0\ 0\ 0\ 1\ 2\ -1\ 1\ 0$$

Assumptions:

It is a generalized version of the lot-size model with uniform replenishing rate (Model 11), since shortage is allowed here. There is a continuous demand with a constant rate r. The amount ordered is delivered also with a constant rate p which must be no smaller than r. An order is given when the inventory decreases to reorder level s. The order interval and the lot-size is constant. When shortage occurs, there is a backorder. Unsatisfied demands are met when the ordered goods are received.

134

Objective:

The total cost per unit time is

$$C(S, q) = c_1 \frac{\left(S - \frac{qr}{p}\right)^2}{2q\left(1 - \frac{r}{p}\right)} + c_2 \frac{(q-S)^2}{2q\left(1 - \frac{r}{p}\right)} + c_3 \frac{r}{q} \quad \left(\frac{qr}{p} \leq S \leq q\right),$$

which has to be minimized.

Solution:

By differentitation of the objective function, using the necessary condition of optimality, the explicit solution can be given in the form

$$S_0 = \frac{\sqrt{2r\bar{c}c_3}}{c_1}, \quad q_0 = \frac{\sqrt{2r\bar{c}c_3}}{\bar{c}};$$

where

$$\bar{c} = \frac{1 - r/p}{1/c_1 + 1/c_2}.$$

I.2.2. Shortage with Lost Sales

23 and 24. *Reorder-level, lot-size systems with lost sales*

Main codes:

$$1\ 1\ 0\ 0\ 0\ 1\ 2\ -1\ 2\ 0$$

Assumptions:

There is a continuous demand with a constant rate r. An order is given, when the inventory decreases to the reorder level s. There is an immediate delivery. The order interval and the lot-size are both constants. If demand is not satisfied (due to a shortage), then there is no backorder, the demand is lost for the supplier. The unit costs (c_1 for inventory carrying, c_2 for shortage and c_3 for ordering) have the usual dimensions.

Objective:

The total cost per unit time is

$$C(S, q) = c_1 \frac{S^2}{2q} + c_2 \frac{(q-S)r}{q} + c_3 \frac{r}{q} \quad (0 \leq S \leq q),$$

which has to be minimized.

Solution:

Setting $k = S/q$ and applying a standard differentiation process, the optimum is expressed in the form

$$q_0 = \sqrt{\frac{2rc_3}{c_1}}, \quad S_0 = \sqrt{\frac{2rc_3}{c_1}} \quad \text{in the case of} \quad c_2 \geq \sqrt{\frac{2c_1c_3}{r}};$$

10*

and

$$q_0 = \infty, \quad S_0 = 0 \quad \text{in the case of} \quad c_2 \leqq \sqrt{\frac{2c_1 c_3}{r}}.$$

25. The marginal return rate of capital invested in inventories

Main codes:

$$1\ 1\ 0\ 0\ 0\ 1\ 2\ -1\ 0\ 0$$

Assumptions:

The assumptions of the model are the same as for the classical EOQ (or EBQ) model, but in this case we are to maximize the rate of return on our batch stock investment. The concept of a marginal rate of return is roughly that we ignore all investment decisions other than those on the batch stock of the item in question. The symbols in the model are the following:

R_m = marginal rate of return or (profit/year) value of batch stock;
p = profit per piece;
S = the cost of placing an order;
D = the demand per year in pieces;
c = the cost per piece.

Objective:

The marginal rate of return is to be maximized:

$$R_m = \frac{\left(p - \dfrac{S}{n}\right)D}{\dfrac{1}{2}n\left(c + \dfrac{S}{n}\right)}.$$

Solution:

The optimal batch size (or lot-size) can be expressed in an explicit form by differentation of R_m:

$$n_m = \frac{S}{p}\left[1 + \sqrt{1 + \frac{p}{c}}\right].$$

26. Influence of discounting on the optimal lot-size

Main codes:

$$1\ 1\ 0\ 0\ 0\ 3\ 2\ -1\ 0\ 1$$

Assumptions:

There is a continuous demand with a constant rate λ. The inventory holding cost is IC per unit quantity and unit time, where C is the purchasing cost and $0 < I < 1$ is a constant. The cost of replenishment is c_3 for each replenishment action. The total cost of inventory for a prescribed time horizon can be calculated on two different ways:

1. discounted costs are considered with a discount factor i,
2. without discounting the future costs.

136

Objective:

The difference between the optimal lot-size for discounted and undiscounted cost functions has to be evaluated. Their respective values are:

$$K(q) = \frac{\lambda}{q}c_{\mathfrak{a}} + \lambda C + \frac{c}{2}(I+i)(q-1) \quad \text{without discounting,}$$

and

$$F(q) = \frac{1}{(1-e^{iq/\lambda})}\left\{A + Cq + \frac{IC}{i}\left[q - \frac{1-e^{-iq/\lambda}}{1-e^{-1/\lambda}}\right]\right\}$$

for discounting.

Solution:

For different values of the cost factors c_1 and c_2 the optimal lot-size are tabulated for different discount factors i.

II. Deterministic Input and Demand, Fixed Replenishment Period, (t, q) and (t, S) Policies

This group consists of the different variations of the optimal lot-size model, similarly to group I. Considering the fact that the different inventory policies (ordering rules) for such simple deterministic models can be transformed into each other, there is no important difference among the operation of models in these two groups. The differences are rather in the special assumptions of the models.

The two main subgroups are distinguished by the characters of their ordering policy, so we have (t, q) and (t, S) policy models. Inside the first main subgroup further classifications may be formed. In the first sub-subgroup, the models make a special assumption concerning demand, in the second on purchase price, and in the third group there are the models for the control of perishables that can be stored only for limited time. A model on equipment stocking policy has been classified into this group too.

In the family of $(t; S)$ policy models two subgroups may be clearly distinguished: in the first subgroup a constant demand rate, while in the second a variable demand rate is assumed.

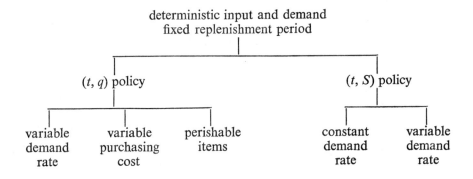

II.1. Deterministic Input and Demand, (t, S) Policy

II.1.1. Constant Demand Rate

27. *Order-level system with backlogging*

Main codes:

$$1\ 1\ 0\ 0\ 0\ 1\ 1\ \ -1\ 1\ 0$$

Assumptions:

The basic model of order-level systems has the following properties:

— The demand is deterministic at a constant rate of r quantity units per unit time.
— The replenishment period is a prescribed constant t_p.
— The replenishment size raises the inventory at the beginning of each period to order level S. Demand is backlogged. The optimal order level has to be determined.
— The replenishment rate is infinite, i.e. each order is delivered in one lot.
— The lead time is zero, however, for the case of positive lead time all the decisions may be easily transformed.
— The unit carrying and shortage cost are given constants denoted by c_1 and c_2 respectively, the replenishment costs usually need not be considered since there are prescribed replenishment periods and one order takes place in each period.

From these assumptions we can conclude that the number of replenishments per unit time will be $I_3 = 1/t_p$, a constant, while the lot-size $q_p = rt_p$ is also a pre-fixed constant.

The average amount of inventory equals

$$I_1(S) = \begin{cases} 0 & \text{if } S \le 0, \\ S^2/2q_p & \text{if } 0 \le s \le q_p \\ S - q_p/2 & \text{if } S \ge q_p. \end{cases}$$

The average shortage is given by

$$I_2(S) = \begin{cases} q_p/2 - S & \text{if } S \le 0, \\ (q_p - S)^2/2q_p & \text{if } 0 \le S \le q_p, \\ 0 & \text{if } S \ge q_p. \end{cases}$$

Objective:

The total cost of the system equals

$$C(S) = \frac{c_2 S^2}{2q_p} + \frac{c_2(q_p - S)^2}{2q_p},$$

when $0 \le S \le q_p$: it is easy to see that the latter relations are valid for the optimal value of S.

Solution:

The necessary condition for optimality can be expressed as

$$0 = \frac{dC(S)}{dS} = \frac{c_1 S}{q_p} - \frac{c_2(q_p - S)}{q_p};$$

this yields $S_0 = q_p \dfrac{c_1}{c_1 + c_2}$. S_0 is in fact the minimizer of the cost function, since the second derivative of the cost function in the point $S = S_0$ is positive, which is a sufficient optimality condition. The minimum total cost belonging to S_0 is

$$C_0 = \frac{1}{2} \cdot q_p \cdot \frac{c_1 c_2}{c_1 + c_2}.$$

28. Order level system with lost sales

Main codes:

$$1\,1\,0\,0\,0\,1\,1 \quad -1\,2\,0$$

Assumptions:

There are the same assumptions as for the previous model (Model 27), except for the following:

If demand cannot be immediately satisfied due to a shortage, then there are no backorders, i.e. the demand is lost for the supplier. At the end of the order period the amount of shortage is $q_p - S$.

The shortage cost is proportional only to the amount (and not the time) of shortage and is a linear function with unit cost factor c_2.

Objective:

The total cost of the system equals

$$C(S) = \begin{cases} c_2(q_p - S)/t_p & \text{if } S \leq 0, \\ c_1 S^2/2q_p + c_2(q_p - S)/t_p & \text{if } 0 \leq S \leq q_p, \\ c_1(S - q_p/2) & \text{if } S \geq q_p, \end{cases}$$

which has to be minimized.

Solution:

The optimal order level S_0 lies in the range $0 \leq S \leq q_p$. Differentiating the objective function we obtain

$$S_0 = \frac{r c_2}{c_1}.$$

This holds only in the case if $r c_2/c_1 \leq q_p$, i.e. if $c_2 \leq c_1 t_p$. Otherwise the optimal solution is equal to q_p.

29. Order level system with continuous replenishment

Main codes:

$$1\,1\,0\,0\,0\,1\,0\,0\,1\,0$$

Assumptions:

There is a continuous demand with a known rate r. The amount ordered periodically is delivered also continuously with a rate λ (where $\lambda > r$), until the whole amount is received in a period. The fixed order cost c_3 can be interpreted also as the setup cost of a production lot. The inventory carrying cost factor c_1 and the shortage cost factor c_2 have the usual dimensions, they are proportional to time and amount. The order level S and the length of the ordering period T is to be determined.

140

Objective:

The total expected annual cost

$$L(T, S) = \frac{1}{T}\left[c_3 + \frac{(c_1+c_2)\lambda S^2}{2r(\lambda-r)}\right] + \frac{c_2 r}{2\lambda}(\lambda-r)\cdot T - c_2$$

is to be minimized.

Solution:

Differentiating the objective function by T and S, the optimal parameters can be expressed as the solution of a system of two equations.

II.1.2. Variable Demand Rate

30. *Order level for constant replenishment periods and linearly increasing rate*

Main codes:

$$1\ 1\ 0\ 0\ 0\ 1\ 0\ -1\ 0\ 0$$

Assumptions:

The system operates only during a prefixed period, the duration of which is H time units. During this period there is a total demand of D units. The rate of demand r changes linearly with time T, i.e., $r = a \cdot T$. The constant $a > 0$ can be determined from the relation

$$D = \int_0^H r\,dT = \frac{aH^2}{2}.$$

Hence $r = 2DT/H^2$. The replenishment period t is constant and has an optimal value $t = H/m$ with some integer m, which is the number of replenishment orders. The model works with a (t, S_i) ordering rule, it means: the S_i order levels change from order to order in consequence of the increasing demand rate. No shortage is allowed.

Objective:

The sum of inventory holding and replenishment order cost is

$$C(m) = c_1 \frac{D \cdot h(m)}{m} + c_3 \cdot \frac{m}{H},$$

where $h(m) = \frac{2m}{3} - \frac{\sum\limits_{i=1}^{m-1}\sqrt{i}}{\sqrt{m}}$,

$m = \frac{H}{t}$ the number of replenishments, $(m = 1, 2, 3, ...)$.

Solution:

For the optimal integer value m_0 of m we have the $C(m_0+1) \geq C(m_0)$ and $C(m_0-1) \geq C(m_0)$ conditions. We can get the optimal solution from the next formula:

$$F(m_0-1) \leq \frac{c_1 DH}{c_3} \leq F(m_0),$$

where $F(m) = \dfrac{m(m+1)}{(m+1)h(m) - m \cdot h(m+1)}$.

The value of m_0 gives $t_0 = \dfrac{H}{m_0}$. S_{i0} are determined on these values.

31. Order level for variable replenishment period and linearly increasing demand rate

Main codes:

$$1\ 1\ 0\ 0\ 0\ 1\ 0\ \ -1\ 0\ 0$$

Assumptions:

The only significant difference, in comparison with the assumptions of the previous model (Model 30), is that this model works on the basis of a (t_i, S_i) policy. Thus both the replenishment periods t_i and the order levels S_i depend on the number of replenishments m and on the order points T_i, $(T_m = H)$. The length of replenishment periods form a monotonically decreasing sequence, while the order levels are still monotonically increasing in consequence of the increasing demand rate. The total demand D in a period with length H has to be satisfied without shortage.

Objective:

The total cost of the system

$$C(m, T_i) = \frac{c_1}{H}\left[\frac{2}{3}DH - \frac{D}{H^2}\sum_{i=2}^{m}(T_i^2 - T_{i-1}^2)T_{i-1}\right] + \frac{c_3 m}{H}$$

has to be minimized.

Solution:

An approximate solution is given. The optimal number of replenishments m_0 is approximated by the solution of the previous model (fixed length of replenishment periods), then the optimal timing of replenishment orders is approximated by

$$T_{i0} = \frac{1}{2}H(i/m_0 + \sqrt{i/m_0}).$$

The order levels are given by

$$S_i = D(T_i^2 - T_{i-1}^2)/H^2 \quad \text{for} \quad i = 1, 2, \ldots, m_0.$$

32. Order level for deterministically varying discrete demand

Main codes:

$$1\ 1\ 0\ 0\ 0\ 1\ 1\ \ -1\ 1\ 0$$

Assumptions:

142

Each planning period H is divided into N scheduling periods with fixed length t_p. The demand x_i occurs at the beginning of each scheduling period i. The x_is may vary for different periods but they are known quantities.

A constant and prefixed lot-size $q_p = \sum_{i=1}^{N} x_i/N$ corresponding to the average demand is ordered which is delivered at the beginning of each scheduling period. First the possible backorders will be fulfilled.

The only variable subject to control is the inventory level S at the beginning of each planning period H. The unit holding cost is c_1 and the unit shortage cost is c_2. They have the usual dimensions.

Objective:

The cost of the system

$$C(S) = (c_1+c_2)\left(mS - \sum_{i=1}^{m} R_i\right) - c_2 NS + c_2 \sum_{i=1}^{N} R_i$$

is to be minimized, where R_i ($i=1, 2, ..., N$) denotes the nondecreasing sequence formed by ordering the numbers

$$R_i' = \sum_{j=1}^{i} x_j - iq_p.$$

Solution:

The minimization of $C(S)$ leads to the following result: if for an integer m_0 the inequality

$$m_0 - 1 \leq N\frac{c_2}{c_1+c_2} \leq m_0$$

holds, then $S_0 = Rm_0$, and further, if $Nc_2/(c_1+c_2)=m_0$, then any $R_{m_0} \leq S_0 \leq R_{m_0+1}$ is a solution for the optimal order level.

33. Order level system with power demand pattern

Main codes:

$$1\ 1\ 0\ 0\ 0\ 1\ 1\ \ -1\ 1\ 0$$

Assumptions:

All the assumptions concerning the order level system with backlogging (Model 27) are unchanged. The only difference is that the demand has not a uniform rate, but it has a power demand pattern. It means that for an order level S and total demand q_p during a replenishment period with length t_p the amount in inventory at time T can be expressed in the form

$$Q(T) = S - q_p \sqrt[n]{T/t_p},$$

where the exponent n may have any real number value. This is the demand pattern index by which the time-dependence of the demand can be characterized.

Objective:

The total cost is

$$C(S) = (c_1+c_2)\frac{S}{n+1}\left(\frac{S}{q_p}\right)^n - c_2 S + c_2 \frac{q_p \cdot n}{n+1}$$

for the value $0 \leq S \leq q_p$: this interval has to contain the optimal S.

Solution:

By the usual minimization technique (by differentiating the objective function) the optimal order level can be expressed as

$$S_0 = q_p \sqrt[n]{\frac{c_2}{c_1+c_2}}$$

and the optimal cost is

$$C_0 = c_2 q_p \left[1 - \sqrt[n]{\frac{c_2}{c_1+c_2}}\right] \cdot \frac{n}{n+1}.$$

II.2. Deterministic Replenishment and Demand, (t, q) Policy

II.2.1. Models for Deterministic, Variable Demand

34. *Deterministic (t, q) policy model with variable intensity of demand*

Main codes:

$$1\ 1\ 0\ 0\ 0\ 1\ 4\ \ -1\ 0\ 0$$

Assumptions:

In the period $[O, H]$ there is a continuously changing demand intensity which can be described by a known function $f(t)$ of time; $f(t)$ is assumed to be a logarithmically concave function (i.e. $\ln f(t)$ is concave). The initial and the closing stock is 0. An order for the supply of goods has a fixed cost c_3, the unit holding cost is assumed to be $c_1 = 1$. The number of orders n and the times of orders T_i ($i = 1, ..., n$) are to be optimized.

Objective:

The total cost in $[O, H]$ is

$$C(n, T_i) = nc_3 + \sum_{i=0}^{n-1} \int_{T_i}^{T_{i+1}} (t - T_i) f(t)\,dt,$$

which has to be minimized.

Solution:

A general recursive solution method is proposed by the author for the optimal ordering times T_i, but it is not detailed. The optimality of the stationary points is ensured by the logconcave property of $f(t)$. The optimal ordered amounts are given by

$$q_i = (T_i - T_{i-1}) f(T_i) \quad i = 1, 2, ..., n.$$

144

35. A generalized deterministic model with varying demand

Main codes:
$$1\ 1\ 0\ 0\ 0\ 1\ 4\ \ -1\ 0\ 0$$

Assumptions:

This is a generalized version of the previous model (Model 34), in the following sense:

— The variable demand can be described by any continuous nonnegative function. (The logconcavity assumption is omitted.)

— Three cost components are considered. The unit purchase price is c_0, the fixed cost of an order is c_3, the unit cost of inventory holding is c_1. Shortage is not allowed.

Objective:

The sum of the above three cost components equals

$$C(n, T_i) = c_0 \int_0^H f(u)\, du + nc_3 + c_1 \sum_{i=0}^{n-1} \int_{T_i}^{T_{i+1}} (t-T_i) f(t)\, dt,$$

which has to be minimized.

Solution:

A general recursive solution method, similar to the solution of the previous model, can be formulated. In this general case, however, the optimality of the stationary points cannot be ensured.

36. Deterministic model with linear trend in demand

Main codes:
$$1\ 1\ 0\ 0\ 0\ 1\ 4\ \ -1\ 0\ 0$$

Assumptions:

The model is similar to the previous ones. The significant difference is that the demand has a linear trend in the time period $[0, H]$. Thus, it is a special case of the previous model with $f(t) = a + bt$.

Objective:

The sum of the purchase, ordering and inventory holding costs is

$$C(n, T_i) = c_0 \int_0^H (a + bt)\, dt + nc_3 + c_1 \sum_{i=0}^{n-1} \int_{T_i}^{T_{i+1}} (t-T_i)(a + bt)\, dt,$$

which has to be minimized.

Solution:

The general solution method of the previous models has been specified for the linear case. An analytic way of solution using the first-order optimality condition based on the derivative of the cost function is proposed. Different tables containing

145

function values for the numerical computations are included which make the solution method more effective. The unique optimality of the solution is ensured. The application of the method is illustrated by some numerical examples.

37. Deterministic model with time proportional demand

Main codes:

$$1\ 1\ 0\ 0\ 0\ 1\ 4\ -1\ 0\ 0$$

Assumptions:

These are similar to those of the previous models, the demand increases in time according to a fixed rate $(a>0)$, i.e. at time t the demand is at. An order is placed when the inventory level decreases to 0. The unit holding cost is c_1, the fixed ordering cost is c_3. The number m of replenishment orders and the times T_i $(i=1, ..., m)$ $(T_m=T)$ of the orders have to be chosen to ensure cost minimum.

Objective:

The total cost of the time period $(0, T)$ is $c_1 Y(m)+c_3 m$, where the total inventory for fixed m equals

$$Y(m) = \frac{aT^3}{3} - \frac{a}{2} \sum_{i=1}^{m-1} T_i(T_{i+1}^2 - T_i^2) \quad \text{if} \quad m \geq 2,$$

and

$$Y(1) = \frac{a}{3} T^3.$$

Solution:

For fixed values of m, the optimal sequence of T_i $(i=1, ..., m)$ is calculated: all these subproblems have a unique solution for every m. This will be substituted for $Y(m)$, which is a decreasing convex function of m. Thus for the optimal value of m we have the relations

$$Y(m)-Y(m-1) \leq \frac{c_3}{c_1} \leq Y(m-1)-Y(m).$$

38. Deterministic model with variable demand and no fixed ordering cost

Main codes:

$$1\ 1\ 0\ 0\ 0\ 1\ 4\ -1\ 0\ 0$$

Assumptions:

Similar to the previous models, but the usual fixed ordering cost is omitted. The optimal time sequence t_i $(i=1, ..., n)$ of ordering and the optimal ordered quantities $y(t_i)$ have to be determined. The demand at time t is $f(t)$ which is a known positive function. The initial stock is 0.

Objective:

The total cost on the time-interval $[t_0, T]$ is

$$C(t_1, ..., t_n) = \sum_{i=1}^{n+1} \int_{t_{i-1}}^{t_i} [y(t_i)+g(t_{i-1})-g(t)+y_z(t_i-_1)]\, dt,$$

146

where

$$g(t) = \int_{t_0}^{t} f(t)\, dt; \quad g(t_i) = \int_{0}^{t_i} f(t)\, dt;$$

furthermore, the inventory at time t_i is given by

$$y_z(t_i) = y_z(t_{i-1}) + y(t_i) - \int_{t_{i-1}}^{t_i} f(t)\, dt; \quad y_z(t_0) = 0$$

Solution:

The dynamic programming algorithm is inefficient here, because of the great computational effort required. For the analytical method it is assumed that the order equals the demand for each period:

$$y(t_i) = \int_{t_{i-1}}^{t_i} f(t)\, dt.$$

This simplifies the determination of the optimal order times t_i, as in this case the function

$$\sum_{i=1}^{n+1} \int_{t_{i-1}}^{t_i} [g(t_i) - g(t)]\, dt \quad \text{has to be minimized.}$$

It is proved that this is equivalent to the minimization of the function

$$\sum_{i=1}^{n+1} g(t_i)(t_i - t_{i-1}),$$

which is an easy task.

39. *A production-inventory model with variable demand*

Main codes:

$$1\ 1\ 0\ 0\ 1\ 1\ 9\ -10\ -1$$

Assumptions:

The production-inventory system is controlled on the time interval $[0, T]$, during which there is a continuous demand $D(t)$ which depends on the actual price $p(t)$ and other time-dependent parameters $a(t)$ and $b(t)$ in the way $D(t) = a(t) - b(t)p(t)$. Here $a(t)$ and $b(t)$ are known functions. A decision has to be made about the price $p(t)$ and about the quantity $q(t)$ of production for each time moment $t \in [0, T]$. The production cost of $q(t)$ is $f[q(t)]$ which is assumed to be a strictly convex, nonnegative, increasing function. The inventory holding cost for a unit of quantity during a time unit is c_1. No shortage is allowed. The inventory amount at time t is

$$I(t) = q(t) - [a(t) - b(t)p(t)]$$

Objective:

The total profit during the time $[0, T]$ is given by the function

$$G[p(t), q(t)] = \int_0^T \{p(t)[a(t)-b(t)p(t)]-c_1 I(t)-f(q(t))\}\, dt,$$

which has to be minimized by the appropriate choice of the functions $p(t)$ and $q(t)$.

Solution:

The results of control theory are applied to derive the solution. First the Lagrange-function is determined, then the optimality conditions are achieved which give a method for determining the optimal functions of $p(t)$ and $q(t)$. Different possible cases are considered separately. Formulas are derived for $p(t)$ and $q(t)$, and references to known solutions algorithms are given. Some examples are also included for the illustration of the methods. The optimal solution is proved to be unique.

40. *Inventory model with deterministically changing demand pattern and backordering*

Main codes:

$$1\ 1\ 0\ 0\ 1\ 1\ 9\ 1\ 1\ 0$$

Assumptions:

There are T time intervals with fixed length in which different amounts d_t of demands occur $(t=1, 2, ..., T)$. During each period there may be an unsatisfied demand which can be satisfied at a later date, however, by the end of the last period all demands must be satisfied. The amount of inventory delivered in period i and used up in period j is denoted by c_{ij}.

The amount ordered (q) may only be an integer multiple of the package size, i.e., it is a discrete variable. The unit inventory holding cost is h_t and the unit shortage cost is g_t, which cost factors may be different for different periods t. The cost is calculated at the end of each period and depends only on the amount of stock or shortage at the end of that period. The unit purchasing cost is c_t, the fixed ordering cost is A.

Objective:

The total cost for a time period $[0, H_k]$ is

$$C[N(H_k), n_k] = G[I_{J^*(k)}|N(H_k), n_k]+A\delta(n_k)+c_k n_k q,$$

where $I_{J^*(k)}$ means the stock level in the period $J^*(k)$ and $N(H_k)$ is the set of deliveries in $[0, H_k]$. The number of batches ordered in period k is n_k. If $n_k=0$, then $\delta(n_k)=0$, while for $n_k>0$ $\delta(n_k)=1$.

Solution:

A dynamic programming algorithm is given.

148

41. *Price and inventory model for profit maximization*

Main codes:

$$1\ 1\ 0\ 0\ 0\ 1\ 9\ -1\ 0\ 0$$

Assumptions:

The demand changes from period to period as a function of time and of the unit selling price p. Suppose that $Q_t(p)=\alpha_t+\beta_t q(p)$, is the demand in period t $(t=1, 2, ..., N)$, where $\beta_t>0$ and $\dfrac{dq(p)}{dq}\leq 0$. No shortage is permitted. $s(t, p')$ denotes the period in which an order was given under price p' in order to satisfy the demand of period t. In period t the unit purchasing cost is c_t, the fixed ordering cost is K_t, the inventory holding cost depends on the stock level at the end of the period t and has a unit cost h_t.

Objective:

The ordering times and the constant selling price are to be determined so as to attain a maximal total profit. The income in period t is $R_t(p)=p\cdot Q_t(p)$. The total cost of N periods is

$$C(p', p) = \sum_{i=1}^{N}\left[C_{s(t, p')}+\sum_{j=s(t, p')}^{t-1} h_j\right]Q_i(p)+\sum_{t=1}^{N}[K_t|s(t, p')=t]$$

and the total profit is

$$\sum_{i=1}^{N} R_t(p)-C(p_{i-1}, p_i).$$

Solution:

A complicated procedure is given for calculating the optimal ordering times and selling price. The finiteness of the proposed method of solution is proved.

II.2.2. Stocking Policy in the Case of Price Increase

42. *Optimal lot-size policy by increasing purchase price*

Main codes:

$$1\ 1\ 0\ 0\ 0\ 1\ 4\ 0\ 0\ 0$$

Assumptions:

The conditions of the classical lot-size model are modified in such a way that the purchasing price b will be increased by a given amount k after a previous announcement. A constant demand rate r and a fixed ordering cost c_3 is considered. The cost of inventory holding is proportional to the purchasing price with a unit factor f. Thus the optimal lot-size before price increase is $q^*=\sqrt{2rc_3/fb}$ and after price increase $\bar{q}^*=\sqrt{2rc_3/f(b+k)}$. Two cases are analysed, the new price may be valid (a) immediately after the announcement, (b) after a time delay t_a. Two feasible strategies are compared. In the first one, there is no extra order at the price increase. In the second one, an extra amount

$$q_1^* = q^*(1+k/b)+kr/fb-rt_r$$

is ordered before the price increase, where the last term denotes the stock level at the moment of price increase. The order is delivered instantaneously, and no shortage is allowed.

Objective:

The costs of the above two strategies are compared. The benefit of purchasing an extra amount before a price increase is expressed as a quadratic function of q_1: $G_{1a,b}(q_1)=K_{1a}(q_1)-K_{1b}(q_1)$.

Solution:

The second ordering strategy is optimal if, for the inventory level rt_r at the time of price increase, the inequality

$$kr/bf+\sqrt{\frac{2rc_3}{(b+k)f}}\frac{b+k}{b}-\sqrt{\frac{2rc_3}{bf}} \geq rt_r$$

holds. Otherwise, the first strategy is optimal. The results are generalized for case (b) too. Here also an explicit expression is derived when it is optimal to order an additional amount before price increase. In both cases, the inventory is raised to the same order level

$$D^* = q^*\left(1+\frac{k}{b}\right)+\frac{kr}{bf}.$$

II.2.3. Inventory Decision for Perishable Goods

43. *An optimal policy for blood-bank systems*

Main codes:

$$1\ 1\ 0\ 0\ 0\ 1\ 9\ \ -1\ 9\ 7$$

Assumptions:

The model is based on the inventory problem of the central blood-bank system of a large hospital. The value of the blood can be characterised by a decreasing step function of storage time. There are M different age groups. The unit value of the blood belonging to group j is V_j in the case that it has been stored no longer than p_j days but at least p_{j-1} days, where $p_0=1$. For the group j there exists a demand D_{ij} in period i ($i=1, 2, ..., n$) which is known in advance. It can be fulfilled with blood from the group $k \leq j$. The unit price is V_j, regardless of the value of k. The inventory level of group j in period i is denoted by I_{ij} and the initial stock is I_{oj}. The cumulative value of the inventory until period i is denoted by R_i and

$$R_0 = \sum_{j=1}^{M} V_j I_{oj}.$$

150

The shortage in period i in group j is S_{ij}. This demand is normally backlogged, however, the lost sales case is also examined. The lot-sizes of deliveries are known.

Objective:

The total amount of inventory has a potential capability in satisfying demand. It can be expressed for period i in the following form:

$$R = R_{i-1} + \sum_{j=1}^{M} V_j[D_{ij} + I_{ij} - I_{i-1,j}],$$

that has to be maximized over the total time horizon $i = 1, 2, ..., M$.

Solution:

Two different policies are examined. For the FIFO (first in first out) policy, the demands are satisfied always with the material of the oldest group on stock which is still appropriate, while with the LIFO (last in first out) policy demands are satisfied by the youngest appropriate group. The following results are derived:

— independent of the specific values of V_j the FIFO policy is optimal, if the demand in the case of shortage is backordered,
— the unsatisfied demand of the time periods, and the lost demand (in the case for where there is no backorder), can also be minimized by the FIFO policy,
— the amount of demand satisfied with the best (youngest) bload group is minimized by the FIFO policy, thus one can optimally keep the age of the blood stock.

44. Economic packaging frequency for perishable, jointly replenished items

Main codes:

$$1\ 1\ 0\ 0\ 0\ 1\ 4\ \ -1\ 2\ 0$$

Assumptions:

The time between the subsequent ordering moments is determined by the model. The goods can be ordered in containers of m different capacities. The ordering cost S_j and the inventory holding cost h_j depend on the type of the container used at delivery. The demand D_j and the profit P_j of unit selling depend also on the volume of the container. The goods can be stored only for a limited time. The product can be used in a time horizon L, afterwards it is completely useless. The type of the container, K_j, and the order frequency T ($T > L$) which maximize the profit are determined by the model. K_j is the ratio between the frequency of set-ups for the product and the frequency of packaging set-ups for the item.

Objective:

The profit maximum is achieved by minimizing the total annual cost, $Z_j(T, K_j)$ being the net value for particular values of T and K_j.

$$Z_j(T, K_j) = \min\left\{D_j P_j; \frac{LD_j P_j}{TK_j}\right\} - \frac{S_j}{TK_j} - \min\left\{\frac{TD_j h_j K_j}{2}, \frac{L^2 D_j h_j}{2TK_j}\right\}.$$

Solution:
An iteration procedure is proposed with the starting solution

$$T_1 = \sqrt{2\left(M + \sum_{i=1}^{m} S_j\right) / \sum_{j=1}^{m} D_j h_j}.$$

The procedure will be stopped in the case where $K_j(T_r) = K_j(T_{r-1})$ holds (approximately). The fast convergence of the procedure is assured.

45. Order level for perishable goods

Main codes:

$$1\,1\,0\,0\,0\,1\,1 \quad -1\,1\,0$$

Assumptions:
There is a continuous demand with a constant rate r. The portion θ of the goods on inventory perishes per unit time. The length of the order period T is fixed. The order level is S which is the initial stock of the periods, since there is no lead-time. During period $[0, t_1]$ there is stock on hand and in $[t_1, T]$ there is a shortage which is backordered and is satisfied at the beginning of the next period.

Objective:
The sum of the value of the perished goods, inventory holding costs and shortage costs will be minimized by the optimal choice of t_1:

$$K(t_1) = \frac{C}{T} D(t_1) + c_1 I_1(t_1) + c_2 I_2(t_2),$$

where C means the unit purchasing cost,

$$D(t_1) = \frac{r}{\theta}[(1-\theta)^{-t_1} - 1] - r \cdot t_1$$

is the amount of perished goods in the period $[0, t_1]$,

$$I_1(t_1) = \frac{r}{\theta(T+1)}\left[\frac{(1+\theta)^{-t_1}-1}{\theta} - t_1\right]$$

is the average inventory on hand,

$$I_2(t_2) = \frac{r}{2(T+1)}(T-t_1)(T-t_1+1)$$

is the average shortage.

Solution:
For the optimal value of t_1 we have

$$M(t_1-1) \leq \frac{c_2 \theta T}{c_1 T + C\theta(T+1)} \leq M(t_1),$$

152

where

$$M(t_1) = \frac{(1-\theta)^{-(t_1+1)}-1}{T-t_1}.$$

Using the optimal value of t_1, the optimal order level is

$$S = \frac{r}{\theta}[(1-\theta)^{-t_1}-1].$$

46. Deterministic inventory and equipment replacement model

Main codes:

$$1\ 1\ 0\ 0\ 1\ 1\ 9\ -1\ 0\ 0$$

Assumptions:

The operation and maintenance costs of a given piece of equipment increase as a function of working hours, thus after a certain time it is economic to substitute it with new equipment held on inventory. During the interval $[0, T]$ there are n purchasing orders for the equipment, which are characterized by the time moments t_i and amounts x_i $(i=1, 2, ..., n)$ of the orders. The cumulative amount ordered and received by time t_i is given by $X(i) = \sum_{j=1}^{i} x_j$. The time intervals between two consecutive replacements of the equipment are denoted by τ_1, τ_2, ..., $\tau_{X(n)}$. The purchase price of v equipments at time t_i is $c(v, t_i)$. $F(y, s, r)$ denotes the additional costs incurred on the interval $[s, s+r]$. It represents the cost of holding $y=0, 1, 2, ...$ equipments in inventory during $[s, s+r]$, the cost of operating an equipment installed at time s during $[s, s+r]$, the cost of replacing the previously operating equipment at time s, the cost of maintenance and repair during $[s, s+r]$ and the salvage value (as a negative cost) at time $s+r$.

Objective:

The feasible policies are denoted: $W = \{w | w = (n, t, x, r)\}$. The total cost for a policy $w \in W$ is

$$C(w) = \sum_{i=1}^{n} [c(x_i, t_i) + \sum_{j=X(i-1)+1}^{X(i)} F(X(i)-j; \sum_{k=1}^{j} \tau_k, t_j)].$$

Solution:

In the general case when the cost factors are time-dependent, the existence conditions of an optimal decision for W are shown. An algorithmic solution is given only for the case when the cost factors are fixed over the whole interval $[0, T]$.

III. Deterministic Replenishment, Stochastic Demand, (t, q) Policy

In this group of models single- and multi-period models can be distinguished. The former ones deal with a single decision and the subgroups can be specified on the basis that the stock on hand at the end of the period may or may not be sold in the future. The multi-period models have also two subgroups: one for the models of perishable goods, and the other for the goods which may be stored without the loss of their value.

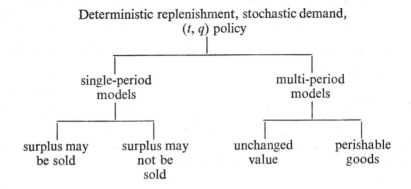

Deterministic replenishment, stochastic demand, (t, q) policy

- single-period models
  - surplus may be sold
  - surplus may not be sold
- multi-period models
  - unchanged value
  - perishable goods

47. Single-period model with time-proportional costs

Main codes:

$$1\ 1\ 0\ 1\ 0\ 4\ 9\ 0\ 2\ 0$$

Assumptions:

An order for the amount h is given before the beginning of the considered time period. At the beginning of the period with length T an initial stock h exists. The demand is random and follows the Poisson distribution with mean demand rate λ. The expected value of demand is μ ($\mu=\lambda T$). The following notations are introduced

$$p(a; b) = \frac{b^a}{a!}\, e^{-b} \quad \text{and} \quad P(a; b) = \sum_{i=a}^{\infty} p(i; b)$$

Then at any time moment t the stock level is

$$\sum_{x=0}^{h} (h-x)p(x; \lambda t) = h - \lambda t + \lambda t P(h; \lambda t) - h P(h+1; \lambda t),$$

154

since for arbitrary $\mu > 0$ we have

$$\sum_{j=0}^{r}(r-j)p(j;\mu) = r-\mu+\mu P(r;\mu)-rP(r+1;\mu).$$

At time moment t the shortage equals

$$\sum_{x=h}^{\infty}(x-h)p(x;\lambda t) = \lambda t P(h;\lambda t)-hP(h+1;\lambda t),$$

since for the Poisson distribution the following equations are valid:

$$\sum_{j=r}^{\infty}(j-r)p(j;\mu) = \mu P(r;\mu)+(\mu-r)P(r+1;\mu)$$

and

$$\mu P(r;\mu)-rP(r+1;\mu) = \mu P(r-1;\mu)-rP(r;\mu).$$

The unit inventory carrying cost is IC, where C is the unit purchase cost. The time-dependent shortage cost is denoted by $\hat{\pi}$ and the amount-dependent shortage cost is denoted by π_0. The unit selling price is S. The stock remained at the end of the period can be sold on a unit price L.

The expected annual carrying cost is

$$ICD(h) = ICT\left\{h-\frac{\lambda T}{2}+\frac{\lambda T}{2}P(h;\lambda t)-hP(h+1;\lambda t)+\frac{h(h+1)}{2\lambda T}P(h+2;\lambda T)\right\}.$$

The time dependent shortage cost is $\hat{\pi}B(h)$, where

$$B(h) = \int_{0}^{T}[\lambda t P(h;\lambda t)-hP(h+1;\lambda t)\,dt] =$$

$$= \frac{\lambda T^2}{2}P(h;\lambda T)-hTP(h+1;\lambda T)+\frac{h(h+1)}{2\lambda}P(h+2;\lambda T).$$

The total expected annual cost can be expressed as

$$C(h) = (S-L)\mu-(C-L)h-(S+\pi_0-L)\sum_{x=h}^{\infty}(x-h)p(x;\lambda T)+$$

$$+ICT\left(h-\frac{\lambda T}{2}\right)+(IC+\hat{\pi})\cdot B(h).$$

Objective:

The objective is to maximize the expected profit:

$$G(h) = (S-L)\mu-(C-L-ICT)h-\frac{1}{2}IC\lambda T^2-$$

$$-\mu\left[S-L+\pi_0-\frac{T}{2}(\hat{\pi}+IC)\right]P(h;\lambda T)+$$

$$+h[S-L+\pi_0-T(\hat{\pi}+IC)]P(h+1;\lambda T)+\frac{IC+\hat{\pi}}{2\lambda}h(h+1)P(h+2;\lambda T).$$

155

Solution:

The optimal order amount is the largest positive integer h for which

$$\Delta(h) = -(C-L-ICT)+\left[S-L+\pi_0+\left(\frac{h}{\lambda}+T\right)(\hat{\pi}+IC)\right]P(h;\lambda T)-$$

$$-\frac{h}{\lambda}(\hat{\pi}+IC)p(h;\lambda T)$$

is still positive.

III.1. Single-Period Models

III.1.1. Models for the Case When Surplus Stock Can Be Sold

48. The "Christmas-tree" problem

Main codes:

$$1\,1\,0\,1\,0\,1\,4\,0\,2\,0$$

Assumptions:

There is one single possibility of ordering. The price per unit purchase is C, the price per unit sale is S. When demand arises, if there is no stock on hand, it causes not only a loss of profit, but a shortage cost is also to be paid. The unit shortage cost is c_2 per unit quantity. If at the end of the selling period there remains stock on hand (surplus stock), then it can be sold for the unit price L, which is smaller than the unit purchase price $(L<S)$. The demand is not known at the time of ordering, it is a random variable. The probability of demand x is $p(x)$, the mean (expected demand) value is denoted by μ. The order amount is an integer.

Objective:

The expected total profit for an order amount h is

$$G(h) = (S-L)-(C-L)h-(S+c_2-L)\sum_{x=h}^{\infty}(x-h)p(x),$$

which has to be maximized.

Solution:

The optimal value of h is the largest integer for which

$$\sum_{j=h}^{\infty}p(j) > \frac{C-L}{S+c_2-L}.$$

49. Single-period model with changing demand rate

Main codes:

$$1\,1\,0\,1\,0\,4\,9\,0\,2\,0$$

156

Assumptions:

There is only one order for a single amount h. There is a stochastic demand with a time-varying mean demand rate. The cumulated demand until time t has a Poisson distribution with mean value $D(t)$. The length of the selling period T is prefixed. It is subdivided into m sub-periods, all of which have constant demand rates, maybe with different constant λ_i for different periods $[t_{i-1}, t_i]$ $i=1, 2, ..., m$, where $t_0=0$ and $t_m=T$. The inventory holding cost and a part of the shortage cost is only time-dependent, it does not depend on the amount of inventory or shortage. The respective cost factors are denoted by c_1 and c_2. There exists another type of shortage cost which depends on the amount of shortage and has a unit cost factor c_4. The price per unit sale is S. The stock remaining at the end of the period can be sold for a unit price L, which is smaller than the price per unit purchase C. The expected value of demand is denoted by μ.

Objective:

The expected total profit in the case of an initial stock h is

$$G(h) = (S-L)\mu - (C-L-c_1T)h - \frac{c_1}{2}\sum_{i=1}^{m}\lambda_i(t_i^2 - t_{i-1}^2) -$$

$$-\mu(S-L+c_4)P(h; \lambda T) + h(S-L+c_4)P(h+1; T) +$$

$$+(c_3+c_1)\sum_{i=1}^{m}[R(h, t_i, \lambda_i) - R(h, t_{i-1}, \lambda_i)],$$

where

$$R(h, t, \lambda) = \frac{\lambda t^2}{2}P(h; \lambda t) - htP(h+1; \lambda t) + \frac{h(h+1)}{2\lambda}P(h+2; \lambda t)$$

($P(h; \lambda t)$ was defined above; see Model 47.)

Solution:

The optimal value of h can be calculated by applying the necessary condition of the optimality. Considering the complex form of this equation, the authors recommend the use of tables in practice.

50. Single-period model with variable demand and random length of period

Main codes:

$$1\ 1\ 0\ 1\ 0\ 4\ 9\ 0\ 2\ 0$$

Assumptions:

This model is a generalized version of the previous model in the sense that the length of the selling period is not known at the time of ordering. The length of the period is T_j with a probability w_j ($j=1, ..., n$).

Objective:

The expectation of the total profit in this case can be expressed by using the objective $G(h)$ of the previous model. We substitute T_j for T and λ_{ij} for λ; μ_j

for μ. If this revised form is denoted by $G(h, T_j)$, then the objective of the generalized model is

$$G(h) = \sum_{j=1}^{m} w_j G(h, T_j) \quad j = 1, 2, ..., m.$$

Solution:

The method of solution is again similar to that of the previous model, with a minor increase in complexity. (Therefore the use of a computer is recommended for calculating the optimal decision.)

51. *Single-period model with time-dependent costs and random length of period*

Main codes:

$$1\ 1\ 0\ 1\ 0\ 1\ 9\ 0\ 2\ 0$$

Assumptions:

This model is a generalized version of Model 48. Here the inventory holding cost is proportional to the length of time during which a unit remains in inventory, and a stockout cost which is proportional to the length of time between the moment when the demand occurs and the end of the time period. The cost per unit time of keeping the item in stock will be denoted by IC, where C is the purchasing cost of the item, and $\hat{\pi}$ will be the cost per unit time of a stockout. Including these costs it is necessary to introduce a distribution for the demand from the beginning of the period up to any moment t of the period. The demand is supposed to follow a Poisson distribution over any time period. The mean rate of demand r is constant over time. The length of the period may be a random variable. It has the value T_j with probability w_j. Any units remaining at the end of the period can be sold at a unit price L ($L < C$).

Objective:

If h units are ordered for a period with length T_j, then the expected total profit is:

$$G(h, T_j) = (S-L)\mu_j - (C-L-ICT_j)h - \frac{1}{2}ICrT_j^2 - \mu_j \times$$

$$\times\left[S - L + \pi_0 - \frac{T_j}{2}(\hat{\pi}+IC)\right]P(h, rT_j) + [h(S-L+\pi_0 - T_j(\hat{\pi}+IC)]P(h+1; rT_j) +$$

$$+ \frac{IC+\hat{\pi}}{2r} h(h+1)P(h+2, rT_j),$$

where $\mu_j = rT_j$, S is the selling price per unit and π_0 is the fixed stockout cost per unit shortage. $p(a, b) = \dfrac{b^a}{a!}e^{-b}$ and $P(a, b) = \sum\limits_{i=a}^{\infty} p(i, b)$ is by definition. The expected total profit is

$$G(h) = \sum_{j=1}^{n} w_j G(h, T_j).$$

Solution:

The largest h for which $\Delta G(h)$ is positive, will be optimal, where

$$\Delta G(h) = \sum_{j=1}^{n} \Delta G(h, T_j),$$

and

$$\Delta G(h, T_j) = -(C - L - ICT_j) + \left[S - L + \pi_0 + \left(\frac{h}{r} - T_j\right)(\hat{\pi} + IC)\right] \times$$

$$\times P(h, rT_j) - \frac{h}{r}(\hat{\pi} + IC)p(h, rT_j).$$

For numerical solution a straightforward tabulation of the above expression is suggested.

52. Spare-parts ordering policy, when surplus stocks can be sold

Main codes:

$$1\ 1\ 0\ 1\ 0\ 1\ 4\ 0\ 1\ 7$$

Assumptions:

The spare part considered plays an important role in production, as its shortage may involve a great loss. There is only a single decision on the amount y ordered which will be on stock at the beginning of the production period. The demand occurs when a part breaks down, i.e., it is a random variable. The demand for quantity r has the probability P_r, where r is an integer. $P(r \leq y) = \sum_{r=0}^{Y} P_r$. The purchasing cost is C per quantity unit and the shortage cost is U per quantity unit. The $(y-r)$ stock remaining at the end of the production period can be sold for a unit price V, $(V < U)$.

Objective:

The total cost which is to be minimized over the production period equals

$$\overline{W}(y) = \sum_{r=0}^{\infty} W_{ry} \cdot P_r = Cy + U \sum_{r=y+1}^{\infty} (r-y)P_r - V \sum_{r=0}^{y} (y-r)P_r,$$

since

$$W_{ry} = \begin{cases} Cy - (y-r)V & r \leq y \\ Cy + (r-y)U & r > y \end{cases}$$

Solution:

The optimal amount of order y_0 is defined by the inequalities

$$\sum_{r=0}^{y_0} P_r \geqq \frac{U-C}{U-V} \geqq \sum_{r=0}^{y_0-1} P_r.$$

159

53. *Single period model with continuous distribution of demand*

Main codes:

$$1\ 1\ 0\ 1\ 0\ 1\ 4\ 0\ 1\ 7$$

Assumptions:

This is the continuous version of the previous model. Here the demand of the production period is random and has a continuous distribution with a probability density function $f(x)$. The profit of sale per unit item is G. The maximum amount of any single order is N.

Objective:

The expected total income minus the costs are expressed by

$$g(y) = (G+C-V)\int_0^y xf(x)\,dx - (C-V)y\int_0^y f(x)\,dx -$$

$$-U\int_y^N xf(x)\,dx + Uy\int_y^N f(x)\,dx,$$

which has to be maximized.

Solution:

In this simple case, y can be easily calculated, applying the analytical optimality conditions. The solution method is illustrated with an example, where the density function has the following type: $f(x)=a-bx$.

III.1.2. Models for the Case When Surplus Stock Cannot Be Sold

54. *Spare-parts ordering, when surplus stock cannot be sold*

Main codes:

$$1\ 1\ 0\ 1\ 0\ 1\ 4\ 0\ 1\ 7$$

Assumptions:

This model considers spare parts, which are important from the point of continuous production. The demand is stochastic and has a value r with probability P_r. Both the C purchasing cost and the U shortage cost depend on the quantity of stock and shortage at the end of the production period. The only difference to Model 52 is that there is no possibility to sell out the surplus stock at the end of the production period.

Objective:

The expected total cost of the production period for any given order y is equal to

$$\overline{W}(y) = \sum_{r=0}^{\infty} W_{ry}P_r = Cy + U\sum_{r=y+1}^{\infty}(r-y)P_r,$$

where
$$W_{ry} = \begin{cases} Cy & \text{if } r \leqq y, \\ Cy+(r-y)U & \text{if } r > y. \end{cases}$$

Solution:

The optimal order amount y_0 is defined by the inequalities

$$\sum_{r=0}^{y_0} P_r \geqq \frac{U-C}{C} \geqq \sum_{r=0}^{y_0-1} P_r,$$

55. *A model for items with a short selling period*

Main codes:
$$1\ 1\ 0\ 1\ 0\ 1\ 4\ 0\ 7\ 7$$

Assumptions:

There is a single order possibility for the item considered. The amount S ordered for the short selling period is on hand at the beginning of the period. The profit after selling a unit is G. If it cannot be sold, then the unit loss is U. The demand of the period is random: $p(S)$ is the probability of the event that the demand exceeds S.

Objective:

The expected total profit of the period

$$E(S) = Gp(S) - U[1-p(S)]$$

has to be minimized.

Solution:

The optimal S is the maximal value for which the inequality $p(S) > \dfrac{U}{G+U}$ holds.

III.2. Multi-Period Models

III.2.1. Models with Unchanged Price

56. *A multi-period model with random demand and fixed leadtime*

Main codes:
$$1\ 1\ 0\ 1\ 0\ 1\ 4\ 0\ 1\ 1$$

Assumptions:

There is a periodic order with a fixed length T of each time-period; the periods have a random demand with density function $f(x)$. The ordered amount will be delivered after a fixed number k of periods. Let q_i be the order which has been given $(k-i)$ periods earlier and thus it will be delivered in period i. The initial

161

stock of the first period is z. The inventory holding cost is s_T, the shortage cost is p_T, both are the function of the amount on hand or the shortage at the end of each period. The ordering cost of q_k quantity is $c(q_k)$.

Objective:

The minimum of the expected total cost with a discounting factor α can be expressed in the form

$$L_\infty^*(z, q_1, q_2, ..., q_{k-1}) = L_T^*(z) + \min_{q_k \geq 0} \left[c(q_k) + \alpha \int_0^\infty L^*(z-x, q_1, ..., q_k) f(x)\, dx \right],$$

where

$$L_T(z) = \begin{cases} \int_0^z s_T(z-x)f(x)\, dx + \int_z^\infty p_T(x-z)f(x)\, dx & \text{if } z > 0, \\ \int_0^\infty p_T(x-z)f(x)\, dx & \text{if } z \leq 0. \end{cases}$$

Solution:

The optimal order at period n is

$$q_n^*(z) = \begin{cases} \hat{Y}_n - z & \text{if } z < \hat{Y}_n, \\ 0 & \text{if } z \geq \hat{Y}_n, \end{cases}$$

supposing that $c(u) = c \cdot n$ and $\hat{Y}_n$ is the unique solution of the equation

$$c + \alpha \int_0^\infty L_{(n-1)T}^{*\prime}(\hat{Y}_n - x) f(x)\, dx = 0,$$

where L' is the first differential function of L.

57. Multi-period model with urgent ordering in the case of shortage

Main codes:

$$1\,1\,0\,1\,0\,1\,4\,0\,0\,1$$

Assumptions:

We have the same assumptions as for the pervious model, the only difference is that in the case of a shortage an urgent ordering is placed for the amount of the shortage. This will be delivered immediately, but with extra costs which will be considered.

Objective:

The minimum of the total discounted costs is

$$L_\infty^*(z, q_1, q_2, ..., q_{k-1}) = \min_{q_k \geq 0} \{ c(q_k) + L_T(z) +$$

$$+ \alpha L_\infty^*(q_1, ..., q_k) \int_z^\infty f(x)\, dx + \alpha \int_0^z L_\infty^*(z + q_1 - x, q_2, ..., q_k) f(x)\, dx \},$$

where the notations are the same as in the previous model.

162

Solution:

Only an approximate solution is given for the case, when the lead-time is one period:

$$q(z) = \begin{cases} \beta(\hat{Y}-z) & \text{if} \quad z < \hat{Y} \quad (0<\beta<1), \\ 0 & \text{if} \quad z \geq \hat{Y}. \end{cases}$$

where for the β constant $(0<\beta<1)$ holds.

58. Multi-period model with random demand

Main codes:

1 1 0 1 0 1 4 1 2 1

Assumptions:

Demands occur at equidistant time points. At time t there is a demand Z_t $(t=1, 2, ..., T)$, which is random and may have one of the possible values $z_1, z_2, ..., z_p$. For the demands Z_t, which are not independent, we assume that they depend on the past demands indirectly only through the previous demand. Thus they form a Markov chain, and the conditional probability $P(Z_{t+1}=z_j|Z_t=z_i)=r_{ij}$ is independent of the time parameter t. The inventory level of the considered item at time moment t is characterized by a vector $s(t) = (s_1(t), ..., s_m(t))$ where $s_i(t)$ is the number of those items which are on stock already through i time periods. $s_i(t)$ is the initial stock at time t period. An item remains in stock from period i until period k with a probability q_{ik}, but $q_{im}>0$. The input amount is $f(t)$ at period t and

$$s_{k+1}(t+1) = \begin{cases} f(t+1) & \text{for} \quad k = 0, \\ q_{ik}s_k(t) & \text{for} \quad k = 1, 2, ..., m-1. \end{cases}$$

Objective:

The expected weighted squared error between the supply and demand is to be minimized. Let w_j $(j=1, ..., n)$ be positive weights, their sum equals to 1. Thus the objective is

$$E^i[\underline{s}(t), f(t+1)] = \sum_{j=1}^{n} r_{ij} w_j [z_j - d_{j1}f(t+1) - \sum_{k=1}^{m-1} q_{ik} d_{ik} s_k(t)]^2.$$

Here the coefficients d_{ik} denote the fraction of the stock which at demand level i can be used to service for the length of k periods. (In the original paper the type of stock has been specified for manpower considering the qualification.)

Solution:

Using the principle of dynamic optimality a functional equation is derived for the optimal solution. This yields a convex quadratic function of s and f for each j. When it is differentiated with respect to f a linear equation results and the unique minimizer f can be expressed as a linear function of s. The parameters of this linear function depend on i and T in such a way that it can be specified by an $n\times(m+1)$ matrix.

163

The general question of convergence is difficult to answer; in special cases, however, it has been proved. Some numerical examples are included for the illustration of the solution procedure.

59. A model with normal and emergency ordering possibilities

Main codes:

$$1\ 1\ 0\ 1\ 1\ 1\ 4\ 2\ 1\ 9$$

Assumptions:

There is a periodic ordering possibility. At the beginning of each period two different orders may be placed:

— an urgent order with immediate delivery, which depends on the initial stock x_i of the ith period and has an upper bound m_n: $0 \leq m_{n-i+1}(x_i) \leq m_n$. The urgent ordering cost per unit is c_0.

— a normal order $z_{n-i+1} > 0$ which is delivered at the end of the period. The normal ordering cost per unit is c ($c < c_0$).

The demands of the periods are independent, nonnegative random amounts ξ with the same φ probability distribution, which is known.

Objective:

The discounted average of the total cost for n periods is to be minimized by the appropriate orders of each periods. Beside the above-mentioned costs of ordering and shortage, the costs of inventory holding and those of capital investment are considered.

$f_n(x)$ denotes the expected total discounted loss of the n-period model:

$$f_n(x) = \inf_{z \geq 0} \left[f^{(1)}(x; m, z) + \alpha \int_0^\infty f_{n-1}(x+m+z-\xi)\varphi(n)\,dn \right],$$

where $f^{(1)}(x; m, z)$ is the one-period expected loss if the initial stock level is x, an emergency order of size m and a regular order of size z are issued.

Solution:

The convexity of the objective function is proved and some special properties of the optimal ordering policy are shown which enhance the efficiency of the numerical optimization procedure.

60. A stochastic production-inventory system

Main codes:

$$1\ 1\ 0\ 1\ 1\ 1\ 9\ 1\ 1\ 0$$

Assumptions:

There is a demand for an item produced and stored in n subsequent time periods. The demand of each period is stochastic. The cumulated demand of the first i periods Q_i has a densitiy function $f_i(Q_i)$. In period i an amount z_i is produced which is not larger than the given production capacity c. The cumulated produc-

164

tion of the first i periods is denoted by y_i while the initial stock is x. The production, inventory holding and shortage costs are functions depending on the amount of production, inventory level or shortage of the respective periods: these functions are denoted by $r(x)$, $q(x)$ and $p(x)$. The demand in case of shortage is backordered.

Objective:

The total discounted cost of n periods (with a discount factor δ) equals

$$G(y_1, \ldots, y_n) = r(y_1) + \int_{-\infty}^{y_1+x} q(y_1+x-Q_1) f_1(Q_1) \, dQ_1 +$$

$$+ \int_{y_1+x}^{\infty} p(Q_1-y_1-x) f_1(Q_1) \, dQ_1 +$$

$$+ \delta \left[r(y_2-y_1) + \int_{-\infty}^{y_2+x} q(y_2+x-Q_2) f_2(Q_2) \, dQ_2 + \int_{y_2+x}^{\infty} p(Q_2-y_2-x) f_2(Q_2) \, dQ_2 \right] + \ldots$$

$$+ \delta^{n-1} \left[r(y_n-y_{n-1}) + \int_{-\infty}^{y_n+x} q(y_n+x-Q_n) f_n(Q_n) \, dQ_n + \right.$$

$$\left. + \int_{y_n+x}^{\infty} p(Q_n-y_n-x) f_n(Q_n) \, dQ_n \right].$$

The values of the cumulative production y_i are to be determined, which minimize the above cost function.

Solution:

The solution is given for linear cost functions $q(x)=ux$ and $p(x)=vx$ using the necessary optimality condition. As a result, the optimal $(y_1, \ldots, y_n)$ is the solution of the system of the following equations:

$$\int_{-\infty}^{y_1+x} f_1(Q_1) \, dQ_1 = \frac{v-r'(y_1)+\delta r'(y_2-y_1)}{u+v},$$

$$\int_{-\infty}^{y_i+x} f_i(Q_i) \, dQ_i = \frac{v-r'(y_i-y_{i-1})+\delta r'(y_{i+1}-y_i)}{u+v}, \quad (i = 2, 3, \ldots, n-1)$$

$$\int_{-\infty}^{y_n+x} f_n(Q_n) \, dQ_n = \frac{v+r'(y_n-y_{n-1})}{u-v}.$$

In the special case, when the production cost is also a linear function with cost factor k per unit, these equations can be written as

$$\int_{-\infty}^{y_i+x} f_i(Q_i) \, dQ_i = \frac{v-k(1-\delta)}{u+v} \quad (i = 1, 2, \ldots, n-1),$$

$$\int_{-\infty}^{y_n+x} f_n(Q_n) \, dQ_n = \frac{v-k}{u+v}.$$

61. *Joint optimization of multi-period orders*

Main codes:

$$1\ 1\ 0\ 1\ 0\ 1\ 4\ 1\ 7\ 0$$

Assumptions:

There is a stochastic demand during N considered time periods. The demands in different periods are not independent of each other, there is a stochastic dependence among them. The assumption is that the joint (N-dimensional) density function of the demands is logarithmically concave, i.e., the logarithm of the density function is a concave function. This assumption is valid for some practically important density functions, e.g., for normal or uniform distribution. The amount of the order placed for the period i is denoted by z_i. The total cost of ordering (or producing) and inventory holding can be expressed as a function $g(z_1, ..., z_N)$, which is convex by assumption. The unit cost of shortage at period i is q_i. In the case of shortage the demand is backordered.

Objective:

For an initial stock z_0 the expected total cost is

$$g(z_1, ..., z_N) + \sum_{i=1}^{N} q_i \int_{z_0+...z_i}^{\infty} (x - z_0 - ... - z_i) f_i(x)\, dx$$

where $f_i(x)$ denotes the density function of the cumulated demand of the first i periods.

Solution:

The problem of finding the optimal order amounts is reduced to the solution of a convex programming problem, for which known numerical methods are available.

III.2.2. Perishable Items

62. *A model for perishable items with variable demand distribution*

Main codes:

$$1\ 1\ 0\ 1\ 0\ 1\ 4\ 1\ 1\ 0$$

Assumptions:

The demands in different periods are independent random amounts D_i with a probability density function f_i ($i=1, 2, ...$) which may be also different but known functions. The length of the periods is given. At the beginning of each period—after reviewing the inventory—an order is placed. After immediate delivery the demand of the period will be satisfied. The items can be stored only for m periods during which time there is no loss of value but after the nth period they must be rejected with a cost θ per unit. The purchase price, the inventory holding cost and the shortage cost are all proportional to the amount and have respective unit cost factors c, h and r. There is backorder in the case of shortage.

166

The initial stock of each period is characterized by the vector $\underline{x}=(x_{m-1}, ..., x_1)$ where x_i is the amount of items which can be stored in i subsequent periods. The demand is always satisfied by the oldest items on stock. The order amount $y=x_m$ is the decision variable.

Objective:

The expected total cost of the nth period is

$$L_n(\underline{x}, y) = cy + h \int_0^{x+y} (x+y-t) f_n(t)\, dt +$$

$$+ r \int_{x+y}^{\infty} (t-x-y) f_n(t)\, dt + \theta \int_0^y G_{m,n}(t, \underline{x})\, dt$$

where $G_{m,n}(t, \underline{x})$, as a function of the demands in the next m periods, expresses the expected number of rejected items and $x = \sum_{i=1}^{m-1} x_i$.

Solution:

The existence of a unique optimal solution $y_n(\underline{x})$ is proved, when for $t>0$ the relation $f_n(t)>0$ holds and $r>c>0$ is valid. Here $y_n(\underline{x})$ is the optimal quantity to be ordered if $\underline{x}$ is the inventory on hand and n periods remain in the horizon. It is the solution of the equation

$$\frac{\partial B_n(\underline{x}, y)}{\partial y}\bigg|_{y=y_n(\underline{x})} = 0,$$

where the left-hand side is continuously differentiable. The optimal ordering policy is the following: if $\sum_{i=1}^{m-1} x_i \leq \bar{x}$, then order $y_n(\underline{x})$ else do not order. Here $\bar{x}$ is the unique positive solution of the equation

$$C(1-\alpha) + hF(\bar{x}) - r[1-F(\bar{x})] = 0.$$

$$B_n(\underline{x}, y) = L_n(\underline{x}, y) + \alpha \int_0^{\infty} C_{n-1}[\underline{s}(y, \underline{x}, t)] f(t)\, dt$$

where

$$\underline{s}(y, \underline{x}, t) = \begin{cases} y-(t-x)^+ & \text{if demand is backordered,} \\ [y-(t-x)^+]^+ & \text{if demand is lost,} \end{cases}$$

and $F(x)$ is the probability distribution function of the demand of a period, α is discount factor $(0<\alpha<1)$, $C_n(\underline{x})$ is the minimum expected discounted cost if $\underline{x}$ is on hand and n periods remain.

63. A model of perishable items with uniform demand distribution

Main codes:

$$1\ 1\ 0\ 1\ 1\ 1\ 9\ 1\ 1\ 0$$

12*

Assumptions:

This is a simplified version of the previous model. Here the demand in each period has the same distribution generated by the probability density function f. The other notations are the same as with the previous model.

Objective:

The expected total cost of a period is

$$L(\underline{x}, y) = cy + \sqrt{h} \int_0^{x+y} (x+y-t)f(t)\,dt + r \int_{x+y}^{\infty} (t-x-y)f(t)\,dt +$$

$$+ Q \int_0^y G_{m,n}(t, \underline{x})\,dt, \quad \text{where} \quad x = \sum_{i=1}^{m-1} x_i;$$

the function $G_{m,n}$ expresses the expected number of items which have to be rejected. For a stock state $\underline{x}$, the minimal discounted cost (with discount factor α) in the case when n orders are still left can be expressed using the idea of dynamic programming in the form $C_n(\underline{x}) = \inf_{y \geq 0} B_n(\underline{x}, y)$, where

$$B_n(\underline{x}, y) = L(\underline{x}, y) + \alpha \int_0^{\infty} C_{n-1}[\underline{s}(y, \underline{x}, t)]f(t)\,dt \quad (n \geq 1).$$

The vector $\underline{s}$ expresses the initial stock state of the next period, the ith component of which means the amount of items which can be stored in subsequent i periods. This is a function of the initial stock state $\underline{x}$, the order y and the demand t of the actual period. $C_0(x) = cx$, since the surplus stock at the end of the time horizon can be sold.

Solution:

The existence of a unique optimal solution is proved under certain circumstances, similarly to the previous model.

64. Model for perishable items in case of urgent ordering possibility

Main codes:

$$1\ 1\ 0\ 1\ 1\ 1\ 9\ 1\ 2\ 0$$

Assumptions:

The demand is random and in each period has the same probability density function f. The order is placed at the beginning of the period. In the case of a shortage an urgent order has to be given to meet the excess demand. It is delivered immediately, but with an extra cost $p(z)$. The items can be stored only for m periods during which time there is no loss of value but after the mth period they must be rejected with the cost $v(z)$. The normal unit cost of purchasing is c, where $c < p'(z)$ and $cz + v(z) \geq 0$. The inventory holding cost $h(z)$ is a function of the rest of the stock at the end of the period. The profit per unit after satisfying a demand unit is r. The future costs are discounted by a factor α. The demand is always satisfied by the oldest items on stock.

168

Objective:

For an initial stock y (after delivery) the expected cost of inventory holding and urgent ordering of a period is

$$L(y) = \int_0^y h(y-x)f(x)\,dx + \int_y^\infty p(x-y)f(x)\,dx.$$

The expected cost of rejecting the items perished in m subsequent periods is

$$V(z) = \int_0^z (z-x)f(x)\,dx.$$

Solution:

The case when the functions $L(y)$ and $V(z)$ are convex and differentiable will be discussed. For $m=1$, i.e. if the surplus stock cannot be used in the next period, all the periods can be analysed independently, thus we have a single-period problem. The optimal amount of order in period r is x_r, the unique solution of the equation $c+L'(x_r)+V'(x_r)=0$.

For $m \geq 2$ the periods cannot be treated separately, however, the stock state vectors $w=(w_1, w_2, \ldots, w_{m-1}) \geq 0$ form a Markov process. Using the idea of dynamic programming, the solution can be formulated in the form of a recursive system of equations. The component w_i of vector $\underline{w}$ is the number of the items on stock which are stored since i periods. The stock state is considered at the beginning of the period when the perished items have been rejected, but the new order has not yet been received.

65. Optimal lot-size model for a special inventory problem of production components

Main codes:

$$1\ 1\ 0\ 1\ 0\ 1\ 4\ 7\ 0\ 0$$

Assumptions:

The model is based on a special inventory control problem. The enterprise considered produces on order, since many different types of components are built into the products. Thus when the demand for a type of component is known (n), an order $x \geq n$ placed. After receiving the items ordered, a quality test will be made for each piece. The number of pieces failing the test is denoted by j.

If $j \leq x-n$, then the demand of production can be satisfied, and $x-n-j$ pieces are surplus, they are sold for a salvage price v which is smaller than the purchasing price. if $j > x-n$, then an urgent order with a fixed cost K is requested. Because of this order we have the same decision problem, since the new delivery has defective pieces again. The probability that a delivery of x pieces of the item contains j defective pieces is $p_x(j)$. The demand of the production must be at last satisfied with n good pieces, thus not only one urgent order may be necessary.

Objective:

The optimal policy ensures the total satisfaction of demands with minimum total costs. The cost components are the following:

— The expected purchase-cost considering the defective pieces is

$$c\left[x - \sum_{j=0}^{x} j p_x(j)\right];$$

— The expected income from selling the unnecessary pieces in case of $j \leq x-n$ is

$$v \sum_{j=0}^{x-n} (x-n-j) p_x(j).$$

— The urgent ordering in the case of $j \geq x-n$ has the fixed cost K and the purchasing price of the urgent ordering. Besides, the loss of selling the unnecessary pieces from the urgent order of volume $n-x-j$ or the cost of a repeated urgent order must be considered again. Thus the expected total cost can be determined in a successive way, applying the idea of dynamic programming:

$$C(n) = \min_{x \geq n}\left\{c\left[x - \sum_{j=0}^{x} j p_x(j)\right] - v \sum_{j=0}^{x-n}(x-n-j)p_x(j) + \right.$$

$$\left. + \sum_{j=x-n+1}^{x} [K + C(n-x+j)p_x(j)]\right\}$$

for $n = 1, 2, 3, \dots$. The optimal decision for the given n is the value of x which gives the minimum of $C(n)$.

Solution:

The solution method is illustrated in the paper by a numerical example for the case when the probability of false pieces has a binomial probability distribution with a given parameter $0 < p < 1$

$$p_x(j) = \frac{x!}{j!(x-j)!} p^j (1-p)^{x-j} \quad j = 1, 2, \dots, x.$$

170

IV. Deterministic Replenishment, Stochastic Demand, (t, S) Policy

The model in this family assume periodic ordering for identical time periods of length t. An order should be placed in such a way that the amount in inventory be brought to a level of S quantity units. Two subgroups in the family may be distinguished. In the first one, the order is received immediately; in the second one, there is a positive leadtime which is the interval between placing an order and its addition to inventory. In this group we consider only the models with known leadtime, the random case will be treated in the groups VII and XII.

The models without leadtime are classified on the basis of the considered time-horizon. Single and multi-period models are distinguished. The models with leadtime are classified on the basis of their behaviour with respect to shortage. There are models in which the demand in the case of shortage is backordered and is satisfied at the next delivery. In another case, if the demand occurs when there is no stock on hand, then it will not wait until the next delivery, but will be lost for the system: this is the lost sales case. In the third group, there are the models which have constraints on the volume or time shortage.

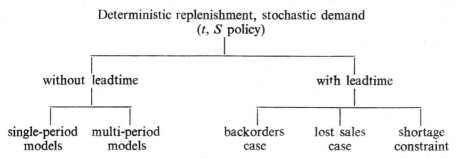

Deterministic replenishment, stochastic demand
(t, S policy)

without leadtime — with leadtime

single-period models — multi-period models — backorders case — lost sales case — shortage constraint

IV.1. Models without Leadtime

IV.1.1. Multi-Period Models

66. The (t, S) policy for random demand

Main codes:

$$1\ 1\ 0\ 1\ 0\ 1\ 0\ 1\ 1\ 0$$

Assumptions:

The inventory level is increased by an order to S in each time period (t). There is an immediate order arrival at the beginning of the periods and backorders in the case of shortage which will be first satisfied when an ordered quantity arrives. Both the inventory holding cost and the shortage cost are proportional to the time and amount of inventory or shortage. The unit cost factors are c_1 and c_2. There is a fixed order cost c_3 per each order.

The demand is random, during a time unit it has the probability density function $f(x, 1)$. The demands of the different time units are independent by assumption, thus the total demand of t subsequent time units $f(x, t)$ can be calculated by convolution, i.e.,

$$f(x, t) = \int_0^x f(x-y, t-1)f(y, 1) \, dy,$$

where $f(x-y, t-1)$ can be expressed the same way using $f(x-y, t-2)$ etc. For example, in the case of exponentially distributed demand with $f(x, 1)=e^{-x}$, the convolution can be explicitly given in the form of

$$f(x, t) = \frac{x^{t-1}}{(t-1)!} e^{-x}.$$

In the case of discrete distribution when the demand of a time unit is y with probability $P(y, 1)$ for $y_{min} \leq y \leq y_{max}$ then the total demand of t periods is x $(t y_{min} \leq x \leq t y_{max})$ with probability

$$P(x, t) = \sum_{y=y_{min}}^{x} P(x-y, t-1) \cdot P(y, 1),$$

which can be successively calculated for $t=2, 3$, etc.

Objective:

The expected total cost for given t and S is

$$C(t, S) = c_1 \int_0^S \left(S-\frac{x}{2}\right) f(x, t) \, dx + c_1 \int_S^\infty \frac{S^2}{2x} f(x, t) \, dx +$$

$$+ c_2 \int_S^\infty \frac{(x-S)^2}{2x} f(x, t) \, dx + \frac{c_3}{t},$$

which has a similar form for discrete demands except that the integrations are replaced by the respective sums and the density function $f(x, t)$ by the probability $P(x, t)$. The cost function has to be minimized by the optimal choice of t and S.

Solution:

For any fixed value t^*, the optimal $S_0(t^*)$ is the solution of the following equation:

$$\int_0^{S_0(t^*)} f(x, t^*) \, dt + \int_{S_0(t^*)}^\infty \frac{S_0(t^*)}{x} f(x, t^*) \, dx = \frac{c_2}{c_1+c_2}.$$

Therefore $S_0(t)$ can be calculated for different values of t. The optimal pair t^*, $S_0(t^*)$ minimizes the cost function $C(t, S)$. For the calculation of this pair no

172

algorithmic solution is given, only simple trials with different values of t are proposed. In the case of discrete demand distribution, the optimal value of $S_0(t^*)$ for a given t^* is determined by the inequalities

$$M[S_0(t^*) - u] \leq \frac{c_1}{c_1 + c_2} \leq M[S_0(t^*)],$$

where

$$M(S, t) = \sum_{x=0}^{S} P(x, t) + \left(S + \frac{u}{2}\right) \sum_{x=S+u}^{\infty} \frac{P(x, t)}{x}$$

(u is the amount of demand and x and S are constrained to discrete units: they can be equal to $0, u, 2u$, etc.)

67. Optimal ordering period for fixed order level

Main codes:

1 1 0 1 0 1 8 1 0 0

Assumptions:

Suppose that the demand is a random variable, while the maximal amount o) the possible demand during a period t is known: $x_{\max}(t) = rtA(t)$, where $A(t$- is a given function of t and r is the average demand rate. No shortage is allow ed. This implies to fix the order level S_p for an order period t, as $S_p(t) = = x_{\max}(t)$.

The inventory holding cost is proportional to the amount and time of stocking and has the unit cost factor c_1. The cost of an order is c_3.

Objective:

The total cost for a given t is

$$C(t) = c_1 \left[S_p(t) - \frac{rt}{2} \right] + \frac{c_3}{t} = c_1 \left[A(t) - \frac{1}{2} \right] rt + \frac{c_3}{t},$$

which has to be minimized.

Solution:

The explicit solution is given for two special functions $A(t)$. For $A(t) = k$ we have $t_0 = \sqrt{2c_3/[c_1 r(2k - 1)]}$, while for $A(t) = 1 + \frac{b}{t}$ $t_0 = \sqrt{2c_3/(rc_1)}$ is the optimal length of the order periods. In the first case the ratio of the maximum demand to the average demand during any period t is assumed to be a constant k. In the second case this ratio depends on t.

68. Optimal order level for fixed ordering frequency

Main codes:

1 1 0 1 0 1 1 1 1 0

Assumptions:

The probability density function of the random demand x during the prefixed order period t_p is denoted by $f(x)$. This demand is assumed to occur following

173

a uniform pattern, i.e. the inventory level of a period decreases linearly. The inventory holding cost and the shortage cost depends on the amount and time of stocking and shortage. The unit cost factors are c_1 and c_2. Because of the fixed ordering frequency, the ordering cost does not influence the optimal order level. In the case of shortage the demand is backordered.

Objective:

The expected total inventory holding and shortage cost for a period is equal to

$$C(S) = c_1 \int_0^S \left(S - \frac{x}{2}\right) f(x)\,dx + c_1 \int_S^\infty \frac{S^2}{2x} f(x)\,dx + c_2 \int_S^\infty \frac{(x-S)^2}{2x} f(x)\,dx.$$

Solution:

To find the optimal level S_0, we have to differentiate $C(S)$ with respect to S and set the derivative equal to zero. This leads to the equation

$$\int_0^{S_0} f(x)\,dx + \int_{S_0}^\infty \frac{S_0}{x} f(x)\,dx = \frac{c_2}{c_1 + c_2},$$

which generally cannot be solved in explicit form; however, numerical methods are available for its solution. In the special case, when the demand follows the distribution $f(x) = xe^{-x/b}/b^2$ (with properly chosen $b > 0$), the explicit solution is $S_0 = b \ln [(c_1 + c_2)/c_1]$. In the case of a discrete demand distribution the following inequalities determine the optimal S_0:

$$M(S_0 - u) \leq \frac{c_2}{c_1 + c_2} \leq M(S_0),$$

where

$$M(S) = \sum_{x=0}^S P(x) + \left(S + \frac{u}{2}\right) \sum_{x=S+u}^\infty \frac{P(x)}{x},$$

and $P(x)$ is the probability of a demand x and u is the minimal demand (expressed in the given unit).

69. The order-level system with instantaneous demands

Main codes:

$$1\,1\,0\,1\,0\,1\,1\,1\,1\,0$$

Assumptions:

The probability density function of the random demand x during the prefixed order period t_p is $f(x)$. The demand occurs at the beginning of each period after the receipt of the order which has been placed for the supply of the period. Thus, first the stock is filled up to the order level S and in the next moment it is decreased

174

to $S-x$ by satisfying the actual demand. If the demand of a period is greater than S, there is a shortage; the demand is backlogged and satisfied in the next period. The inventory holding cost and the shortage cost are proportional to the amount of stock or shortage. The costs per unit are c_1 and c_2, respectively.

Objective:

The sum of expected cost of the inventory holding and shortage is

$$C(S) = c_1 \int_0^S (S-x)f(x)\,dx + c_2 \int_S^\infty (x-S)f(x)\,dx.$$

Solution:

By differentiating $C(S)$ with respect to S and setting it equal to zero, the solution of the equation

$$\int_0^{S_0} f(x)\,dx = \frac{c_2}{c_1+c_2}$$

gives the optimal order level S_0. In the case of a discrete demand distribution, the optimality criteria are

$$F(S_0-u) \leq \frac{c_2}{c_1+c_2} \leq F(S_0)$$

with

$$F(S) = \sum_{k=0}^S P(x),$$

where $P(x)$ is the probability of a demand x, and u is the demand unit.

70. Order level with power demand pattern

Main codes:

$$1\ 1\ 0\ 1\ 0\ 1\ 0\ 1\ 1\ 0$$

Assumptions:

There is a given order period with length t_p. At the beginning of each period the inventory level is S. The total demand x of a period is a random variable and has a known probability density $f(x)$ in each period. Demand occurs during a period with a so-called power demand pattern which means that the inventory at the time T is

$$Q(T) = S - x\sqrt[n]{T/t_p}$$

where n is the demand-pattern index. In the special case $n=1$ we have the linear (or uniform) decrease of inventory and for $n=\infty$ we have the case of instantaneous demand, which cases are handled in the previous models. The other assumptions are the same as in these models (66—69).

Objective:

The total expected cost of inventory holding and shortage is

$$C(S) = c_1 \int_0^S \left(S - \frac{n}{n+1}\,x\right) f(x)\,dx + c_1 \int_S^\infty \frac{S\left(\frac{S}{x}\right)^n}{n+1}\,f(x)\,dx +$$

$$+ c_2 \int_S^\infty \left[\frac{nx + S\left(\frac{S}{x}\right)^n}{n+1} - S\right] f(x)\,dx,$$

which has to be minimized for S.

Solution:

By differentiating $C(S)$, analogously to the previous models, the optimal value can be expressed as the solution of the equation

$$\int_0^{S_0} f(x)\,dx + \int_{S_0}^\infty \left(\frac{S_0}{x}\right)^n f(x)\,dx = \frac{c_2}{c_1 + c_2}.$$

For discrete demand distribution when the demand x has a probability $P(x)$, the optimality criteria are

$$M_n(S_0 - u) \le \frac{c_2}{c_1 + c_2} \le M_n(S_0),$$

where

$$M_n(S) = \sum_{x=0}^S P(x) + \sum_{x=S+u}^\infty \frac{(S+u)^{n+1} - S^{n+1}}{u(n+1)x^n}\,P(x)$$

and u is the demand unit.

71. Centralized inventory control system

Main codes:

$$1\ 1\ 0\ 1\ 0\ 1\ 1\ 1\ 1\ 0$$

Assumptions:

A central supplier controls a store according to a (t_p, S) inventory policy. Both the inventory holding and shortage costs are proportional to the (possibly negative) inventory level S_0 at the end of the period. The unit cost factors are denoted by C_I and C_S, respectively. The purchasing unit price is C_v. No fixed ordering cost is considered. The demand of a period is a random variable with the probability density function $f(y)$.

Objective:

The expected total cost of a period is

$$C(S) = C_v(S - S_0) + C_I \int_0^S (S - y) f(y)\,dy + C_S \int_S^\infty (y - S) f(y)\,dy \quad (S \ge S_0).$$

176

Solution:

The derivative of the cost function is zero at the optimal value. This nonlinear equation can be solved generally by numerical methods. No specific solution method is given in the paper.

72. Minimax ordering policy for finite time-horizon

Main codes:

1 1 0 1 0 1 1 1 1 0

Assumptions:

An order $y=\delta_n(x)$ is placed at the beginning of each time period (all periods have the same length t_p). x is the initial level of stock at the beginning of periods. n periods are considered, this is the time-horizon of optimization. The demand of a period is a random variable: it has an expected value μ and variance σ^2 which are known, but the distribution is not known. The purchase price per unit is c. No fixed ordering cost is considered. The unit holding cost is h and the unit shortage cost is p. Future costs are discounted by a factor α. $\pi_n(\delta_1, \delta_2, ..., \delta_n)$ is the ordering policy, $\delta_n(x)=y$ is the ordering rule for the nth period.

Objective:

The ordering policy minimizes the maximum expected discounted cost over the n periods. The maximum is examined over all the probability distributions which have the expected value μ and variance σ^2.

$$G_n(y) = cy + p(\mu - y) - \alpha[c(y-\mu) - G_{n-1}(y^*_{n-1})] + T_n(y),$$

where

$$T_n(y) = \max_{\sigma^2/(\sigma^2+\mu^2) \le \pi < 1} \{(p+h)[\pi(y-\mu) + \sigma\sqrt{\pi(1-\pi)}] +$$

$$+ \alpha\pi H_{n-1}(y - \mu + \sigma\sqrt{(1-\pi)/\pi})\}$$

and

$$H_{n-1}(u) = \begin{cases} 0 & \text{if } u \le y^*_{n-1}, \\ G_{n-1}(u) - G_{n-1}(y^*_{n-u}) & \text{if } u > y^*_{n-1}. \end{cases}$$

Solution:

It is proved that a limiting base stock policy characterizes the infinite model. The author shows some conditions under which a myopic minimax ordering policy exists for nonstationary distributions.

73. Minimax ordering policy for infinite time-horizon

Main codes:

1 1 0 1 0 1 1 1 1 0

Assumptions:

This is a version of the previous model. In this model the infinite time-horizon is considered (i.e., $n \to \infty$, using the notations of the previous model).

Objective:

The expected total cost is maximized for all possible demand distributions F, which have an expected value μ and variance σ^2 (the family of these distribution functions is denoted by $\Gamma_+(\mu, \sigma)$). The optimal order level is the value y_0, which minimizes the maximal cost, where the maximum is taken on the distribution class $\Gamma_+(m, b)$ (for any particular y). Thus the objective function is a minimax problem which can be formulated using the idea of dynamic programming.

It is supposed that the cost functions of the n-period time-horizon $C_n(y)$ are monotonously and uniformly convergent as n tends to infinity. For this asymptotic cost function $C(y)$ the following equation is valid:

$$C(x) = \min_{y \geq x} \max_{F \in \Gamma_+(\mu, \sigma)} \left\{ c(y-x) + L(y, F) + \alpha \int_0^\infty G(y-t)\, dF(t) \right\} = \min_{y \geq x} [G(y) - cx],$$

where

$$G(y) = cy + L(y, F) + \alpha \int_0^\infty G(y-t)\, dF(t)$$

and

$$L(y, F) = \begin{cases} \int_0^y h(y-t)\, dF(t) + \int_y^\infty p(t-y)\, dF(t) & \text{if } y \geq 0, \\ \int_0^\infty p(t-y)\, dF(t) & \text{if } y < 0. \end{cases}$$

Solution:

The $C(x)$ is convex and the sequence y_n^* contains convergent subsequences. Any limit point y of the sequence $\{y_n^*\}$ minimizes $G(y)$. The optimal ordering policy for the infinite horizon problem is a stationary policy $\pi(\delta^*, \delta^*, ...)$ that is characterized by a single critical number y. In this case it is to order up to $\bar{y}$ whenever the inventory level x is less than $\bar{y}$ and not to order otherwise: $\delta^* = \max(\bar{y}-x, 0)$, where $\bar{y}$ is the smallest minimizing solution of $c(1-\alpha)y + p(\mu-y) + (p+h)T_1 y$.

74. Order-level system for spare parts

Main codes:

$$1\ 1\ 0\ 1\ 0\ 1\ 1\ 1\ 1\ 0$$

Assumptions:

For a working process, k pieces of a component part are necessary. During the work they may break down and have to be replaced by spare parts. The demand for the spare parts is random with a known probability distribution. The stock is increased in each period (of identical length) to the order level S. If the number of perfect parts is less than k, then a shortage cost appears which is proportional to the amount of shortage and has a unit cost p. If the number of perfect parts is greater than k, then there is an inventory holding cost h, which is proportional to the excess stock.

178

Objective:

The expected total cost of a period is

$$C(S) = E\{cD(S)+h[S-D(S)-k]^+ + p[k-S+D(S)]^+\},$$

where E denotes the expected value operator, $(y)^+ = \max\{0, y\}$ and $D(S)$ is the expected demand of a period.

Solution:

In general the optimal order level S can be calculated from the inequalities which result for the first difference of the objective function from the necessary condition of the optimality. For special cases, when the demand of a period has a binomial or normal distribution, explicit solutions are derived.

75. Production smoothing by safety stock

Main codes:

$$1\ 1\ 0\ 1\ 0\ 1\ 1\ 1\ 1\ 0$$

Assumptions:

In the time period $(k-1, k)$ an amount u_k is produced. The demand r_k of this period is random and follows a normal distribution. The demand of the different periods are independent random amounts. The inventory level of period k is $I_k = I_{k-1} + u_{k-1} - r_k$. The optimal amounts of production have to be determined. The changes in the production go with costs: p is the cost factor for an increase per unit and q is the factor for a decrease of production per unit from a period to the next one. Both the inventory holding and the shortage cost are proportional to the amount of stock and of shortage, having unit cost factors h and v, respectively. Since r_k has a normal distribution for all k, we can suppose that the random amounts I_k and u_k have also normal distributions; their parameters are denoted by (μ_I, δ_I) and (μ_u, δ_u).

Objective:

The average expected cost of a period is

$$F = \frac{1}{\delta_I \sqrt{2\pi}} \int_{-\infty}^{\infty} L(I) \exp\left\{-\frac{1}{2\delta_I^2}(I-\mu_I)^2\right\} dI +$$

$$+ \frac{1}{\delta_u \sqrt{2\pi}} \int_{-\infty}^{\infty} P(u) \exp\left\{-\frac{1}{2\delta_u^2} u^2\right\} du,$$

where $P(u) = pu^+ + qu^-$ and $L(I) = hI^+ + vI^-$ with $y^+ = \max\{y, 0\}$ and $y^- = -\min\{y, 0\}$.

Solution:

The optimal decision u_k depends on the deviation of the actual inventory level I_k from the safety stock μ_I. This safety stock μ_I will be held fixed during the entire production smoothing process. The optimal μ_I is calculated by an ordinary static optimization procedure which is the first step of the solution. The second one is the linear dynamic optimization problem which has an explicit solution. For the optimal solution we have

$$u_k = -\frac{\alpha}{\alpha+\beta}(I_k - \mu_I),$$

where

$$\alpha = \exp\left(-\frac{1}{2}x^{*2}\right), \quad x^* = \varphi^{-1}\left(\frac{v}{v-h}\right)$$

and

$$\beta = \frac{p-q}{h-v}.$$

76. A model for discrete demand, occurring at random times in random amounts

Main codes:

1 1 0 1 1 1 9 1 2 0

Assumptions:

The t_ξ time between the ξth and $(\xi+1)$-th discrete demands is random, and has a $p_T(.)$ probability distribution function with mean τ and maximum value M. The amount of ξ-th demand d_ξ is a random integer value with known discrete probability distribution, which is denoted by $p_D(.)$. The maximal possible demand B is also given. The order is placed depending on the actual stock level after a demand occurrence and on the time interval since the occurrence of the previous demand. There is an immediate order receipt and in the case of shortage the sales are lost. The capacity of the store is N units. The cost factors: h the holding cost per unit of stock per unit time, π the penalty cost per unit of lost sales, K the fix ordering cost, c the purchasing cost.

The problem can be formulated by a Markovian renewal process where the different states are the possible values of the stock level. The δ policy can be characterized by a (γ, v, P) vector, where γ is the expected single transition cost, v the mean single state duration, P is the matrix of transition probabilities.

Objective:

The order level S is fixed and the time between the orders may change. The objective is to minimize the expected total cost for a unit per time. The explicit cost expression is not given in the paper.

Solution:

For fixed S, the time until the next order is optimized. The solution is a vector of $S-1$ dimensions. Its components give the optimal time until the next order for different stock levels 1 to $S-1$. After this step the value of S is optimized. The solution method can be based on linear programming or on some other iterative procedure.

77. Production planning for stochastic demand process

Main codes:

1 1 0 1 1 1 9 0 1 −1

Assumptions:

The production of an item is considered on the finite time interval $[0, T]$. The optimal production rate $r(t)$ is to be determined as a continuous function of time. The demand is described by a continuous stochastic process. The net inventory level at time t—which may also be negative in case of shortage—is denoted by $I(t)$. There is a backorder for shortages. The production cost $F(r(t))$ and the sum of inventory holding and shortage costs $h(I(t))$ is considered.

Objective:

The expected total cost for the time-interval $[O, T]$ has to be minimized by selecting an appropriate production rate function $r(t)$:

$$\min_{r(t), 0 \leq t \leq T} E\left\{ \int_0^T \{F[r(t)] + h[I(t)]\}\, dt \right\}$$

where E denotes the expected value operator.

Solution:

The method is based on the stochastic variation principle and control theory. The stochastic problem is first reduced to an equivalent deterministic problem for which an algorithmic solution is given. The existence of a unique solution $r(t)$ is also proved.

IV.1.2. Single-Period Models

78. *Order-level model, surplus can be sold*

Main codes:

1 1 0 1 0 1 1 0 1 0

Assumptions:

The total demand in the period $[O, T]$ considered is a random amount x with a known probability density function $f(x)$. The inventory is filled up by order to the level S at the beginning of the period. Before filling up there is an initial stock z. Both the inventory holding and shortage costs depend on the amount of stock or shortage at the end of the period. The respective costs per unit are S_T and P_T. The purchase price per unit is c. No fixed order cost is considered. The surplus stock at time T can be sold for the price per unit $r (r < c)$.

Objective:

The total expected cost of the period $[O, T]$ is

$$L_T(S) = \int_0^S [S_T(S-x) - r(S-x)] f(x)\, dx + \int_S^\infty P_T(x-S) f(x)\, dx + c(S-z),$$

which has to be minimized.

Solution:

The existence of an optimal solution is proved under different assumptions.

79. Nonlinear cost model, surplus can be sold

Main codes:

$$1\ 1\ 0\ 1\ 0\ 1\ 1\ 1\ 1\ 0$$

Assumptions:

During a given period t, the total demand is a random amount x with probability density function $f(x)$. The initial stock level is h. After the receipt of the amount q ordered the inventory level becomes $S = q + h$. The inventory holding cost C_c and the income C_r from the sale of the surplus stock are both functions of the stock level at the end of the period. The purchase price per unit and the cost of sale per unit during the period are c and z. The unit shortage cost is denoted by b.

Objective:

The total expected cost of the period is

$$g(S) = cS + \int_0^S [C_c(S-x) - C_r(S-x) - zx] f(x)\, dx + \int_S^\infty [b(x-S) - zS]\, f(x)\, dx,$$

which has to be minimized for S.

Solution:

Differentiating the cost function with respect to S and setting it equal to 0, the optimal order level S can be determined by solving the equation

$$c + \int_0^S [C_c(S-x) - C_r(S-x)] f(x)\, dx - \int_S^\infty [b(x-S) + z] f(x)\, dx = 0.$$

80. Stochastic single-period model without leadtime

Main codes:

$$1\ 1\ 0\ 1\ 0\ 1\ 1\ 0\ 2\ 0$$

Assumptions:

An interval with given length T is investigated during which time a random demand ξ occurs. The initial stock is x. The decision variable is the amount to be ordered z which is on hand at the beginning of the time interval. The inventory holding and shortage costs are proportional to the net inventory at the end of the time interval: they are expressed by a function L of the maximal inventory level. The unit of purchasing price is denoted by c. No fixed ordering cost is considered.

Objective:

The expected total cost of the period is to be minimized, i.e., we have to find the solution of the problem:

$$\min_{z \geq 0} C(z) = \min_{z \geq 0} \{cz + L(x+z)\}.$$

Solution:

In general, for an arbitrary known continuous demand distribution function, the necessary condition for optimality can be expressed by differentiating the objective function. It leads usually to the solution of the nonlinear equation $c+L'(\bar{x})=0$ by some numerical method.

The amount to be ordered is determined by

$$z = \begin{cases} \bar{x}-x & \text{if } x < \bar{x} \\ 0 & \text{if } x \geq \bar{x}, \end{cases}$$

where $\bar{x}$ is the order level.

For a discrete demand distribution an algorithmic solution is derived (by use of the first differences) and illustrated on Poisson distributed demand.

81. A minimax ordering policy for a single period

Main codes:

$$1\ 1\ 0\ 1\ 0\ 1\ 1\ 1\ 1\ 0$$

Assumptions:

A fixed time interval is considered during which a random demand occurs. The expected value μ and the variance σ of the demand is known, but its F distribution function is unknown. The purchase price and the unit inventory holding and shortage costs are denoted by c, h and p, respectively. All these costs are linear and proportional to the amount ordered, or obtained at the end of the period as stock or shortage. The inventory is filled up to the level S at the beginning of the period according to an order. The stock left over at the end of the period can be salvaged with a return of γ times the initial purchase cost c ($\gamma < 1$).

Objective:

For any fixed value of S, the total discounted cost of the time-period (with discount factor α) is:

$$G(S) = \max_{F} [c(1--\alpha\gamma)S + L(S, F) + c \cdot \alpha \cdot \gamma \cdot \mu]$$

where L is the expected holding and shortage costs incurred during the period. The objective function can be written as follows:

$$G(S) = c\alpha\gamma\mu + c(1-\alpha\gamma)S + p(\mu-S) + (p+h)T_1(S)$$

where

$$T_1(S) = \begin{cases} [\sqrt{\sigma^2+(S-\mu)^2}+(S-\mu)]/2 & \text{if } S \geq (\sigma^2+\mu^2)/2\mu, \\ \sigma^2 S/(\sigma^2+\mu^2) & \text{if } 0 \leq S < (\sigma^2+\mu^2)/2\mu, \\ 0 & \text{if } S < 0. \end{cases}$$

(The cost $G(S)$ is maximal for all possible demand distributions F which have expected value μ and variance σ^2.)

Solution:

The cost in the case of the most unfavourable demand distribution is expressed by $G(S)$, which has to be minimized for S. The optimal solution of this minimax optimization procedure is

$$S_0 = \begin{cases} 0 & \text{if } \sigma^2/(\sigma^2+\mu^2) \geq \theta \\ \dfrac{\mu+\sigma(2\theta-1)}{\sqrt{2\theta(1-\theta)}} & \text{otherwise} \end{cases}$$

where

$$\theta = \frac{p-c(1-\alpha\gamma)}{p+h}.$$

82. The "newsboy problem"

Main codes:

$$1\ 1\ 0\ 1\ 0\ 1\ 1\ 1\ 2\ 0$$

Assumptions:

The demand of the period considered is an integer random variable, which has the value x with probability $P(x)$ and may have values between x_{min} and x_{max}. The purchasing price per unit is b and the selling price per unit is d $(d>b)$. The surplus stock at the end of the period can be sold only for a unit price a $(a<b)$. The demand is lost in the case of shortage.

Objective:

In the case of an initial stock level S, the total expected profit of the period is

$$C(S) = d \sum_{x=x_{min}}^{S} xP(x)+d \sum_{x=S+1}^{x_{max}} SP(x)+a \sum_{x=x_{min}}^{S-1} (S-x) P(x)-bS,$$

which has to be maximized.

Solution:

Differentiating $C(S)$ with respect to S, we obtain that the optimal value of the initial stock S is the integer for which the following inequalities hold:

$$F(S_0-1) \leq \frac{d-b}{d-a} \leq F(S_0),$$

where

$$F(S) = \sum_{x=x_{min}}^{S_0} P(x).$$

IV.2. (t, S) Models with Fixed Leadtime

IV.2.1. Demand Is Backordered

83. Order level for random demand with uniform rate

Main codes:

$$1\ 1\ 0\ 1\ 0\ 1\ 1\ 1\ 1\ 1$$

Assumptions:

The length t_p of the order periods is prefixed. The demand of each order period is a random amount x which has the known probability density function $f(x)$. The leadtime between ordering and receipt is a fixed time interval with length L. The demand of the leadtime is v which has the probability density $h(v)$. At the beginning of each order period an order is placed for the amount $S-y$, where S is the order level and y is the actual inventory level which may also be negative (in the case, when shortage has been backordered). Having received the order after the leadtime L, first the backorder is satisfied, if there is any. Both the inventory holding cost and the shortage cost are proportional to time and amount of stock level or shortage and have the costs per unit c_1 and c_2.

In this version of the model the demand is assumed to occur with a uniform rate, i.e. the inventory level of the period decreases linearly.

Objective:

The purchase price and ordering cost do not influence the optimal order level S. The total expected cost of the inventory holding and shortage is for continuous values of x, v and S:

$$C(S) = \int_0^S \int_0^{S-v} c_1 \left(S-v-\frac{x}{2}\right) f(x) h(v)\, dx\, dv +$$

$$+ \int_0^S \int_{S-v}^{\infty} \left[c_1 \frac{(S-v)^2}{2x} + c_2 \frac{(x-S+v)^2}{2x} \right] f(x) h(v)\, dx\, dv +$$

$$+ \int_S^{\infty} \int_0^{\infty} c_2 \left(v-S+\frac{x}{2}\right) f(x) h(v)\, dx\, dv.$$

If the demand is measured in discrete units, then the integrals are replaced by the respective sums and the probability density by the discrete probability distribution.

Solution:

By differentiating the cost function and setting it equal to zero, we obtain the following equation from which the optimal order level (S_0) can be determined:

$$\int_0^{S_0} \left[\int_0^{S_0-v} f(x)\, dx + \int_{S_0-v}^{\infty} \frac{S_0-v}{x} f(x)\, dx \right] h(v)\, dv = \frac{c_2}{c_1+c_2}.$$

185

In the discrete case (when x, r and S may have values $0, u, 2u, ...$) the solution is given by the inequalities

$$K(S_0-u) \leqq \frac{c_2}{c_1+c_2} \leqq K(S_0),$$

where

$$K(S) = \sum_{v=0}^{S} \left[\sum_{x=0}^{S-v} P(x) + \left(S-v+\frac{u}{2}\right) \sum_{x=S-v+u}^{\infty} \frac{P(x)}{x} \right] H(v).$$

84. Order level for instantaneous random demand

Main codes:

$$1\,1\,0\,1\,0\,1\,1\,1\,1\,1$$

Assumptions:

This is a modified version of the previous model. All the notations are the same. The only difference is that the total amount of the demand of one period occurs instantaneously at the beginning of each period, once in each period.

Objective:

In the case when x, v and S are continuous, the expected cost of inventory holding and shortage is

$$C(S) = \int_0^S \int_0^{S-v} c_1(S-v-x)f(x)h(v)\,dx\,dv + \int_0^S \int_{S-v}^{\infty} c_2(v-S+x)f(x)h(v)\,dx\,dv +$$

$$+ \int_S^{\infty} \int_0^{\infty} c_2(v-S+x)f(x)h(v)\,dx\,dv.$$

In the case of discrete demand the integrals and density functions have to be replaced by the respective sums and discrete distributions.

Solution:

By differentiation of $C(S)$ we obtain that the optimal order level is the solution of the equation

$$\int_0^S \int_{S-v}^{\infty} f(x)h(v)\,dx\,dv = \frac{c_2}{c_1+c_2}.$$

In the discrete case the solution is given by

$$K(S_0-u) \leqq \frac{c_2}{c_1+c_2} \leqq K(S_0)$$

where

$$K(S) = \sum_{v=0}^{S} \sum_{x=0}^{S-v} p(x)H(v).$$

186

85. *The probabilistic scheduling-period order-level system with leadtime*

Main codes:

$$1\ 1\ 0\ 1\ 0\ 1\ 0\ 1\ 1\ 1$$

Assumptions:

The scheduling period is the time between two consecutive orders: it is also subject to control in this model together with the order level. Let $f(x, t)$ be the probability density function of demand x during a scheduling period with length t. Let $h(v)$ be the probability density of demand v during the lead time L. Beside the cost of inventory holding c_1 and the cost of shortage c_2 the ordering cost has to be considered which is a fixed amount c_3 for each order.

Objective:

The expected total cost for a given value of t and S is

$$C(t, s) = c_1 \int_0^S (S-v)k(v, t)\, dv + c_2 \int_S^\infty (v-S)\, h(v, t)\, dv + \frac{c_3}{t},$$

where

$$k(v, t) = \int_0^v \int_{v-r}^\infty \frac{f(x, t)}{x} h(v)\, dx\, dv.$$

Solution:

For any given t, the corresponding optimal order level $S_0(t)$ can be found from the equation

$$\int_0^{S_0(t^*)} k(v, t)\, dv = \frac{c_2}{c_1 + c_2}.$$

The corresponding minimum cost can then be shown to be

$$C_0(t) = \frac{c_2 r t^*}{2} + (c_1 + c_2) \int_0^{S_0(t^*)} vk(v, t^*)\, dv + \frac{c_3}{t^*},$$

where r is the mean rate of demand per unit of time. Hence, the value of t_0 which minimizes $C_0(t)$ has to be found. No general explicit solution can be obtained, however. In order to solve the problem we have to apply some numerical technique.

In the case of a discrete demand distribution the solution can be found similarly as in the previous models (83—84).

86. *The inventory-bank system*

Main codes:

$$1\ 1\ 0\ 1\ 0\ 1\ 1\ 1\ 1\ 1$$

Assumptions:

The ordering period has a fixed length t_p. The amount of demand is supposed to be a discrete random variable, i.e., during this period demand S occurs with

187

probability $P(S)$. In an ordering period the demand rate is assumed to be constant. The ordering rule of the inventory bank system is the following. At the beginning of each ordering period the average demand of the previous M periods $\bar{S}$ is calculated. The so-called inventory-bank level is $N \cdot \bar{S}$, where N is an integer. If the actual inventory level is less than the inventory bank level then the difference is ordered. If there is a leadtime, the ordered, but so far not received, amount has to be added to the actual physical stock (i.e., this inventory level has to be considered). In the case of a shortage, demand is backordered and will be satisfied at the next receipt. The order cost is a fixed value c_3. The inventory holding and shortage have the cost factors c_1 and c_2 per unit quantity and time.

Objective:

For the values of the decision variables N and M, the expected total cost can be approximated for a large value of M in the case of zero lead-time by

$$C(M, N) = c_1 \sum_{S=0}^{NS} \left(N\bar{S} - \frac{S}{2}\right) P(S) + c_1 \sum_{S=NS+u}^{\infty} \frac{(N\bar{S})^2}{2S} P(S) +$$

$$+ c_2 \sum_{S=NS+u}^{\infty} \frac{(S - N\bar{S})^2}{2S} P(S) + c_3 \sum_{S=u}^{\infty} P(S),$$

where u means the unit amount.

For arbitrary M and N, the exact cost expression for the zero leadtime case is

$$C(M, N) = c_1 \sum_{Q=0}^{Q_{max}} \left[\sum_{S=0}^{Q} \left(Q - \frac{S}{2}\right) P(S) + \sum_{S=Q+u}^{\infty} \frac{Q^2}{2S} P(S) \right] G(Q, M, N) +$$

$$+ c_2 \sum_{Q=0}^{Q_{max}} \left[\sum_{S=Q+u}^{\infty} \frac{(S - Q)^2}{2S} P(S) \right] G(Q, M, N) +$$

$$+ c_3 \sum_{Q=0}^{Q_{max}} \sum_{S=0}^{S_{max}} P(S) G(Q, M, N) B(Q, S),$$

where $G(Q, M, N)$ is the probability distribution of the initial stock level Q, which has the maximal possible value Q_{max}; finally, $B(Q, S) = 1$ if an order is placed, while $B(Q, S) = 0$ if no order is placed.

Solution:

The cost function is very complicated, direct optimization procedures cannot be applied. The author indicates two different ways for optimization: the first applies Markov chains, the second relies on simulation techniques. Some concepts, concerning the computer realization of the latter method are also described.

87. (t, S) policy shortage cost proportional to the amount of shortage

Main codes:

$$1\ 1\ 0\ 1\ 0\ 1\ 0\ 0\ 1\ 1$$

Assumptions:

A (t, S) policy with stochastic demand is considered. The demand is backordered in the case of shortage. The sum of the inventory reviewing and ordering costs at each order is a fixed amount L. The shortage cost is proportional to the amount of shortage, while the inventory holding cost is proportional both to the time and amount of inventory on hand. The respective unit cost factors are denoted by π and IC, where C is the unit cost of purchasing. The average demand rate is r. The leadtime is random with known distribution. The leadtimes of the different orders are assumed to be independent, and the deliveries are assumed to occur in the same sequence as the orders were given.

Objective:

The expected total annual cost is

$$K(S, t) = \frac{L}{t} + IC\left(S - \mu - \frac{\lambda t}{2}\right) + \pi E(S, t),$$

where μ is the expected demand during the leadtime and $E(S, t)$ is the expected annual shortage.

Solution:

For a given value of t, the optimal S can be derived by differentiating the objective function with respect to S and setting it equal to zero. If the value of t is also a subject of optimization, then the partial derivatives have to be equal to zero: the corresponding system of two equations can be solved numerically e.g. by the Newton-method.

88. (t, x, nQ) *policy with Poisson demand*

Main codes:

1 1 0 1 0 1 0 0 1 1

Assumptions:

The inventory reviewing period is of length t. At each review the following policy is applied: if the inventory level is above the reorder level s, then no order is given. If the inventory level is below s, then an order for an amount nQ is given where n is the smallest positive integer for which $nQ \geq s$. The demand occurs in discrete amounts (one by one) and for any time interval the total demand has Poisson distribution with mean demand rate r. The inventory reviewing cost is J at each review, the fixed order cost is A. There are two types of shortage costs: one is proportional to the amount and the other is proportional to the time of shortage. The unit cost factors are denoted by π and $\hat{\pi}$. The inventory holding cost has the usual dimension $\frac{[\$]}{[Q][T]}$ and the unit cost factor is denoted IC.

Objective:

The expected total annual cost (as a function of Q, s, t) is

$$K = \frac{J}{t} + \frac{A}{t}P_{Qr} + IC\left(\frac{Q}{2} + \frac{1}{2} - \mu - \frac{rt}{2}\right) + \pi E(Q, s, t) + (\hat{\pi} + IC)B(Q, s, t),$$

189

where $E(Q, s, t)$ is the expected annual amount of shortage, $B(Q, s, t)$ is the expected annual time of shortage and P_{Or} is the probability of ordering in a given period.

Solution:

A numerical procedure is proposed for the minimization of the objective, however, this method may yield some local minimum which is higher than the global minimum of the objective function. (A similar case is obtained, if the Poisson distribution is approximated by a normal distribution.)

89. (t, s, S) policy with stochastic demand

Main codes:

$$1\ 1\ 0\ 1\ 0\ 1\ 0\ 0\ 1\ 1$$

Assumptions:

In a period with length t the total demand is a discrete random amount x with probability $p(x, t)$. For n subsequent periods the total demand can be expressed using the n-fold convolution of the probability distribution denoted by $p^{(n)}(x, t)$. $P(x, t)$ is the complementary cumulative of $p(x; t)$. The demand of the different periods is assumed to be independent. At each review the following policy is applied: if the inventory level x is below the reorder level s, then the amount $S-x$ is ordered. There is a constant leadtime. The review cost is J, the fixed order cost is A. The sum of the inventory holding and shortage costs for a period is a function of the initial stock $r+j$ and the review period t, and is denoted by $H(r+j; t)$.

Objective:

The expected total annual cost (as a function of the parameters S, s, t) equals

$$K(S, s, t) = \frac{J}{t} + \frac{A + \sum_{n=0}^{\infty} \sum_{j=1}^{S-s} p^{(n)}(S-s-j, t) H(s+j, t)}{t \sum_{n=1}^{\infty} \sum_{j=1}^{S-s} n p^{(n-1)}(S-s-j, t) P(j, t)}.$$

Solution:

The explicit cost function is derived for the case when the demand is generated by a Poisson-process and also for the continuous case; a sophisticated numerical solution method is also suggested.

90. Order-level system with stochastic process demand

Main codes:

$$1\ 1\ 0\ 1\ 0\ 1\ 1\ 1\ 1\ 1$$

Assumptions:

The demand of an interval with length s has the probability distribution $G(x, s)$. Thus the demand is generated by a stochastic process which is homogeneous in time. The mean demand rate is r. The demands of disjunct time intervals are

190

assumed to be independent, thus the stochastic process of the demand has in-dependent increments (it is a stationary process). The order periods have fixed length t, the leadtime L is also constant. The inventory holding, shortage and order costs have the usual dimensions and respective unit costs c_1, c_2 and c_3.

Objective:

The expected total annual cost, as a function of the order level S, can be expressed in the following form:

$$K(S) = \frac{c_3}{t} + (c_1 + c_2) B - c_1 \left(rL + \frac{rt}{2} \right),$$

where

$$G(x) = \frac{1}{t} \int_L^{t+L} G(x, s)\, ds \quad \text{and} \quad B = \int_S^{\infty} x\, dG(x) - S[1 - G(S)].$$

Solution:

The optimal order level is the solution of the equation $G(S) = c_2/(c_1 + c_2)$. Its solution is studied in the case, when the demand is a Wiener process, i.e.,

$G(x, s) = \Phi \left(\dfrac{x - rs}{\sigma \sqrt{s}} \right)$, where Φ denotes the standard normal distribution function:

hence the optimal S_0 can be determined by the numerical solution of a nonlinear equation.

91. Stochastic periodic model with leadtime

Main codes:

$$1\,1\,0\,1\,0\,1\,1\,1\,1\,1$$

Assumptions:

There is a fixed order period with length t. The order of period i is delivered in period $i + L$, thus the leadtime is Lt. The purchasing unit cost is c, the fixed ordering cost is c_3 at each order. The inventory carrying cost and the shortage cost are both functions denoted by h and p. The demand is random and it is backordered in the case of shortage.

Objective:

$$C(x) = -cx + \min_{y \geq x} \{c_3 \delta(y - x) + G(y)\}, \quad \delta(z) = \begin{cases} 1 & z > 0 \\ 0 & z = 0 \end{cases}$$

where

$$G(y) = cy + h(y) + p(y).$$

If the inventory holding and shortage costs are linear with unit cost factors c_1 and c_2, then

$$G(y) = \begin{cases} cy + c_1 \displaystyle\int_0^y (y - \varrho)\, dF^{(L+1)}(\xi) + c_2 \int_y^{\infty} (\xi - y)\, dF^{(L+1)}(\varrho) & \text{for} \quad y > 0, \\[4mm] cy + c_2 \displaystyle\int_0^{\infty} (\xi - y)\, dF^{(L+1)}(\xi) & \text{for} \quad y \leq 0, \end{cases}$$

where $F(x)$ is the demand distribution and $F^{(L)}(x)$ denotes the L-fold convolution.

Solution:

For convex cost functions $h(y)$ and $p(y)$, the total cost is a convex function, too. In the case of continuous demand the single solution can be calculated from the derivative of the objective function. In the case of linear inventory holding and shortage costs, the optimal order level S_0 is the solution of the equation

$$F^{(L+1)}(S_0) = \frac{p-c}{p+h}.$$

IV.2.2. Lost Sales Case

92. Stochastic order period and order level model

Main codes:

$$1\ 1\ 0\ 1\ 0\ 1\ 0\ 0\ 2\ 1$$

Assumptions:

A (t, S) policy is considered with stochastic demand. The probability density function of the demand during an order period is denoted by $h(x; t)$. The expected mean demand rate is r. The inventory review cost is J at each review (it can be considered as a fixed order cost). The unit purchasing price is C, the unit inventory holding cost is IC. The shortage cost is proportional to the amount of shortage and has a cost factor π.

Objective:

The expected annual total cost is

$$K(t, S) = \frac{J}{t} + IC\left[S - \mu - \frac{rt}{2}\right] + \left[IC + \frac{\pi}{t}\right] \int\limits_{S}^{\infty} (x - S)h(x; t)\, dx,$$

where μ is the expected demand during the leadtime.

Solution:

The optimal order level S is the solution of the equation $\int\limits_{S}^{\infty} h(x; t)\, dx = \dfrac{IC \cdot t}{\pi + IC \cdot t}$ for fixed t. To determine the optimal t value the simultaneous solution of the above equation and the equation $\dfrac{\partial K(t, S)}{\partial t} = 0$ is suggested. For this, for example, Newton's method can be applied. The solution technique is illustrated by an example.

93. Stochastic model with a one period leadtime

Main codes:

$$1\ 1\ 0\ 1\ 0\ 1\ 1\ 1\ 2\ 1$$

192

Assumptions:

There is a fixed order period with length t. The leadtime is equal to the length of the order period, thus there is a one-period delay in the system. The demand is random and in each period has the same probability distribution. In the case of shortage the demand is lost for the supplier. The purchasing unit price is c, the fixed ordering cost is c_3 at each order. The inventory carrying cost and the shortage cost are supposed to be functions denoted, by h and p, respectively.

Objective:

The expected total cost is

$$C(x) = -cx + \min_{y \geq x} \{c_3 \delta(y-x) + cy + L(y, x)\}, \quad \delta(z) = \begin{cases} 1 & z > 0 \\ 0 & z = 0 \end{cases}$$

where $L(y, x) = h(y, x) + p(y, x)$, which, for linear inventory holding and shortage cost with cost factors h and p, can be written as

$$\int_0^x \left[h \int_0^{y-x_1} (y-x_1-x_2) \, dF(x_2) + p \int_{y-x_1}^{\infty} (x_1+x_2-y) \, dF(x_2) \right] dF(x_1) +$$

$$+ \int_x^{\infty} \left[h \int_0^{y-x_1} (y-x_1-x_2) \, dF(x_2) + p \int_{y-x_1}^{\infty} (x_1+x_2-y) \, dF(x_2) \right] dF(x_1);$$

where x_1 and x_2 denote the demand of the first and second period. Both random quantities have the same distribution function $F(x)$.

Solution:

No general solution method is given for the minimization of the above cost function.

IV.2.3. Models with Shortage Constraints

94. Scheduling period model with leadtime and no shortage

Main codes:

$$1\ 1\ 0\ 1\ 0\ 1\ 8\ 1\ 0\ 1$$

Assumptions:

The demand x is random and has the probability density function $f(x, t)$ during any time period with length t. The mean demand rate is r, and there is a maximal rate of demand $x_{\max}$. The probability density of the demand during the leadtime with length L is $h(v)$. At each time period t an order is given. The length of the order period has to be optimized. The order level $S_p = (L+t) x_{\max}$ is prescribed, since no shortage is allowed. The fixed ordering cost is c_3. The unit inventory holding cost is c_1.

Objective:

The expected total annual cost is

$$C(t) = c_1\left(S_p - rL - \frac{rt}{2}\right) + \frac{c_3}{t} = c_1\left[(L+t)A(L+t) - \left(L+\frac{t}{2}\right)\right]r + \frac{c_3}{t},$$

where

$$A(t) = \frac{x_{max}}{rt}.$$

Solution:

The derivative of the objective function has to be set equal to zero. This equation serves to determine the optimal value of t. In general no explicit solution can be given, but for special cases of $A(t)$ explicit solutions are derived. For example,

$$A(t) = k: \quad t_0 = \sqrt{2c_3/[c_1 r(2k-1)]},$$

and for

$$A(t) = 1 + b/t: \quad t_0 = \sqrt{2c_3/(rc_1)}.$$

95. Order level, when the probability of the shortage is constrained

Main codes:

1 1 0 1 0 1 1 1 1 1

Assumptions:

There is a fixed order period with length t. The leadtime L is also constant. The demand of the period with length $t+L$ has the probability distribution function $F(x|t+L)$ and the probability density function $f(x|t+L)$. The probability of the shortage is constrained, the continuous supply has to be ensured with some probability $1-\alpha$ (which is supposed to be near 1). This is the required service reliability level.

Objective:

The expected stock on hand at the end of a period is

$$B(S|t+L) = \int_0^S (S-x)f(x|t+L)\,dx$$

in the case of an order level S. Thus the inventory holding cost is $htB(S|t+L)$ which has to be minimized under the mentioned probabilistic constraint.

Solution:

The Lagrange-function of the above constrained optimization problem is

$$H(S, G) = htB(S|t+L) + G[1 - F(S|t+L) + (1-\alpha)]$$

with the Lagrange-multiplier G. By minimization the optimal order level S can be expressed: $S = F^{-1}(\alpha|L+t)$, where F^{-1} denotes the inverse function of F.

194

96. *Order level when the expected value of shortage is constrained*

Main codes:

$$1 1 0 1 0 1 1 1 1 1$$

Assumptions:

The random demand during the prefixed order period t plus leadtime L is described by the distribution function $F(x|t+L)$. The density function is denoted by $f(x|t+L)$. The mean demand rate is r. The expected value of the shortage at the end of the order periods is constrained. It has to be smaller than or equal to the prescribed service level β.

Objective:

The expected value of shortage at the end of an arbitrary period equals

$$D(S|t+L) = \int_0^S (x-S)f(x|t+L)\,dx.$$

The expected inventory holding cost $htB(S|t+L)$ is minimized under the shortage constraint $D(S|t+L) \leq \beta$.

The expected value of inventory at the end of an arbitrary period is expressed in the form

$$B(S|r+L) = \int_0^S (S-x)f(x|t+L)\,dx.$$

Solution:

The Lagrange-function of the above constrained optimization problem is

$$H(S, G) = htB(S|t+L)+G[D(S|t+L)-(1-\beta)\,tr],$$

where G is the Lagrange-multiplier. By minimization of $H(S, G)$, the optimal order level can be expressed as

$$S=D^{-1}[(1-\beta)\,tr|t+L]$$

where D^{-1} denotes the inverse function of D.

97. *Order level with periodical withdrawals*

Main codes:

$$1 1 0 1 0 1 0 1 7 1$$

Assumptions:

The demand consists of a deterministic and a random part. A quantity P is demanded at each p unit of time. At every unit of time there is a random demand with known distribution and with mean demand rate d. Thus the expected total demand of a time unit is $\lambda=P/p+d$. The length of the order period is t units of time, which is a function of p. Two cases are considered: (a) $t=k \cdot p$; (b) $t=p/k$, where k is a positive integer.

195

The leadtime is L. The distribution function of the demand during the time $t+L$ is $F(x)$. The order can be revoked before delivery with a unit cost K proportional to the amount revoked. The order level S means the sum of the stock on hand and on order minus the withdrawals. The fixed order cost is c_3, the inventory carrying cost is c_1.

Objective:

The total expected cost for a time unit is

$$C(S) = \frac{c_3}{t} + \frac{c_1}{2}(2S - \mu - dL) + \frac{K}{t} \cdot b(S),$$

where $b(S)$ is the expected amount of withdrawals and $\mu = d(L+t)$.

Solution:

No exact solution method is given, only an approximation using heuristic arguments is proposed.

V. Deterministic Replenishment, Stochastic Demand, (s, q) Policy

This group consists of the so-called reorder point models. When the stock level decreases to or below the level of the reorder point (s) an order with given lot-size (q) is placed. The further classification of these models is based on the behaviour of the demand in case of shortage and on the stock review system:

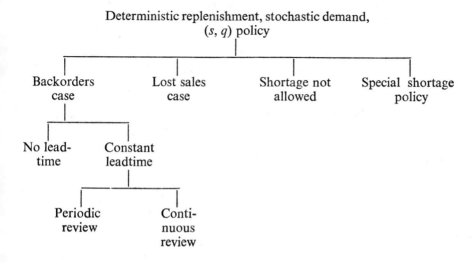

V.1. Backordered Demand

V.1.1. Models with No Leadtime

98. *Stochastic reorder point lot-size system*

Main codes:

$$1\ 1\ 0\ 1\ 0\ 1\ 2\ 1\ 1\ 0$$

Assumptions:

There is prescribed reviewing period with length w_p. During this time the demand x is a random amount with a probability density function $f(x)$ in each reviewing period, where the demand intensity $\bar{x}$ is assumed to be constant. At each review an order is given, if the stock level is not above the reorder point s. The order amount is an integer multiple of a constant lot-size, i.e., it equals nq, where n is the smallest integer for which the actual stock level minus backorders plus nq is greater than s. The amount ordered is delivered instantly and first the possible backordered demands are fulfilled.

Under general circumstances, the initial stock of the review periods have a uniform distribution in the interval $[s, s+q]$. The fixed order cost is c_3. The inventory carrying cost and the shortage cost is proportional to the time and amount of stock level or shortage. The unit cost factors are denoted by c_1 and c_2.

Objective:

a) If the demand x and the control variables s and q are continuous, then the expected total annual cost is

$$C(s, q) = \frac{c_1+c_2}{q} \int_s^{s+q} T(Q) \, dQ + c_2 \left(\frac{\bar{x}}{2} - \frac{q}{2} - s\right) + \frac{c_3[1-v(q)/q]}{w_p}$$

where

$$T(Q) = \int_0^Q \int_0^z m(y) \, dy \, dz \quad \text{with} \quad m(y) = \int_0^\infty \frac{f(x)}{x} \, dx \quad \text{and}$$

$$v(q) = \int_0^q \int_0^y f(x) \, dx \, dy.$$

b) For discrete variables with unit u

$$C(s, q) = (c_1+c_2) \frac{u^2}{q} [N(s+q-u) - N(s-u)] +$$

$$+ c_2 \left(\frac{\bar{x}}{2} - \frac{q+u}{2} - s\right) + \frac{c_3[1-uv(q-u)/q]}{w_p}$$

where

$$N(B) = \sum_{S=0}^{B} \sum_{z=0}^{S} \sum_{y=0}^{z} K(y)$$

with

$$K(y) = \begin{cases} P(0) + \dfrac{u}{2} \displaystyle\sum_{x=u}^{\infty} \dfrac{P(x)}{x} & \text{for} \quad y = 0 \\[3mm] \dfrac{u}{2} \dfrac{P(y)}{y} + u \displaystyle\sum_{x=y+u}^{\infty} \dfrac{P(x)}{x} & \text{for} \quad y = u, 2u, \ldots, \end{cases}$$

and $P(x)$ is the probability of demand x during a review period.

$$v(s) = \sum_{z=0}^{S} \sum_{x=0}^{z} P(x).$$

198

Solution:

a) For any given q, the optimal s_0^q can be determined by partial differentiation, as the solution of the equation $\partial C(s, q)/\partial s=0$. Similarly, for given s the optimal q_0^s can be fixed. The overall optimal pair s_0, q_0 is the solution of a system of two non-linear equations which can be evaluated using iterative numerical solution techniques.

b) For the discrete case, the equations are to be replaced by a system which consists of two pairs of inequalities which can be solved also by some iterative numerical method.

99. Stochastic reorder-point system

Main codes:

$$1\ 1\ 0\ 1\ 0\ 1\ 2\ 1\ 1\ 0$$

Assumptions:

They are basically similar to those of the previous model. The only difference is that the lot-size q_p is prescribed. Thus the order cost does not influence the optimal value of the reorder point.

Objective:

With the notations used in the previous model, the expected annual total cost (without ordering costs) can be expressed as follows.

a) In the continuous case we have

$$C(s) = (c_1+c_2)\frac{K(s+q_p)-K(s)}{q_p}+c_2\left(\frac{\bar{x}}{2}-\frac{q_p}{2}-s\right),$$

where

$$K(S) = \int_0^S T(Q)\,dQ$$

b) In the discrete case with unit u we have

$$C(s) = (c_1+c_2)\frac{u^2}{q_p}N(s+q_p-u)-N(s-u)]+c_2\left(\frac{\bar{x}}{2}-\frac{q_p+u}{2}-s\right).$$

Solution:

a) The optimal s_0 is the solution of the equation

$$\frac{T(s+q_p)-T(s)}{q_p} = \frac{c_2}{c_1+c_2}.$$

b) In the discrete case, the inequalities

$$g(s_0-u) \leq \frac{c_2}{c_1+c_2} < g(s_0)$$

14*

199

give the necessary and sufficient condition for the optimality of s_0 with the notation

$$g(s) = \frac{W(s+q_p)-W(s)}{q_p/u},$$

where $W(s) = \sum_{z=0}^{s} \sum_{y=0}^{s} k(y)$.

V.1.2. Models with Constant Leadtime

V.1.2.1. Periodic Inventory Review Models

100. Stochastic reorder-point lot-size system with leadtime

Main codes:

$$1\ 1\ 0\ 1\ 0\ 1\ 2\ 1\ 1\ 1$$

Assumptions:

This model is a generalized version of the reorder-point lot-size system (Model 98). The difference is that a leadtime L is assumed between placing and receiving an order. The notation z is replacing s, meaning the sum of inventory on hand and on order at the beginning of the review period.

Objective:

a) In the continuous case we have

$$C(z, q) = (c_1+c_2)\frac{K(z+q)-K(z)}{q}+c_2\left(\frac{\bar{x}}{2}+\bar{v}-z-\frac{\bar{q}}{2}\right)+\frac{c_3}{w_p}\left[1-\frac{v(q)}{q}\right]$$

where the notations are the same as used in connection with Models 98 and 99, except the influence of the leadtime changes the earlier meaning of $m(y)$ to

$$m(y) = \int_0^y d(y-v)e(v)\,dv,$$

where $d(w) = \int_w^{x_{max}} \frac{f(x)}{x}\,dx$ and $e(v)$ is the probability density function of the demand v during the leadtime. Its expected value is denoted by $\bar{v}$.

b) In the discrete case we can write

$$C(z, q) = (c_1+c_2)\frac{u^2}{q}[N(z+q-u)-N(z-u)]+$$

$$+c_2\left(\bar{v}+\frac{u-\bar{x}}{2}-z-\frac{q}{2}\right)+\frac{c_3}{w_p}\left[1-\frac{uv(q-u)}{q}\right],$$

200

where we apply the notations of Models 98 and 99, except that $R(y)$ should be replaced in the expression of $N(y)$ by $D(y)$, which is defined by

$$R(y) = \sum_{v=0}^{y} D(y-v)E(v)$$

where $E(v)$ is the probability of demand v during the leadtime and $v(k) = \sum_{j=0}^{k} \sum_{x=0}^{j} P(x)$. The expected value of v is denoted by $\bar{v}$.

Solution:

The same principles should be applied as in the case of model 98 and 99.

101. Stochastic reorder-point system with leadtime

Main codes:

$$1\ 1\ 0\ 1\ 0\ 1\ 2\ 1\ 1\ 1$$

Assumptions:

We have the same assumptions as in the previous model, the only difference is that the lot-size q_p is a prefixed value. Thus it is a special case of the previous model.

Objective:

Both in the continuous and discrete case the objective function is the same as in the previous model, but it has only the variable s to control, the value of q is replaced by q_p. The order cost does not influence the optimal value of s, thus it may be omitted.

Solution:

In the continuous case, the optimal s_0 is the solution of the equation

$$\frac{T(z+q_p)-T(z)}{q_p} = \frac{c_2}{c_1+c_2},$$

where $T(Q) = \int_0^Q \int_0^z m(y)dy \; dz$ with $m(y) = \int_0^y \int_{y-v}^{\infty} \frac{f(x)}{x} dx \, e(v)dv$. In the discrete case, the optimal s_0 is defined by the inequalities

$$g(z_0-u) \leq \frac{c_2}{c_1+c_2} \leq g(z_0),$$

where

$$g(z) = \frac{W(z+q_p)-W(z)}{q_p} \quad \text{with} \quad W(S) = \sum_{j=0}^{S} \sum_{y=0}^{j} R(y)$$

and $R(y)$ defined in connection with the previous model.

102. (s, q) policy with constant leadtime

Main codes:

1 1 0 1 0 1 2 2 1 1

Assumptions:

The mean value of the shortage is assumed to be small, when related to the lot-size q, moreover, the time interval in which no stock is on hand is considerably smaller than the average intervals between the deliveries. The assumptions are satisfied if we have a relatively high shortage cost factor c_2 relative to the inventory holding cost factor c_1. The leadtime L is constant: during this time there is a random demand x with probability density function $f(x)$. The mean demand rate is denoted by r. The fixed order cost is c_3.

Objective:

The total expected annual cost is

$$C(s, q) = \frac{r}{q}\left[c_3 + c_2 \int_s^\infty (x-s)f(x)\, dx\right] + c_1\left(\frac{q}{2}+s\right).$$

Solution:

The optimal value of the reorder-point s and lot-size q have to satisfy the following system of equations:

$$q = \sqrt{\frac{2r\left[c_3 + c_2 \int_s^\infty (x-s)f(x)\, dx\right]}{c_1}}$$

and

$$\int_s^\infty f(x)\, dx = \frac{c_1}{rc_2}\cdot q,$$

which can be solved by an iterative procedure starting with the initial trial

$$q_0 = \sqrt{2rc_3/c_1}.$$

103. (s, q) system with normal demand distribution

Main codes:

1 1 0 1 0 1 2 2 1 1

Assumptions:

The demand of the constant lead-time l has a normal distribution with mean D and standard deviation σ. The usual cost factors C_h, C_s and C_0 are considered for inventory holding, shortage an fixed order cost.

Objective:

For given reorder point z and lot-size Q, the expected total annual cost is

$$C_T = \frac{D}{Q}C_0 + \left[\frac{Q}{2} + z\sqrt{l}\sigma_d\right]C_h + \left(C_h + \frac{D}{Q}C_s\right)\sqrt{l}\sigma_d\{f(z) - z[1 - \varphi(z)]\}$$

where φ denotes the probability function of the standard normal distribution and σ_d represents the standard deviation of demand per unit time.

Solution:

From the necessary conditions of the optimum

$$\frac{\partial C_T}{\partial z} = 0 \quad \text{and} \quad \frac{\partial C_T}{\partial Q} = 0$$

the solution cannot be expressed in an explicit form. Applying the parametrization

$$a = \frac{2C_h C_0}{DC_s} \quad \text{and} \quad b = \frac{2C_h \sqrt{l}\sigma_d}{DC_s}$$

we obtain a linear function for $\partial C/\partial z$ for any fixed z. This can be represented on a nomogram from which the optimal value of z can be read for the given a, b. The optimal Q can be determined on a similar way or by solving the equation $\partial C_T/\partial Q = 0$ substituting the optimal z_0 for z. Beside the graphical solution, an iterative numerical procedure is also presented.

104. *The cost of changing the continuous review policy to periodic one*

Main codes:

$$1\ 1\ 0\ 1\ 0\ 1\ 2\ 2\ 1\ 1$$

Assumptions:

The demand of a unit time has a normal distribution with parameters D and σ. The leadtime has a constant length l. The shortage cost depends only on the amount of shortage, while the inventory holding cost depends also on the stocking time. The order cost consists of a part, dependent on the amount ordered, and of a constant part.

Objective:

The model compares two inventory policies: the (s, q) policy with continuous review and the (t, S) policy with periodic review. The expected total cost per unit time for a periodic review system is:

$$Dc_0 + \frac{c_1}{T} + \frac{1}{2}c_2 DT + c_2[z - D(T + l)] + \frac{c_3}{T}\int_z^\infty (X - z)f(X, T + l)\,dX.$$

The cost function of the corresponding lot size reorder level system is:

$$Dc_0 + \frac{Dc_1}{Q} + \frac{1}{2}c_2 Q + c_2(r-Dl) + \frac{Dc_3}{Q} \int_r^\infty (X-r)f(X, l)\,dX,$$

where
c_0 = cost of item per unit;
c_1 = replenishment cost;
c_2 = cost of inventory holding per unit per unit time;
c_3 = unit shortage cost;
D = mean demand per unit time;
σ^2 = variance of demand per unit time;
$f(X, t)$ = probability distribution of the demand X in time t;
l = leadtime;
T = review period;
z = gross inventory (i.e., actual inventory plus balance on order);
Q = lot-size;
r = reorder level.

Solution:

The aim of the analysis is to evaluate the surplus cost in inventory carrying, shortage and order by changing the continuous review to periodic one. This is compared to the surplus cost of the administration of continuous reviewing in order to select the more economic policy.

105. (s, q) policy with general leadtime demand distribution

Main codes:

1 1 0 1 0 1 2 2 1 1

Assumptions:

The demand during the leadtime τ is a random amount with the probability density function $f(x, \tau)$. The mean demand rate is λ.

Objective:

The total expected cost for unit time

$$K(r, Q) = \frac{\lambda}{Q} A + IC \left(\frac{Q}{2} + r - \mu\right)(IC + \hat{p})\frac{\beta(r)}{Q}$$

has to be minimized, where $\mu = \lambda \cdot \tau$ and the cost factors IC, $\hat{p}$, A of the inventory holding, shortage and ordering have the usual dimensions; furthermore,

$$\beta(r) = \int_r^\infty \int_x^\infty (y-x)f(x, \tau)\,dy\,dx.$$

Solution:

A nonlinear system of equations have to be solved which results from the first-order condition of optimality, (the partial derivatives of the cost function are equal to zero at the minimum point which is proved to be an interior point). No specified method of solution is detailed in the paper.

204

106. Stochastic reorder-point lot-size model, with Poisson demand

Main codes:

$$1\ 1\ 0\ 1\ 0\ 1\ 2\ 2\ 1\ 1$$

Assumptions:

An (s, q) policy is investigated (here the reorder point is denoted by r, the lot-size by Q), where shortage is backordered and the demand of the leadtime is generated by a Poisson process: units are demanded at random moments with exponentially distributed inter-arrival times. All the variables are discrete. The average demand rate is λ. Two different shortage cost factors are considered: $\hat{\pi}$ is proportional to the time, while π is proportional to the amount of shortage. The inventory holding cost factor is IC and the constant cost of order is denoted by A.

Objective:

The total expected cost for unit time,

$$K(Q, r) = \frac{\lambda}{Q} A + IC \left[\frac{Q+1}{2} + r - \mu \right] + \pi E(Q, r) + (\hat{\pi} + IC) B(Q, r)$$

has to be minimized, where the expected number of backorders for a time unit is $E(Q, r)$ and the expected number of backorders at a random instant is $B(Q, r)$. Both $E(Q, r)$ and $B(Q, r)$ are expressed by means of the Poisson distribution function in a rather complicated form. The parameter μ denotes the expected demand during the lead-time.

Solution:

Under the assumption that $\mu < r + Q$, a simple iterative procedure can be derived. For the general case a complicated computer program has been developed which yields a local optimum which may be different from the globally optimal solution of the problem.

107. Stochastic reorder-point and lot-size model with normally distributed leadtime demand

Main codes:

$$1\ 1\ 0\ 1\ 0\ 1\ 2\ 2\ 1\ 1$$

Assumptions:

The assumptions are the same as in the previous model; the only difference is that the leadtime demand is approximated by normal distribution.

Objective:

It differs from the previous one only in the expression of $E(Q, r)$ and $B(Q, r)$ (as a result of using the normal distribution).

Solution:

It is suggested by an iterative process.

205

108. *Minimization of the cost induced by the average inventory level*

Main codes:

$$1\ 1\ 0\ 1\ 0\ 1\ 2\ 2\ 1\ 1$$

Assumptions:

The demands occur according to a Poisson process with parameter λ, while the amount of a demand is also random with expected value μ. The leadtime τ is constant. $t^{(j)}$ means the time moment when j-th order is placed.

Objective:

It is not the total expected cost which is considered (as is usually the case), but the average inventory level is calculated and the cost belonging to this level is minimized. The average level of inventory is

$$\bar{y}(t) = Q - \mu\lambda t + Q \sum_j l[t - (t^{(j)} + \tau)]$$

where

$$l(t) = \begin{cases} 0, & \text{if } t \le 0 \\ 1, & \text{if } t > 0. \end{cases}$$

The average level cost of period $[0, T]$ is

$$\varphi(Q, r) = \frac{c_1 + c_2}{2H} R^2 \left(\frac{H}{Q} - 1\right) + \frac{c_1}{2H}(HQ + 2HR - 2QR) + c_3 \frac{H}{Q},$$

where $R = r - \mu\lambda\tau$; $H = \mu\lambda T$; c_1, c_2 and c_3 are the inventory holding, shortage and ordering cost factors.

Solution:

If the partial derivatives of the objective are set to zero, then it leads to a cubic equation with respect to the optimal order amount Q^*. Having determined Q^*, the optimal reorder point r^* can be obtained by substitution.

109. (s, q) *policy with gamma distributed demand*

Main codes:

$$1\ 1\ 0\ 1\ 0\ 1\ 2\ 2\ 1\ 1$$

Assumptions:

The ordered amount q is delivered with a leadtime τ. The average inventory level can be approximated by the expression $q/2 + y - \lambda\tau$, where y is the reorder level and λ is the mean demand rate. The leadtime demand follows gamma distribution with parameters k and μ. The expected shortage equals

$$\Pi(y) = \int_y^\infty (x - y) f(x)\, dx$$

where $f(x)$ denotes the density function of the respective gamma distribution.

If the shortage cost does not depend on the amount of shortage, then

$$\Pi(y) = \int_y^\infty x f(x)\, dx$$

Objective:

The inventory holding and ordering cost factors are s and g. The shortage cost may be constant or may depend on the time and amount of shortage. The shortage cost factor is denoted by d. The expected total cost,

$$L = s\left(\frac{q}{2} + y - \lambda\tau\right) + \frac{\lambda}{q}[g + \pi(y)]$$

is to be minimized.

Solution:

Applying the first-order optimality criterion, the following system of equations have to be solved in an iterative way:

$$q = \bar{x} - y + \sqrt{(y-\bar{x})^2 + \frac{2\lambda}{s}\left[g + \frac{d}{\mu\Gamma(k)}(\mu y)^k e^{-\mu y}\right]},$$

where

$$y = \frac{1}{\mu}\left\{\frac{\Gamma(k)\left(1 - \dfrac{sq}{\lambda d}\right)}{\displaystyle\sum_{n=0}^{\infty} \frac{(-1)^n (\mu y)^n}{n!(k+n)}}\right\}^{1/4}$$

with the starting points $y_0 = \bar{x}$ and $q = \sqrt{\dfrac{2\lambda g}{s}}$. ($\Gamma(k)$ denotes the incomplete gamma function.)

110. Continuous review (s, q) policy with discrete demand

Main codes:

$$1\ 1\ 0\ 1\ 0\ 1\ 2\ 2\ 1\ 1$$

Assumptions:

The demand occurs in stochastic, discrete amounts. There is a constant leadtime L. During this time some demand q_L occurs with probability $P_L(q_L)$. The leadtime demand is supposed to be less than the amount of order q, thus only a single order has to be placed at any moment. For an order placed, a constant ordering cost k and a purchasing cost with unit cost factor c is considered. The shortage cost is proportional to the amount of shortage (controlled right before the delivery of the next order), with a unit cost factor π. The inventory holding cost factor h has the usual dimension (\$/amount/time).

Objective:

The expected total cost per time unit equals

$$c(s, q) = \frac{kM}{Q} + cM + h\left(\frac{Q}{2} - M_L + s\right) + \left(\frac{hM_L}{2Q} + \frac{M\pi}{Q}\right) \sum_{q_L > s} (q_L - s) p_L(q_L)$$

where M means the expected time between two subsequent orders, M_L is the expected leadtime demand, s is the reorder point and Q represents the lot-size.

Solution:

The optimal reorder point s and lot-size Q can be determined by a numerical procedure based on the first order optimality condition. For the case when the random demand is approximated by normal distribution, a simplified procedure is derived and illustrated in the paper.

111. *The service level of an* (s, q) *policy*

Main codes:

$$1\ 1\ 0\ 1\ 0\ 3\ 2\ 2\ 1\ 1$$

Assumptions:

This is a special model for the description of a system. For a fixed lot-size q and reorder point s, the probability of the event that a demand at an arbitrary time moment can be satisfied without shortage is determined. A Poisson-type demand process and constant leadtime L is considered. The shortage is backordered.

Objective:

The service level is the above probability which is expressed for (s, q) policy in the form

$$P_\alpha = 1 - \frac{1}{Q} \sum_{y=r+1}^{r+Q} \sum_{k=y}^{\infty} \varphi(k|\lambda), \quad \text{where} \quad \varphi(k|\lambda) = \frac{(\mu\lambda)^k}{k!} e^{-\mu\lambda} \ k = 0, 1, 2, \ldots$$

r and Q represent the reorder point and reorder quantity respectively, and $\varphi(x|L)$ is the probability distribution function of the leadtime demand.

Solution:

The service level P_α is shown to coincide with the following ratio:

$$\frac{\text{expected directly satisfied demand}}{\text{expected total demand in a delivery cycle}}$$

considered for an order period. The probability of the event "no shortage occurs in a period" is also determined in the paper.

V.2. Models with Lost Sales and Shortage

112. *Lot-size, reorder-point model with general stochastic demand*

Main codes:
$$1\ 1\ 0\ 1\ 0\ 1\ 2\ 2\ 2\ 0$$

Assumptions:

A continuous, stochastic demand is considered with a mean demand rate λ. The usual cost factors IC and A are taken for inventory holding and ordering cost. The shortage cost is proportional to the amount of shortage and has unit cost π. At every time only a single order may be outstanding. The time interval, when shortage occurs, is assumed to be negligable related to the time without shortage.

Objective:

The expected total cost per time unit equals

$$K(r, q) = \frac{\lambda A}{q} + IC\left(\frac{q}{2} + r - \mu\right) + \left(IC + \frac{\pi\lambda}{q}\right)\left[\int_r^\infty xh(x)\,dx - rH(r)\right],$$

where the probability density and distribution function of the leadtime demand are denoted, respectively, by $h(x)$ and $H(x)$, and μ is the expected leadtime demand.

Solution:

The optimal value of the lot-size q and reorder point r are the solution of a pair of nonlinear equations, which can be obtained, calculating the partial derivatives of the objective function.

113. *Stochastic lot-size, reorder-point model with Poisson demand*

Main codes:
$$1\ 1\ 0\ 1\ 0\ 1\ 2\ 2\ 2\ 1$$

Assumptions:

An (s, q) policy is analyzed here, supposing that the demand is generated by a Poisson process: some discrete amounts are demanded at random moments. The number of units demanded at a time interval follows a Poisson distribution. Thus the model can be described by a Markov process, which has a steady state. The mean demand rate is λ, at shortage the demand is lost. At every time only a single order may be outstanding. The same cost factors are considered as in the previous model.

Objective:

The expected total cost per time unit equals

$$K(r, Q) = \frac{\lambda}{Q + \lambda\hat{T}}\left[A + IC\left(\frac{1}{2\lambda}Q(Q+1) + \frac{Qr}{\lambda} - \frac{Q\mu}{\lambda}\right)\right] + \left(\frac{ICQ}{\lambda} + \pi\right)\times$$

$$\times\left[\mu P(r-1;\ \lambda\tau);\ -\frac{r}{\lambda}P(r;\ \lambda\tau)\right]$$

where $\hat{T}$ is the expected time of a period when the system is out of stock, μ is the expected demand of the constant leadtime L. The probability of demand r during the leadtime is denoted by $P(r; \lambda\tau)$ which means the Poisson distribution with parameter $\lambda\tau$.

Solution:

The solution method is given only for the special case, when the value of $\hat{T}$ is neglected. In many practical cases this serves as a good approximation for the optimal parameters of the general case.

114. *A heuristic model for perishable goods*

Main codes:

$$1\ 1\ 0\ 1\ 0\ 1\ 2\ 2\ 2\ 1$$

Assumptions:

The considered item is perishable at an exponential rate. This means that if the inventory level at time t is equal to $I(t)$ then in u units of time the number of good pieces decreases to

$$I(t+u) = I(t)e^{-\theta u},$$

where θ is the rate of perishing. The mean demand rate is fixed at λ units per unit time. The total demand during any leadtime is a random variable with continuous distribution function $F(.)$ and density $f(.)$. The expected inventory level at time $t+u$ considering the perishability and demand can be expressed in the form of

$$I(t+u) = e^{-\theta u}(I(t)+\lambda/\theta)-\lambda/\theta.$$

Orders are placed according to the (s, q) policy. Beside the cost factors of the previous model h, p and k a purchasing cost of the perished items with price c has to be taken into account. A constant leadtime τ exists.

Objective:

The expected total cost per time unit is equal to

$$A(Q, s) = \left\{\frac{1}{T}K+cQ+\frac{hQ}{\theta}-pw\bar{F}(w)+p\int_w^\infty xf(x)\,dx\right\}-\frac{h\lambda}{\theta}-c\lambda$$

where $\mu=\lambda\tau$, $w=s(1+\theta\tau/2)+\mu$ and T is the expected value of the length of a period between two consecutive orders.

$$T = \frac{1}{\theta}\ln\left(1+\frac{\theta Q}{s\theta+\lambda}\right).$$

The function $\bar{F}(x)$ is the complementary cumulative of the leadtime demand.

Solution:

A heuristic approach was applied and evaluated by computer simulation.

210

V.3. Models in Which No Shortage Is Allowed

115. *Stochastic lot-size system with constrained demand in a review period*

Main codes:

$$1\ 1\ 0\ 1\ 0\ 1\ 3\ 1\ 0\ 0$$

Assumptions:

A stochastic demand is considered which has a density function $f(x)$ for the demand of an inventory review period with a fixed length w_p. The maximal demand of this period is $x_{\max}$, a known value. The demand rate is constant during a review period. No shortage is allowed, thus the reorder point is prescribed: $s_p = x_{\max}$. The order amount is the smallest integer multiple of q, for which the increased inventory level becomes larger, than s_p. An immediate delivery is assumed. The initial stock level of the review periods is uniformly distributed on the interval $[s_p; s_p + q]$. The inventory holding and shortage cost factors have the usual dimensions

$$[c_1] = \frac{[\$]}{[Q][T]}, \quad [c_3] = [\$].$$

Discrete and continuous versions of the model are described.

Objective:

a) For continuous q and x, the expected total cost equals

$$C(q) = c_1 \frac{q}{2} - c_3 \frac{V(q)}{q w_p} + c_1 \left(x_{\max} - \frac{\bar{x}}{2} \right) + \frac{c_3}{w_p},$$

where

$$V(x) = \int_0^x \int_0^y f(z)\, dz\, dy.$$

b) For discrete q and x (with possible values $0, u, 2u, \ldots$), we have

$$C(q) = c_1 \frac{q}{2} - c_3 \frac{u V(q-u)}{q w_p} + c_1 \left(x_{\max} - \frac{\bar{x}+u}{2} \right) + \frac{c_3}{w_p},$$

where

$$V(x) = \sum_{y=0}^{x} \sum_{z=0}^{y} P(z)$$

and $P(z)$, $z = 0, u, 2u, \ldots, x_{\max}$, is the probability distribution of demand.

Solution:

a) For the continuous case, the derivative of the objective function has to be equal to 0 at the optimum point. This is a necessary but not sufficient condition; more than one solution may exist which must be compared with each other to select the optimal one. The case $q=0$ is to be considered separately. The solution is illustrated by examples.

b) For the discrete case, the following necessary conditions have to be fulfilled by the optimal q_0:

$$R(q_0-u) \leq \frac{2c_3}{c_1 w_p} \leq R(q_0),$$

where

$$R(q) = \frac{q(q+u)}{\sum\limits_{x=0}^{q} xP(x)} \qquad q = 0, u, 2u, \ldots$$

The solutions have to be compared with each other in order to select the global optimizer.

116. Constrained demand during review period and leadtime

Main codes:

$$1\ 1\ 0\ 1\ 0\ 1\ 3\ 1\ 0\ 1$$

Assumptions:

This model is an extension of the previous one. Constant leadtime L is assumed during which the maximal amount of the delivery is a known constant v_{max}. No shortage is allowed, thus the reorder point is prescribed by $s_p = x_{max} + v_{max}$. All the other assumptions are the same as in the previous model.

Objective:

a) For continuous values of v, q and for periodic review with length w_p (>0), the expected total cost equals to

$$C(q) = c_1 \frac{q}{2} - c_3 \frac{V(q)}{qw_p} + c_1 \left(x_{max} + v_{max} - \frac{\bar{x}}{2} - \bar{v} \right) + \frac{c_3}{w_p},$$

where $\bar{x}$ and $\bar{v}$ denote the mean demand during a review period and during the leadtime, $V(x)$ is defined by

$$V(x) = \int_0^x \int_0^y f(z)\, dz\, dy.$$

For continuous review ($w_p = 0$) we have

$$C(q) = c_1 \left(v_{max} + \frac{q}{2} - \bar{v} \right) + c_3 \frac{r}{q}.$$

b) For discrete values of q, x and v: $0, u, 2u, \ldots$, in the case $x_{max} > u$ we can write

$$C(q) = c_1 \frac{q}{2} - c_3 \frac{uv(q-u)}{qw_p} + c_1 \left(x_{max} + v_{max} - \frac{\bar{x}+u}{2} - \bar{v} \right) + \frac{c_3}{w_p},$$

where

$$V(x) = \sum_{y=0}^{x} \sum_{z=0}^{y} P(x),$$

212

while in the case $x_{max}=u$ there holds

$$C(q) = c_1 \left(v_{max} + \frac{q}{2} - \bar{v} + \frac{u - rw_p}{2} \right) + c_3 \frac{r}{q}.$$

Solution:

The methods in all of the above cases are similar to the solution method of the previous model, only the cost expressions are somewhat more complicated.

117. Lot-size model for demands at random moments in random amounts

Main codes:

1 1 0 1 0 1 2 2 0 0

Assumptions:

The demand occurs in random amounts at random moments. Both the amounts and the interarrival times of demands may have an arbitrary distribution; the expectations are denoted by μ and λ, respectively. At each demand the inventory is reviewed. When the demand cannot be satisfied from stock, an order is placed for an amount nQ, where n is the smallest integer which ensures the satisfaction of demand after the arrival of the ordering amount. The leadtime is supposed to be negligible. Shortage is not allowed and the reorder point is determined by $s_p = 0$. The inventory holding cost factor h and a fixed cost K of an order are taken into account.

Objective:

The expected value of the sum of holding and ordering cost is:

$$C = (K\mu/\lambda Q) + (hQ/2).$$

Solution:

Applying the theory of Markovian renewal processes, the optimality of the lot size

$$Q_w = \sqrt{\frac{2K\mu}{h\lambda}}$$

is proved.

V.4. Models with Special Shortage Policy

118. Stochastic model with perturbed demand

Main codes:

1 1 0 1 0 1 2 2 9 0

Assumptions:

This is an extended version of Model 97 since the demand rate λ may also be a random variable. The expected value of demand is μ. The demand process may be perturbed and changed from the expected one described by $R(t)$ to

$R'(t)$ as a function of time. In the case of shortage we know that $E(R(t)-$
$-R'(t))=\varphi(t-t_0)$, where t_0 is the time moment when shortage occurred and

$$\int_0^\infty \varphi(t)dt=I.$$

This plays a role when r is determined, as in the cost function the demand is considered as being influenced by customers' reactions.

Objective:

The following notations are used:
M = order size
L = reorder level,
$1-\alpha$ = service level,
H = holding cost per item per period,
C = revenue coefficient per item,
$N(t)$ = Poisson demand process,
T_i^* = time moment when the first stockout occurs.

The expected profit in the backorders case equals to

$$E(P_i) = \lambda_i C - \frac{H(M-L)^2}{2M},$$

where λ_i is the demand rate occurring in the ith cycle (its mean is μ). In the lost sales case, this can be written as

$$E(P_i) = \frac{C(M-L)}{M} \cdot \frac{\mu_0}{1+\alpha I} - \frac{H(M-L)^2}{2M}.$$

Solution:

The optimal values of M and L are determined similarly to the solution procedure outlined in connection with Model 97.

The special case, when $H(t)$ can be described by a Poisson process is detailed. For nonnegative inventory levels, the process $N^*(t)$ is characterized by the binomial distribution with parameters (t/T_i^*) and $(M-L)$. The expected inventory holding cost of period i can be expressed in the form

$$\frac{H}{2}(M-L)T_i^*.$$

The solution is a specific version of the above procedure.

119. Model with the possibility of withdrawals

Main codes:

1 1 0 1 0 1 2 1 7 1

Assumptions:

The demand for the considered spare parts consists of two components:
a) a random demand D occurs at every unit of time which has a probability density function $f_D(d)$ with mean $\bar{d}$ and standard deviation σ_D;

214

b) a fixed demand P occurs at every ϱ units of time; thus the demand rate is $r = P/\varrho + \bar{d}$. The leadtime L is constant, during this interval the demand x has the probability density function $f_x(x)$ with mean $\bar{x}$ and standard deviation δ_x which can be derived from the density function $f_D(d)$.

Beside the usual cost factors of inventory holding c_1 and the constant cost of ordering c_3, a new cost factor is considered: if a unit of order is renounced because it is expected to be not necessary, then a cost K_s appears.

Objective:

The expected total cost for a time unit equals to

$$K(s, q) = \frac{rc_3}{q} + c_1 \left(\frac{q}{2} + s - \mu \right) + \frac{K_s r}{q} \bar{b}(s),$$

where μ is the expected leadtime demand and $\bar{b}(s)$ is the expected value of the renounced orders during a period which is expressed by means of the probability density function $f_D(d)$ in a rather complicated way.

Solution:

The optimal amount to be ordered is determined by differentiating the cost function; this equals

$$q_0 = \sqrt{\frac{2r(c_3 + K_s \bar{b}(s^*))}{c_1}}.$$

In the case when c_1 is much greater than $K_s \bar{b}(s^*)$, this expression can be approximated by the Wilson-formula:

$$q_0 \approx \sqrt{\frac{2rc_3}{c_1}}.$$

For the optimal value of the reorder point s, no explicit formula can be given, therefore an iterative solution procedure is suggested in the paper.

VI. Deterministic Delivery, Stochastic Demand, (*s*, *S*) Policy

These models are based on the reorder-point order-level policy. In other words, an order is placed if the inventory level decreases to or under the reorder point *s*. The amount of order should increase the inventory level to its maximal level *S*.

The models of this group can be classified first by the type of shortage policy, then by the type of the leadtime. It is a strange fact that the type of inventory review does not play an important role in the classification. This is due to the small number of continuous review models, which are typical for (*s*, *q*) policy models. Three deterministic delivery and stochastic demand models of not (*s*, *S*) but other rather special policy, have also been connected to this group of models.

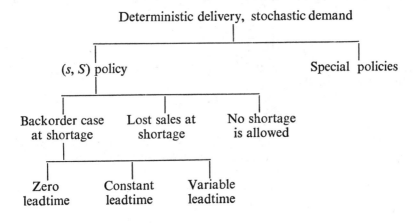

VI.1. Backorder Case at Shortage

VI.1.1. Models without Leadtime

120. *Stochastic reorder-point order-level system*

Main codes:

$$1\ 1\ 0\ 1\ 0\ 1\ 5\ 1\ 1\ 0$$

Assumptions:

A stochastic demand occurs during the fixed w_p reviewing period. The probability density function of the demand $f(x)$ is the same in each reviewing period.

216

The demand intensity is the same in each reviewing period. At shortage the demand is backordered and is satisfied immediately after receiving the next order. The inventory holding and shortage costs are linear with unit cost factors c_1 and c_2 and dimension

$$[c_1] = [c_2] = \frac{[\$]}{[Q][T]}.$$

The constant cost of an order is denoted by c_3. The model considers a single item and a single location. It is a generalized version of the order-level system (Model 143) and of the reorder point model (Model 129), since here both parameters (s and S) are subject to control. The control variables may be discrete or continuous, depending on the type of demand.

Objective:

a) For continuous x, s and S we can write

$$C(s, S) = \frac{c_1 J_1 + c_2 J_2 + c_3}{\bar{t}}$$

where the average stock level equals

$$J_1(s, S) = \begin{cases} w_p \int_0^S (S-y)g(y)\, dy + \int_0^{S-s} J_1(s, m) f(S-s-m)\, dm & \text{if} \quad s \geq 0 \\ w_p \int_0^S (S-y)g(y)\, dy + \int_0^S J_1(s, m) f(S-m)\, dm, & \text{if} \quad s \leq 0; \end{cases}$$

with the notation

$$g(y) = \int_y^{\infty} \frac{f(x)}{x}\, dx.$$

The average shortage level can be expressed by

$$J_2(s, S) = \begin{cases} w_p \int_S^{\infty} (y-S)g(y)\, dy + \int_0^{S-s} J_2(s, m) f(S-s-m)\, dm, & \text{if} \quad s \geq 0, \\ w_p \int_S^{\infty} (y-S)g(y)\, dy + \int_0^S J_2(s, m) f(S-m)\, dm + \\ \qquad\qquad + \int_s^0 B(s, n) f(S-n)\, dn, & \text{if} \quad s \leq 0, \end{cases}$$

where

$$B(s, n) = w_p(-n + \bar{y}) + \int_s^n B(s, r) f(n-r)\, dr,$$

and finally

$$\bar{t}(k) = w_p + \int_0^k \bar{t}(v) f(k-v)\, dv,$$

is the expected length of an order period for a given parameter $k = S - s$.

217

b) For the discrete case, with unit u and probability distribution of demand $P(x)$, the expected total cost of the system is

$$C(s, S) = (c_1 + c_2)u \sum_{Q=s+u}^{S} B(Q-n) H(Q) -$$

$$- c_2\left(\bar{Q} - \frac{\bar{x}}{2}\right) + \frac{c_3}{w_p} H(S)[1 - P(0)],$$

where Q is the inventory level at the beginning of the reviewing period, furthermore,

$$B(Q) = \begin{cases} 0 & (Q < 0), \\ \sum_{y=0}^{Q} G(y) = \sum_{y=0}^{Q} \sum_{r=0}^{y} K(r) & (Q \geq 0); \end{cases}$$

$$K(r) = \begin{cases} P(0) + \dfrac{u}{2} \sum_{x=u}^{\infty} \dfrac{P(x)}{x} & (r = 0), \\ \dfrac{u}{2}\dfrac{P(r)}{r} + u \sum_{x=r+u}^{\infty} \dfrac{P(x)}{x} & (r = u, 2u, \ldots); \end{cases}$$

$$H(Q) = A(S-Q)/ \sum_{y=0}^{S-s-u} A(y) \quad (Q = s+u, s+2u, \ldots, S);$$

$$A(y) = \begin{cases} 1 & (y = 0), \\ \sum_{x=u}^{y} A(y-x) R(x) & (y = u, 2u, \ldots); \end{cases}$$

$$R(x) = \frac{P(x)}{1 - P(0)} \quad (x = u, 2u, \ldots).$$

Solution:

a) For the continuous case no general solution method is given, it is derived only for a special demand distribution. (Namely, when $f(x) = \dfrac{x}{b^2} e^{-x/b}$ $(b>0)$.) For the general case the discretization of the demand distribution and decision variables is suggested.

b) In the discrete case, for given $k = S - s$, the corresponding optimal reorder point s_0^k can be found using the following inequalities which are necessary optimality conditions:

$$N(s_0^k - u) \leq \frac{c_2}{c_1 + c_2} \leq N(s_0^k),$$

where

$$N(s) = \sum_{Q=s+u}^{s+k} G(Q) H(Q)$$

Similar necessary conditions (which may not always be sufficient) exist also for k_0^s—the optimal k, corresponding to some given reorder point s.

The following search procedure is suggested for finding a pair of locally optimal values of s_0^* and S_0^*. For any given reorder point s_i the corresponding locally optimal order level S_{i+1} can be derived. Similarly, for given order level S_{i+1}, the corresponding locally optimal reorder point s_{i+1} can be computed using the above inequalities. The search stops when $S_{i+1} = S_i$ or $s_{i+1} = s_i$ holds (approximately).

121. (s, S) policy for exponentially distributed demand

Main codes:

$$1\ 1\ 0\ 1\ 0\ 1\ 5\ 1\ 1\ 0$$

Assumptions:

We consider an (s, S) policy ($0 \leq s \leq S$), where unsatisfied demand is backlogged, leadtime is negligible. The principal costs are as follows: storage cost (h), shortage cost (p) and ordering cost consisting of a fixed part (K) and a variable part (c). Demand is exponentially distributed: $\varphi(\xi) = e^{-\xi}$.

Objective:

$$C(S, s) = \frac{K}{1+S+s} + c + h \frac{s - 1 + \dfrac{S^2 - s^2}{2}}{1+S-s} + \frac{(h+p)e^{-s}}{1+S-s}.$$

Solution:

Using the variables $\Delta = S - s$ and s and employing the standard procedure of calculus, we obtain as the unique solution to the equations $\partial C/\partial \Delta = 0$; $\partial C/\partial s = 0$

$$\Delta_0 = \sqrt{\frac{2K}{h}} \quad \text{and} \quad e^{-s_0} = \frac{h + \sqrt{2Kh}}{h+p}.$$

These results are valid whenever $\sqrt{2Kh} \leq p$. In the case where $\sqrt{2Kh}$ exceeds p, s_0 should be taken as zero and Δ_0 remains the same. If the demand is $\varphi(\xi) = \dfrac{1}{\mu} e^{-\xi/\mu}$ with mean μ, then

$$\Delta_0 = \sqrt{\frac{2K\mu}{h}} \quad \text{and} \quad e^{-s_0/\mu} = \frac{\sqrt{2K\mu h} + h\mu}{\mu(h+p)}.$$

122. (s, S) policy optimization by the solution of an integral equation using simulation

Main codes:

$$1\ 1\ 0\ 1\ 0\ 1\ 5\ 1\ 1\ 0$$

Assumptions:

The demand of a period is a random amount with a probability density function $\Phi(\xi)$. The leadtime is negligible. The inventory holding and shortage costs are proportional to the stock level or shortage level at the end of a period. The

219

unit cost factors are denoted by h and p. The stock level at the beginning of the period following a decision is x and $Q = S - s$.

Objective:

For given x, the expected cost of inventory holding and shortage is

$$L(x) = h \int_0^x (x-u)\,\varphi(u)\,du + p \int_x^\infty (u-x)\,\varphi(u)\,du.$$

Solution:

For the optimal parameters we have the following relations:

$$m(S, Q) = 0 \quad \text{and} \quad \int_0^Q m(S, x)\,dx = K,$$

where K denotes the constant ordering cost and

$$m(S, x) = -L'(S+x) + \int_0^x m(S, x-u)\,\varphi(u)\,du.$$

For the solution of the above equations, a fast simulation technique was developed using a simulation system called CSMP.

123. *Solution of the (s, S) model by the use of the renewal function*

Main codes:

$$1\ 1\ 0\ 1\ 0\ 1\ 5\ 1\ 1\ 0$$

Assumptions:

A periodic review (s, S) policy is considered, where the demand of a period is random with distribution function $F(x)$, density function $f(x)$ and expected value μ. The sum of the inventory holding and shortage cost of a period is a function $L(x)$ of the initial stock x. L is supposed to be a convex function. The ordering cost consists of a constant K and of another sum which is proportional to the ordered quantity (unit purchasing price, c), and $D = S - s$.

Objective:

The expected total cost for a reviewing period is

$$a(x|s, S) = \frac{K + L(S) + \int_0^D L(S-x)\,m(x)\,dx}{1 + M(D)} + c\mu$$

where $M(x)$ is the renewal function

$$M(x) = \sum_{n=1}^\infty F^{(n)}(x),$$

and the superscript (n) denotes the n-fold convolution. The derivative of $M(x)$ is denoted by $m(x)$.

Solution:

At the optimum point the partial derivatives of the cost function are equal to zero. Hence, we have the following equation for the optimal values of s and S

$$L'(D+s)+\int_0^D L'(D+s-x)\,m(x)\,dx = 0$$

and

$$[1+M(D)]\cdot\left\{L'(D+s)+\int_0^D L'(D+s-x)\,m(x)\,dx+L(s)\,m(D)\right\}-$$

$$-\left\{K+L(D+s)+\int_0^D L(D+s-x)\,m(x)\,dx\right\}\cdot m(D) = 0.$$

124. A stationary solution of the Arrow—Harris—Marschak model

Main codes:

$$1\ 1\ 0\ 1\ 0\ 1\ 5\ 1\ 1\ 0$$

Assumptions:

The stochastic demand of the reviewing period with length w_p is characterized in each period by the same density function $f(x)$. Demand is supposed to occur at the end of the periods in one lot. The inventory holding cost is proportional to the time and amount of stock on hand and has the unit cost factor c_1. In each period, when shortage occurs, a fixed shortage cost c_2 is taken into account. The constant ordering cost is c_3.

Objective:

Applying the notation $k=S-s$, the expected total cost of an ordering period for $s\geq 0$ equals to

$$K(s,k) = c_1(s+k)w_p+c_2\int_{s+k}^\infty f(x)\,dx+c_3\int_k^\infty f(x)\,dx+\int_0^k K(s,v)f(k-v)\,dv;$$

for $s\leq 0$ we have

$$K(s,S) = c_1 Sw_p+c_2\int_S^\infty f(x)\,dx+c_3\int_{S-s}^\infty f(x)\,dx+$$

$$+\int_0^S K(s,v)f(S-v)\,dv+\int_0^{-s} B(v)f(S-s-v)\,dv,$$

where $B(v)$ is the expected total cost during that portion of the ordering period over which only shortages occur. It can be determined from the equation

$$B(v) = c_2+c_3\int_v^\infty f(x)\,dx+\int_0^v B(w)f(v-w)\,dw.$$

221

The expected length of an order period is

$$\bar{t}(k) = w_p + \int_0^k \bar{t}(v) f(k-v)\, dv,$$

thus the expected total cost of the system for time unit equals to

$$C(s, S) = \frac{K(s, S)}{\bar{t}(k)}.$$

Solution:

Only a special demand distribution (exponential distribution with density function $f(x) = \frac{1}{b}\,e^{-x/b}$ is calculated; the method is based on the solution of differential equations, where the specific properties of the demand distribution are utilized. In the original version of the model, formulated by Arrow, Harris and Marschak, a dynamic programming solution approach was used.

125. (s, S) *model with exponential demand distribution*

Main codes:

1 1 0 1 0 1 5 1 1 0

Assumptions:

The demand occurs at the beginning of each reviewing period and has an exponential distribution with density function

$$f(x) = \frac{1}{b}\,e^{-x/b}.$$

Leadtime is zero. At shortage situations, the demand is backordered. The inventory holding, shortage and ordering cost factors have the usual dimensions

$$[c_1] = [c_2] = \frac{[\$]}{[Q][T]} \quad \text{and} \quad [c_3] = [\$].$$

Objective:

$C(s, S)$ is derived from the objective function of Model 120, with simple substitution.

Solution:

The system of equations obtained can be solved by a numerical procedure.

126. *The improved Roberts-approximation*

Main codes:

1 1 0 1 0 1 5 1 1 0

222

Assumptions:

The demand of a reviewing period has a discrete distribution, the probability of the demand q is $p(q)$. The demand may occur in several lots. There is an immediate delivery of the orders which can be placed only at given review points. The inventory holding cost and the shortage cost is proportional to the quantity of net inventory at the end of the period with unit cost factors h and g. The constant ordering cost is K. The notation $S-s=D$ is used.

Objective:

The expected total cost per unit time equals

$$L(S, D) = \frac{1}{1+M(D)} \left[K + L(S) + \sum_{z=0}^{D} L(S-z) \, m(z) \right],$$

where $m(z) = p(z) + \sum_{q=0}^{z} p(z+q) \, m(q)$,

$$M(D) = \sum_{z=0}^{D} m(z),$$

and

$$L(y) = h \sum_{q=0}^{y} (y-q) p(q) + g \sum_{q=y+1}^{\infty} (q-y) p(q), \quad \text{if} \quad y \geq 0.$$

Solution:

An approximation of the optimal solution is given, which is an improved version of the Roberts approximation. The approximation of the optimal D can be expressed as

$$D^* = \left[\sqrt{2m_1 K/h} \right],$$

where m_1 is the average demand of a period. The optimal value of the reorder point s can be calculated using an iterative procedure to determine the greatest value of s which satisfies the inequality

$$\sum_{q=s}^{\infty} (q-s+1) p(q) \geq (D^*+\beta) \frac{h}{h+g}$$

with

$$\beta = \frac{m_1 + m_2}{2m_1},$$

where

$$m_2 = \sum_{q=1}^{\infty} q^2 p(q).$$

127. (s, S) model with reliability constraint

Main codes:

$$1 \ 1 \ 0 \ 1 \ 0 \ 1 \ 5 \ 1 \ 1 \ 0$$

Assumptions:

The demand of a reviewing period has geometric distribution, $p(q)=(1-a)a^q$, where $q=0, 1, 2, \ldots$ $(0<a<1$, a given constant). The demand may occur in several lots within a period. Orders can be placed at every review point, they are delivered instantly. The probability of shortage in a period is constrained by the reliability level P_α. The cost factors h, g and K are the same as in the previous model.

Objective:

The expected total cost per time unit,

$$L(S, D) = \frac{m_1 K + (1+m_1)(S-m_1)h + (D+2S+1)hD/2 + m_1^2(g+h)a^{s-1}}{D+m_1+1},$$

is to be minimized under the reliability constraint

$$P_\alpha = \frac{m_1 a^s}{D+m_1+1},$$

where $D=S-s$, furthermore, $m_1 = a/(1-a)$ is the expected value of demand in a period.

Solution:

The optimal value of D is approximated by the integer part of

$$\sqrt{2m_1 K/h}.$$

The optimal S can be calculated directly from the above-presented reliability equation.

128. (s, S) model with geometric demand distribution

Main codes:

$$1\ 1\ 0\ 1\ 0\ 1\ 5\ 1\ 1\ 0$$

Assumptions:

This model is similar to the previous one, but the demand with geometric distribution occurs at the beginning of each reviewing period. Its expected value is $\mu = \dfrac{q}{1-q}$, as the density function is given by

$$P(\xi) = (1-q)q^\xi \quad (\xi = 0, 1, 2, \ldots).$$

In the case of shortage, the demand is backordered. The cost factors are the same as in the previous model with the additional factor c, the unit price.

Objective:

The expected total cost is expressed as a function of the reorder point s and of $D=S-s$:

$$a(x|s, S) = c\mu + \frac{1}{\mu+D}\left[K\mu + (h+p)\mu^2 q^s + hDs - h\frac{D^2-D}{2} + h\mu(D+s-\mu-1)\right].$$

224

Solution:

The first differences of the objective function are taken with respect to s, while D is fixed. The optimal s is the smallest integer for which this difference is non-negative, i.e.:

$$\frac{\mu q^{s+1}}{\mu+D} < \frac{h}{h+p} \leqq \frac{\mu q^s}{\mu+D}.$$

Similarly, using the first differences with respect to D for the optimal D we have the relation

$$D(D+1) = 2\mu K/h.$$

129. Stochastic reorder-point system

Main codes:

1 1 0 1 0 1 5 1 1 0

Assumptions:

This is a simplified version of Model 120. The order level is a prescribed value $S=S_p$ with $S_p=s+k_p$. In this way the cost of ordering for a unit of time is the same constant value. Thus only the inventory holding cost factor c_1 and the shortage cost factor c_2 has to be considered in the optimization problem.

Objective:

a) In the continuous case, the expected total cost is

$$K(s) = c_1 J_1(s, k_p)+c_2 J_2(s, k_p),$$

where J_1 is the expected total inventory of a period and J_2 is the expected total shortage of a period. They can be expressed in a similar form as it was done in Model 120. (The value of S is to be substituted by $s+k_p$ in the mentioned expressions.)

b) In the discrete case, the expected total cost can be also expressed similarly to that of Model 120 by substituting $S=s+k_p$.

Solution:

a) In the continuous case no general solution is given. For the special demand distribution $f(x)=\dfrac{1}{b^2}\, xe^{-x/b}$, $(x\geqq0)$ the optimal solution is derived.

b) In the discrete case the optimal reorder point s_0 is the solution of the inequality

$$N(s_0-u) \leqq \frac{c_2}{c_1+c_2} \leqq N(s_0),$$

where the notations of Model 120 are used.

130. Inventory policy for slowly moving items

Main codes:

1 1 0 1 0 1 6 1 1 0

Assumptions:

The length of the reviewing period is fixed. A single demand occurs on the average after every period of length p. The amount of demand is random with density function $f(x)$. The case of a normal distribution is described in detail. The order placed at the end of a reviewing period is delivered instantly. An order is placed in each period when demand occurred. The inventory holding cost Lh is proportional to the amount of net inventory. When the demand cannot be met, a C_n shortage cost (cost of an emergency order) arises. The constant ordering cost is C_r, the cost of inventory reviewing is C_s.

Objective:

The order level R is to be determined which minimizes the expected total cost per demand:

$$C_d = RLhp + C_s p + C_r + C_n \int_R^\infty f(x)\, dx.$$

Solution:

The optimal R^* value of R for normally distributed demand with parameters μ and σ can be expressed as

$$R^* = \mu + \sigma \sqrt{2 \ln\left(C_n/\sqrt{2\pi}\sigma Lhp\right)},$$

If $C_n/\sqrt{2\pi}\,\sigma Lhp < 1$ there are no stationary values.

131. Inventory system with periodic and emergency ordering possibilities

Main codes:

$$1\ 1\ 0\ 1\ 0\ 1\ 6\ 1\ 1\ 0$$

Assumptions:

A (t_p, q) ordering system is considered, where each period has a discrete demand with the same probability distribution. The normal order is placed at the end of each period of a fixed length t_p. The order is delivered at the beginning of the reviewing period. The ordered amount (q) is to be optimalized. If the inventory level during the t_p period decreases below a prescribed IMIN level, an emergency order is placed, according to an (s_p, S) policy, namely this order is determined by the order level IOL. On the other hand, when the inventory level after the delivery of the standard (periodic) order arises the inventory level over the capacity of the store IMAX, the excess inventory must be sold out with a loss. The emergency ordering has also an additional cost, while at the normal ordering a price reduction with rate α is given by the vendor. The unit purchasing price is denoted by RC, all the other costs are expressed as related to this price:

CE: unit cost of emergency ordering,
CO: unit selling off cost of excess stock,
CH: inventory holding cost of a unit in a period,
CS: shortage cost of a unit in a period,

where $CH < CE < CS$ and $CH < CO < CS$. In a period only a single emergency ordering is allowed, this gives a lower bound for IOL, as—by supposition—the maximal demand L of a period is a known constant value:

$$\text{IMIN}+L+1 \leq \text{IOL} \leq \text{IMAX}.$$

Moreover, the size of a standing order q is assumed to be equal to or greater than the average demand during the scheduling period.

Objective:

The initial stock of the periods can be described by a Markov process which has a stationary distribution with probability Φ_i for the state i. The transition cost from state i to state j has the expected value $E(c_{ij})=ec_{ij}$. The expected total cost equals to

$$C(q, \text{IOL}) = \sum_{i=\text{IMIN}+1}^{\text{IMAX}} \Phi_i \sum_{j=\text{IMIN}+1}^{\text{IMAX}} ec_{ij} RC.$$

Solution:

The value of the cost function is calculated for all possible pairs of the discrete control parameters q and IOL. The theory of Markov processes is used to derive the stationary distribution Φ_i. The purchasing price of the periodic and emergency ordering has to be added to the optimal value of the cost $C(q, \text{IOL})$ to obtain the optimal value of the total cost. It is shown that, with increasing lead-time, this mixed (t_p, q) and (s_p, S) policy works more efficiently than the stochastic order-level policy with leadtime (Model 83).

132. The sample-size necessary for the simulation of an (s, S) model

Main codes:

$$1\ 1\ 0\ 1\ 0\ 3\ 5\ 1\ 1\ 0$$

Assumptions:

A periodic review (s, S) policy with zero leadtime is considered. The reviewing period has an exponentially distributed random demand with parameter λ. The expected value, the empirical variance and the correlation of stock and shortage at the end of a period are estimated from a random sample generated by simulation method.

Objective:

The minimal sample-size for estimating the expected values of stock and shortage has to be determined which ensures that the difference of the estimated and the exact value will not exceed a prescribed bound with a probability of 0.95. (Of course, both the bound and the chosen probability level may be changed, if desired.)

Solution:

The minimal sample-size, which satisfies the above criterion, was estimated using the inequality of Tschebycheff and the central limit theorem. The analysis was completed by simulation runs, which gave the expected value and standard deviation of the necessary sample size. These results are collected in tables for the different values of $s\lambda$ and $(S-s)\lambda$.

VI.1.2. Models with Constant Leadtime

133. *Stochastic reorder-point order-level system with leadtime*

Main codes:

$$1\ 1\ 0\ 1\ 0\ 1\ 5\ 1\ 1\ 1$$

Assumptions:

The probability density function of the demand during the fixed reviewing period w_p is $k(r)$. The maximal possible demand is denoted by r_{max}. An (s, S) inventory policy is used with backorders in the case of shortage. The dimensions of the usual cost factors are given as follows:

$$[c_1] = [c_2] = \frac{[\$]}{[Q][T]}; \quad [c_3] = [\$].$$

Objective:

a) For continuous x, v, z and Z, we have

$$C(z, Z) = c_1 \int\limits_{z-r_{max}}^{Z} \left[\int\limits_0^{Q'} (Q'-r)k(r)\,dr\right] m(Q')\,dQ' +$$

$$+ c_2 \int\limits_{z-r_{max}}^{Z} \left[\int\limits_Q^{r_{max}} (r-Q')k(r)\,dr\right] m(Q')\,dQ' + c_3 \frac{h}{w_p},$$

where Q' is the initial stock on hand of a reviewing period, a random amount with density function $m(Q')$ and $h = P(Q = S)$.

b) For discrete x, v, s and S with possible values $\ldots -2u, -u, 0, u, 2u, \ldots$ we can write

$$C(s, S) = (c_1 + c_2)\, u \sum\limits_{Q'=s+u}^{S} B(Q'-u)\,M(Q') -$$

$$- c_2 \left(\bar{Q}' - \frac{\bar{x}}{2}\right) + \frac{c_3}{w_p} M(S)[1-P(0)]$$

with the notations used in Models 129 and 143.

Solution:

a) In the continuous case, $\left(f(x) = \dfrac{1}{b} e^{-x/b}\right)$ expression of the total cost is explicitly given and minimized only for the exponential demand distribution.

b) In the discrete case, first the distribution $M(Q')$ is to be determined, then an iterative procedure (similar to the one described in connection with Model 120) can be applied.

134. *Distribution of the inventory level, in case of an (s, S) policy*

Main codes:

$$1\ 1\ 0\ 1\ 0\ 3\ 5\ 1\ 1\ 1$$

228

Assumptions:

The probability distribution function of the demand of a period is $\Phi(\xi)$. It has a continuous density function $\varphi(\xi)$. The demands of the different periods are independent random variables. The lead-time has a length of λ. Thus the demand during the leadtime has the distribution $\Phi^{(\lambda)}(x)$, where the superscript means the λ-fold convolution of $\Phi(x)$.

Objective:

The probability distribution function $H(y)$ of the inventory level at the end of a period is to be determined. This enables the derivation of explicit expressions for the cost function of different models.

Solution:

Applying the usual notation for the renewal function

$$M(x) = \sum_{t=1}^{\infty} \varphi^{(t)}(x)$$

and its derivative $m(x)$, the distribution function in question can be expressed as

$$H(y) = 1 - \frac{\int_s^S \Phi^{(\lambda)}(z-y)\, m(S-z)\, dz + \Phi^{(\lambda)}(S-y)}{1+M(S-s)}.$$

135. Determination of the reliability level of an (s, S) policy

Main codes:

$$1\ 1\ 0\ 1\ 0\ 3\ 5\ 1\ 1\ 1$$

Assumptions:

The demand of the period with length $T+L$ has a discrete distribution $F(k|T+L)$, where T is the length of a reviewing period and L is the lead-time.

Objective:

For given values of the decision parameters s and S, the reliability level (i.e., the probability that no shortage will occur in a period), is to be expressed.

Solution:

The above service level for $D=S-s$ is equal to

$$P_\alpha(S, D) = \frac{1}{1+M(D)}\left[F(S|T+L) + \sum_{k=0}^{D} F(S-k|T+L)\, m(k)\right]$$

applying the notations M, m of the previous model. With the optimal decision parameter S_0— which results the minimal expected cost (having unit cost factors c_1 and c_2 of inventory holding and shortage)—the next inequality holds:

$$P_\alpha(S_0-1, D) < \frac{c_2}{c_1+c_2} \leq P_\alpha(S_0, D).$$

16 Chikán

136. *Necessary sample-size of simulation in case of positive leadtime*

Main codes:

$$1\ 1\ 0\ 1\ 0\ 3\ 5\ 1\ 1\ 1$$

Assumptions:

This is a generalized version of Model 132. A positive leadtime is considered which is an integer multiple of the length of the reviewing period.

Objective:

The minimal sample-size has to be determined which ensures that the deviation of the estimated and the exact value will not exceed a prescribed bound with a probability 0.95. (The value of the bound is changeable.)

Solution:

The expected value and standard deviation of the minimal sample-size, which satisfies the above criterion, is determined by simulation for different lengths of leadtime. The results are collected in tables for different parameter values.

137. *Reliability level for different solutions of the* (s, S) *policy*

Main codes:

$$1\ 1\ 0\ 1\ 0\ 9\ 5\ 1\ 1\ 1$$

Assumptions:

The random demand is forecasted at each inventory reviewing moment. The forecasting error in the ith period $\tau_{i,L}$ has a normal distribution with zero expectation and $\sigma_{i,L}$ standard deviation. The decision parameters s_i and S_i may change from period to period depending on the forecasted demand. A constant leadtime L and backorders case is considered.

Objective:

The aim of the analysis is to compare the reliability levels provided by different decision parameters, which result from different (s, S) policy-based methods. Two different measures of the reliability level are considered:

$$\frac{P}{\beta} = \frac{\text{expected satisfied demand in a period}}{\text{expected total demand in a period}}$$

P_α = probability of no shortage in a period.

Solution:

The reorder point is equal to

$$s_i = \sum_{j=1}^{i+L} \mu_j + k\sigma_{i,L},$$

where μ_j is the forecasted demand of period j and $k>0$ is a safety factor. Hence, $S_i = s_i + D_i$ with a parameter D_i calculated by different methods. The Wilson formula, the dynamic programming algorithm of Wagner and Whitin and a

230

heuristic method called the part-periods-method were used to determine different levels of S_i for which the values of the above reliability level indicators were compared by simulation.

138. Production-inventory control, as an application of the Wiener-filter theory

Main codes:

$$1\ 1\ 0\ 1\ 1\ 1\ 5\ 2\ 1\ 1$$

Assumptions:

A continuous, linear, stochastic production-inventory system is considered. The system can be described by the equations

(1) $dI(t)/dt = v(t) - r(t)$

(2) $v(t) = G_f\{n\}$

(3) $n(t) = -G(I)$

where $I(t)$ is the inventory (state variable), $n(t)$ is the amount of production (control variable), $r(t)$ is the demand. G_f is a linear operator characterizing the changes in the system (in the paper considered this is the production delay), G is also a linear operator of control. Both G_f and G are independent of time.
The possible controls are restricted to the class of linear operators, thus

$$n(t) = -\int_{-\infty}^{t} G(t-t')\,I(t')\,dt',$$

where $G(t)$ is a weight function of G. The demand is a stationary stochastic process with expectation $E\{r(t)\} = 0$ and autocorrelation function $R_{rr}(\tau) = E\{r(t)\cdot r(t+\tau)\}$.

Objective:

A quadratic cost function is applied:

$$Q = C_n E\{n^2\} + C_I E\{I^2\} + C_n E\{w^2\},$$

where $w(t) = G_h(n)$ and C_n, C_I, C_w are empirical weight factors of the costs. The first two terms express the deviation of the amount of production and of the inventory from certain prescribed values with their cost consequences. G_h is the differential operator. Thus $w(t) = dn(t)/dt$ and the last term of Q expresses the cost due to the changes in production rate. The first term describes the overtimes and surplus costs, the second one the inventory holding cost.

Solution:

First the system equations are transformed by the Fourier-transformation, then the Wiener—Hopt decomposition principle is applied.

16*

139. $(S-1, S)$ *inventory system with constant leadtime*

Main codes:

1 1 0 1 0 1 5 1 1 2

Assumptions:

The demand for a spare part is periodic. At the beginning of the first period the initial stock equals Q. The demand δ_j satisfied in this period is also random. The demands of different periods are independent and identically distributed. At the end of each period when demand appeared, an order is placed for an amount equal to the demand, (independently of the fact whether the demand can be satisfied or a shortage occurred). The delivery leadtime has a length of M periods. Whenever possible, demands are filled from current inventory, but no sooner than z periods after the order has been placed.

Objective:

Two different types of costs are considered. H is the purchasing and inventory holding cost for a unit quantity and unit time, $G(M)$ denotes the cost during the leadtime concerning transportation or repair of the spare parts. Hence, the total cost $HQ+G(M)$ has to be minimized.

Solution:

First the number of reorders is determined, then the effectivity measures of the system:

— expected number of reorders,
— expected time of shortage,
— probability of shortage

are calculated.

The solution is illustrated with a numerical example.

VI.1.3. Models with Varying Leadtime

140. $(S-1, S)$ *system with state-dependent leadtime*

Main codes:

1 1 0 1 0 1 5 2 1 3

Assumptions:

A continuous review $(S-1, S)$ inventory policy is considered. The demand has a Poisson distribution with parameter λ. The leadtime is an exponentially distributed random period. It depends on the unsatisfied demand m, which remained after satisfying the previous demands, according to the formula $1-e^{-\mu(m)t}$ (μ is a known function). The service rate m and thus also the leadtime distribution may change only at the moments of starting the service. The unsatisfied demand is backordered.

The following unit cost factors are considered:

C_I inventory holding cost with dimension [$]/[Q]/[T],
C_B shortage cost with dimension [$]/[Q]/[T] and
C_U shortage cost with dimension [$]/[Q].

Objective:

The expected total cost,

$$C(S) = C_I \sum_{n=0}^{S} (S-n) p_n + C_B \sum_{n=S}^{\infty} (n-S) p_n + C_U \lambda \sum_{n=S}^{\infty} p_n$$

is to be minimized, where p_n denotes the probability of n unsatisfied demands at time t.

Solution:

The determination of the probabilities p_n is equivalent to the determination of the steady-state probabilities of a queueing problem with Poisson input. Three different iterative procedures are suggested to find the optimal control parameter S. (Only a single local optimum exists for this model: this advantageous feature is utilized by each of the three methods.)

141. $(S-1, S)$ system with leadtime depending on the actual state

Main codes:

1 1 0 1 0 1 5 2 1 3

Assumptions:

This is a modified version of the previous model. Here the leadtime depends on the actual state of the system, (i.e., how many customers are left unsatisfied). The probability that in an arbitrary small time-interval Δt an order will be satisfied is $\mu(n)\Delta t + o(\Delta t)$, where n is the number of unsatisfied orders. The service rate may change continuously. This is equivalent to the case when the server continuously controls the system and changes the service rate depending on the actual queue length.

All the other assumptions are the same as in the previous model.

Objective:

It has a similar form as in Model 140.

Solution:

An iterative procedure is suggested which is based on the solution of differential equations. Again, one can make profit of the fact that only a single local optimum exists which is then globally optimal.

VI.2. Lost Sales at Shortage

142. *Inventory model for an unknown demand distribution*

Main codes:

$$1\ 1\ 0\ 1\ 0\ 1\ 5\ 1\ 2\ 0$$

Assumptions:

The total demand of a period is a discrete random variable. An amount i is demanded with probability p_i. This is an unknown distribution, estimated only from a sample of earlier periods. K_i denotes the number of the periods when the amount i was demanded. The system is characterized by the vector $K=(K_0, ..., K_N)$ and by the actual inventory level $0 \leq k \leq L$. This serves as the base for selecting the order level denoted by n. Both the demand of a period and the storage capacity are bounded from above by the constant N and L, respectively.

Objective:

The inventory holding cost of m units in a period is σ_m ($m=0, 1, ..., L$), the shortage cost of n units is π_n ($n=1, 2, ..., N$). The cost of ordering l units is w_l ($l=0, 1, ..., L$) thus the total cost of a period can be estimated by

$$C_{K,k}^{(n)} = w_{n-k} + \sigma_n + \sum_{i>n} \pi_{i-n} p_i^{(K)},$$

using the estimation $p_i^{(K)}$ of the demand distribution p_i (based on the sample K).

Solution:

The optimal reorder point and order level are calculated by linear programming method based on a Markov decision process.

VI.3. Models Where Shortage Is Not Allowed

143. *Stochastic order-level system with known maximal demand*

Main codes:

$$1\ 1\ 0\ 1\ 0\ 1\ 6\ 1\ 0\ 0$$

Assumptions:

The random demand of a reviewing period with fixed length w_p has the probability density function $f(x)$ and mean $\bar{x}$. The demand rate is constant during this period and the maximal amount of demand is a known constant x_{max}. There is an immediate delivery. No shortage is allowed, thus the reorder point is prescribed by $s_p = x_{max}$ or $s_p = x_{max} - u$ in a discrete case with unit u. The unit inventory holding cost is c_1 with dimension $[\$]/[Q]/[T]$. The fixed ordering cost is denoted by c_3.

Objective:

The initial stock Q of a reviewing period is a random amount with density function $g(Q)$. After receiving the order, the initial stock increases to S. The

234

probability $h = P\ (Q = S)$ is known. For continuous x and S, the total expected cost for a unit time equals

$$C(S) = c_1 \left[\int_{s_p}^{S} Qg(Q)\, dQ + hS - \frac{\bar{x}}{2} \right] + c_3 \frac{h}{w_p}.$$

For a discrete demand distribution $P(x)$, where $x = 0, u, 2u, \ldots$

$$C(S) = \frac{c_1 \sum\limits_{Q=x_{max}}^{S} QA(S-Q) + \frac{c_3}{w_p}[1-P(0)]}{\sum\limits_{y=0}^{S-x_{max}} A(y)} - \frac{c_1 \bar{x}}{2}$$

where the function $A(y)$ is defined recursively by the equations

$$A(y) = \begin{cases} 1 & \text{if } y = 0, \\ \sum\limits_{x=u}^{y} A(y-x)\dfrac{P(x)}{1-P(0)} & \text{if } y = u, 2u, \ldots. \end{cases}$$

Solution:

For the continuous case, $g(Q)$ cannot be explicitly given in a closed form. The discretization of the demand distribution is suggested, and illustrated in an example. For discrete demand distribution the cost function can be calculated for fixed S. The necessary (but not sufficient) conditions of optimality are

$$M(S_0 - u) \leq \frac{c_3[1-P(0)]}{c_1 u w_p} \leq M(S_0),$$

where

$$M(S) = \frac{[J(S-x_{max})]^2}{A(S-x_{max}+u)} - K(S-x_{max}-u)$$

$$K(r) = \begin{cases} 0, & r < 0 \\ \sum\limits_{i=0}^{r} J(i) = \sum\limits_{i=0}^{r} \sum\limits_{y=0}^{i} A(y), & r = 0, u, 2u, \ldots \end{cases}$$

which may have several solutions: all these solutions have to be compared to find the global optimum.

144. (s, S) model with arbitrary inter-arrival time of demands

Main codes:

$$1\ 1\ 0\ 1\ 0\ 1\ 5\ 2\ 0\ 0$$

Assumptions:

Units are demanded at random moments. The inter-arrival time between two consecutive demands is random with an arbitrary density function $\varphi(x)$. Thus the moments when demands arise form a renewal process. A continuous inventory reviewing is applied. At the inventory level s, an order is placed for an amount $Q = S - s$ which is delivered instantly. The inventory level at a random time t

is $H(t)$ which is known to have a uniform distribution on $(s+1, ..., s+Q)$ as t tends to infinity. No shortage is allowed. The purchasing price per unit of item is c, the constant ordering cost is K. The inventory holding cost has the usual cost factor h with dimension $[\$]/[Q]/[T]$.

Objective:

The expected long-term cost per unit time equals

$$F(s, Q) = \frac{K \cdot \bar{D}}{Q} + c \cdot \bar{D} + h\left[s + \frac{Q+1}{2}\right],$$

where $\bar{D}$ denotes the expected demand per unit time.

Solution:

For the optimal decision parameters we have $s^*=0$ and

$$Q^*(Q^*-1) \leqq \frac{2K\bar{D}}{h} \leqq Q^*(Q^*+1).$$

145. *Order level system for demands at random moments in random amounts*

Main codes:

1 1 0 1 0 1 6 2 0 0

Assumptions:

The time between two consecutive demands is random with expected value λ. The amount of demand is also random, it has an exponential distribution with parameter μ. At each demand the inventory level is reviewed. If the total demand cannot be satisfied, then an order is placed and received immediately. The ordered amount is determined in such a way that after satisfying all of the demands, an inventory level Q has to be remained.

Objective:

The sum of ordering and inventory holding costs for a unit time has the expected value

$$K(Q) = \frac{1}{\mu+Q}\left(\frac{K\mu}{\lambda} + \frac{hQ^2}{2} + hQ\mu\right)$$

with the usual inventory holding and ordering cost factors h and K.

Solution:

The optimal Q^* can be expressed as

$$Q^* = -\mu + \sqrt{\frac{2K\mu}{h\lambda} - \mu^2}.$$

VI.4. Models with Specific Inventory Control

146. *Cost reduction by the introduction of continuous reviewing*

Main codes:

1 1 0 1 0 1 9 9 1 1

Assumptions:

The demand during a time interval $[0, t]$ is described by a Wiener process, i.e., the cumulative demand of a period $[0, t]$ has a normal distribution with parameters $(mt, \sigma\sqrt{t})$. A fixed leadtime L is considered. Two different inventory policies are possible, the periodic review system with (t_p, S) policy and the continuous review system with (s, q) policy.

Objective:

The cost functions of the (t_p, S) and (s, q) policies are compared. The difference of the costs (related to the optimal parameters of both policies) is estimated.

Solution:

The cost saving by changing the periodic review policy (t_p, S) to the continuous review policy (s, q), considering the sum of inventory holding and shortage cost for a unit of time, can be expressed as

$$K_1 - K_2 = (c_1 + c_2)\,\sigma t_p\,\varphi\left(\frac{S - mL}{\sigma\sqrt{L}}\right)\frac{1}{4\sqrt{L + t_p}},$$

where the usual inventory holding and shortage cost factor c_1 and c_2 are used. The expected number of orders is the same in the models. For the optimal parameters $q = t_p/m$.

147. *Optimal spare parts replacing policy*

Main codes:

1 1 0 1 0 1 9 0 0 0

Assumptions:

A single spare part is at disposal to satisfy the demand for replacing the similar n components of a system. The failure rate of the components and of the spare part is known. The whole system works during a time period of length T. The working components may not be changed among each other, they can be only replaced from stock by spare parts.

Objective:

The expected loss, caused by the failure when the components were not replaced immediately, is minimized. The shortage cost of component i per unit time is L_i and $L_1 \geq L_2 \geq \dots \geq L_n$.

$$L = [M(x) + N(\tau_{j+1})]/[1 - F(x)]^{j+1},$$

where

$$M(x) = L_{j+1}(T-x)[1-F(x)]^{J+1} +$$

$$+L_j\left[\int_x^T (T-t)f(t)[1-F(t)]^{j-1}[F(t)-F(x)](1+1/j)\,dt\right];$$

$$N(\tau_{j+1}) = L_{j+1}\left[\int_{\tau_{j+1}}^T (T-t)f(t)[1-F(\tau_{j+1})]^j\,dt - (T-\tau_{j+1})[1-F(\tau_{j+1})]^{j+1}\right] +$$

$$+L_j\int_{\tau_{j+1}}^T (T-t)f(t)\left[(1-F(t))^j(1+1/j)-(1-F(\tau_{j+1})^j/j)\right]\,dt;$$

$$L_{j.} = L_1 + ... + L_j;$$

$F(.) =$ the distribution function of failure;
$f(.) =$ the density function of failure;
$\tau_i \quad =$ the time between the failure and the replacement of component i.

Solution:

First, the optimal replacing policy is derived for the two-component case. If it is possible, at failure, first the more expensive component is to be replaced. At the failure of the cheaper component, it is economic to wait a certain time before replacement. When, during this time, the more expensive component does not fail, the cheaper component is to be replaced. For exponential failure rates and, under certain circumstances, also for concave failure rates the optimal replacement policy for n components is characterized by a series of times $\tau_i^*(x)$ $(i=1, 2, ..., n)$; $\tau_i^*(x)$ means the time when the component i is to be replaced, if it failed at time x, and before time $\tau_i^*(x)$ no component, more expensive than i, failed.

148. *Inventory policy under imperfect asset information*

Main codes:

1 1 7 1 0 1 9 7 2 7

Assumptions:

The demand of the subsequent periods $\xi_1, \xi_2, ...$ are independent, identically distributed random variables: their expectation and the standard deviation is denoted by m and v. The real level of inventory, and that described in books, often do not coincide, due to the failures in recording. The failure is a random amount $\eta_1, \eta_2, ...$ (by supposition, at the end of the subsequent periods): these are also supposed to be independent and identically distributed, with mean value 0 and standard deviation σ. The demand and failure in the same period may depend on each other.

The probability of shortage may not exceed a certain level α. A safety stock is held in inventory in accordance with the given service level. It has a unit holding cost h per unit amount and unit time. At each occasion, the supervision of the records and the inventory level has the cost K.

Objective:

Two different policies are considered.

If the supervision periods have the same length n, the sum of the inventory holding cost of the safety stock and the cost of supervision for a unit of time equals

$$C(n) = \frac{K}{n} + hB(n),$$

where $B(n)$ denotes the amount of safety stock.

If the supervision periods have different lengths depending on the time, when the demand after the previous supervision exceeds the amount t, then the cost function is equal to

$$C[N(t)] = \frac{K}{E[N(t)]+1} + h\tilde{B}(t),$$

where

$$N(t) = \begin{cases} \max(k : \xi_1 + \ldots + \xi_k \leq t) & \text{if } \xi_1 \leq t, \\ 0 & \text{if } \xi_1 > t \end{cases}$$

and $\tilde{B}(t) = m^{-1/2} B(t)$.

Solution:

The cost function is to be minimized under the constraint on the probability of shortage. By approximating the probability distribution by a normal distribution (with the given expected values and standard deviations) an explicit solution is derived. Namely, n^* and t^* are the integer part of the following expressions:

$$n^* = \left\{ 2K/\sigma h \Phi^{-1}\left(1 - \frac{\alpha}{2}\right) \right\}^{2/3}$$

and

$$t^* = m\left\{ 2K/\sigma h \Phi^{-1}\left(1 - \frac{\alpha}{2}\right) \right\}^{2/3}.$$

VII. Models with Stochastic Delivery and Stochastic Demand

In this group of models, both the delivery and demand are considered as random variables. The delivery is usually characterized by the uncertainty of leadtime and the demand is characterized by the uncertainty of the demanded quantity. Two main subgroups are distinguished depending on whether the control parameter of an ordering is the lot size (q) or the order level (S). The lot size models are often based on the results of queuing theory. This is especially characteristic for the case when both delivery and demand have a random character. The group of models with order level are subdivided according to the parameter controlling the frequency of ordering, which may be the order period (t) or the reorder point (s). The specific feature of this last subgroup is that, in many models, an order is immediately placed after every demand.

Models of stochastic delivery and deterministic demand are rather rare: in our sample we have only two such models (158 and 159). Therefore we did not form a separate group for these models but we discuss them in this group.

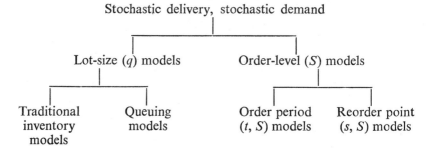

VII.1. Lot-Size Models with Stochastic Delivery and Demand

VII.1.1. Models of Traditional Inventory Theory Approach

149. (s, Q) *model with random leadtime*

Main codes:

$$1\ 1\ 1\ 1\ 0\ 1\ 2\ 2\ 1\ 3$$

Assumptions:

Shortage is permitted. Inventory reviewing is continuous, and the leadtime is a random variable. The demand during the leadtime L is a discretized random quantity r with probability $p(q)$ and expectation M_L. The demand process is stationary, with a mean demand rate M. Demand occurs at random moments. The shortage cost depends on the amount of shortage with unit cost factor π. The expected value of shortage is denoted by I_1, while its second moment is denoted by I_2. The inventory holding cost is proportional to the time and quantity and has the unit cost factor h. The constant ordering cost is K.

Objective:

The method of operators is used to derive the expected total cost for unit time. The exact form is rather complicated, a simplified approximate form is given:

$$C(s, Q) = \frac{Q \cdot h}{2} + sh - M_L h + \frac{hI_2}{2Q} + \frac{KM}{Q} + \frac{M\pi I_1}{Q},$$

where s is the reorder point, Q is the order quantity and

$$I_0 = \sum_{q>s}^{\infty} p(q); \quad I_1 = \sum_{q>s}^{\infty} (q-s)p(q); \quad I_2 = \sum_{q>s}^{\infty} (q-s)^2 p(q);$$

$$M_L = \sum_{q=s}^{\infty} qp(q).$$

Solution:

Setting the partial derivatives of $C(s, Q)$ to zero, one can obtain that

$$Q_0 = \sqrt{I_2 + \frac{2KM}{h} + \frac{2M\pi I_1}{h}}$$

while the value s_0 satisfies the equation

$$\frac{I_1}{Q_0} + \frac{M\pi I_0}{hQ_0} = 1.$$

These parameter values serve a good approximation of the optimal parameters, if the cost function is convex; namely if

$$I_0 + \frac{M\pi p(s)}{h} > 1.$$

150. *A continuous reorder-point, lot-size model*

Main codes:

$$1\ 1\ 1\ 1\ 0\ 1\ 2\ 2\ 1\ 3$$

Assumptions:

The demand of a time period with length t has the probability density function $f(x, t)$. The leadtime is also random with density function $g(t)$. Only one order may be outstanding at any time. Shortage is backordered, the shortage cost is proportional to the amount of shortage with unit cost factor π. The constant ordering cost is A, the inventory holding cost factor is IC with dimension $[\$]/[Q]/[T]$. The reorder point r and lot-size q are positive.

Objective:

The expected total cost per time unit is equal to

$$K(r, q) = \frac{\lambda}{q} A + IC\left(\frac{q}{2} + r - \mu\right) + \frac{\pi\lambda}{q}\left[\int_r^\infty xh(x)\,dx - rH(r)\right].$$

where λ is the mean demand rate,

$$h(x) = \int_0^\infty f(x, t)\,g(t)\,dt,$$

and

$$\mu = \int_0^\infty xh(x)\,dx$$

is the expected demand of the leadtime, finally,

$$H(r) = \int_{-\infty}^r h(x)\,dx.$$

Solution:

Setting the partial derivatives of $K(s, q)$ to zero, we obtain

$$q_0 = \sqrt{\frac{2\lambda[A + \pi \cdot \bar{\eta}(r)]}{IC}}, \quad r_0 = H^{-1}\left(\frac{ICq}{\pi\lambda}\right),$$

where

$$\bar{\eta}(r) = \int_r^\infty xh(x)\,dx - rH(r)$$

is the expected shortage of a period, and H^{-1} denotes the inverse of the distribution function H.

151. *Reorder-point, order quantity model with stochastic leadtime*

Main codes:

$$1\ 1\ 1\ 1\ 0\ 1\ 2\ 2\ 1\ 3$$

Assumptions:

The expected total demand of a planning period is A. The leadtime is a random variable with mean M and variance V. The leadtime demand L has the density function $\varphi(L)$ with mean μ_L and variance $\sigma_L^2 = M\sigma^2 + V\mu^2$, where μ and σ are the mean and standard deviation of demand per unit time. The unit cost of inventory holding H and the unit cost of shortage π have the dimensions

$$H = \frac{[\$]}{[Q][T]}, \quad \pi = \frac{[\$]}{[Q]}.$$

The fixed ordering cost is K.

242

Objective:

The expected total cost (for reorder point R and order quantity Q) of the whole planning period is

$$T(Q, R) = \frac{AK}{Q} + H\left(\frac{q}{2} + R - \mu_L\right) + \pi\,\frac{AS}{Q},$$

where

$$S = \int_R^\infty (L - R)\,\varphi(L)\,dL$$

is the expected value of the shortage in an ordering period.

Solution:

The optimum values of R and Q satisfy the pair of simultaneous equations:

$$\int_R^\infty \varphi(L)\,dL = \frac{HQ}{\pi A},$$

$$Q = \sqrt{\frac{2A(K + \pi S)}{H}}.$$

The sensitivity of the solution is also analyzed with respect to M, V and σ/μ.

152. Reorder-point, order quantity policy with reliability constraint and random leadtime

Main codes:
$$1\ 1\ 1\ 1\ 0\ 1\ 2\ 2\ 1\ 3$$

Assumptions:

Discrete amounts are demanded at random instants, with a mean demand rate λ. The inventory is reviewed continuously and if the sum of the inventory on hand plus on order minus the backordered quantity level is equal to s, the quantity q is ordered. The leadtime is also a random variable with mean τ_L. The shortage is backordered. The expected value of shortage for a unit time may not exceed a prescribed value M.

Objective:

The cost of inventory holding with unit factor h and the ordering cost A are considered. Their dimensions are: $\dfrac{[\$]}{[Q]\,[T]}$ and $\dfrac{[\$]}{[Q]}$, respectively. The expected total cost for a unit time

$$R(s, q) = \frac{\lambda A}{q} + h\left(s + \frac{q+1}{2} + \lambda\tau_L\right)$$

is minimized under a reliability constraint for the expected value of shortage:

$$E(s, q) \leq M,$$

meaning that the expected quantity of shortage must not exceed a predetermined level.

Solution:

A general solution is not given by the author. A solution algorithm is given for the case when the interarrival time between two consecutive demands follows an exponential distribution, the random amount of demands has a geometric distribution and the leadtime is normally distributed.

153. Graphical and numerical estimation of the optimal (s, q) policy

Main codes:

$$1\ 1\ 1\ 1\ 0\ 1\ 2\ 2\ 1\ 3$$

Assumptions:

The demand during the random leadtime is approximated by a normal distribution with mean λ and forecast error σ. (It is the standard deviation between the actual and expected demand during the leadtime.) λ is the annual number of units demanded and Q is the order quantity. The time of shortage is supposed to be relatively small, compared to the time without shortage, thus the average level of the physical stock can be well approximated by the sum of the safety stock and half of the lot-size. The unit inventory holding cost is IC, the constant ordering cost is A. Two different types of shortage costs are considered alternatively:

1. A fixed cost γ of shortage occurrence, independent of time and amount of shortage.
2. A shortage cost depending on the amount of shortage with unit cost factor π.

Objective:

For the two different types of the shortage costs we can write respectively

1)
$$K_1 = A\frac{\lambda}{Q} + IC\left(\frac{Q}{2} + t\sigma\right) + \gamma\varphi\lambda/Q$$

2)
$$K_2 = A\frac{\lambda}{Q} + IC\left(\frac{Q}{2} + t\sigma\right) + \pi\lambda\sigma(\varphi - t\Phi)/Q,$$

where $t = \dfrac{s}{\sigma}$, Φ and φ denote the probability distribution function and the probability density function of the standard normal distribution. C is the unit value of the item.

Solution:

The optimal t is the solution of the respective equations

1)
$$\frac{1}{2}\left(\frac{Q_w}{\sigma}\right)\left(\frac{\gamma}{A}\right) = \frac{1}{\varphi}\left[1 + \left(\frac{\gamma}{A}\right)\Phi\right]^{1/2}$$

2)
$$\frac{1}{2}\left(\frac{\pi\sigma}{A}\right)\left(\frac{Q_w}{\sigma}\right) = \frac{1}{\Phi}\left[1 + \left(\frac{\pi\sigma}{A}\right)(\varphi - t\Phi)\right]^{1/2},$$

244

depending on the type of shortage cost. Furthermore,

$$Q_w = \sqrt{\frac{2\lambda A}{IC}}$$

is the amount of order according to the Wilson formula. For the approximate solution of the above equations a fast iterative procedure is given. Nomograms are also presented for the practitioner.

154. Reorder-point lot size policy with partial backordering

Main codes:

1 1 1 1 0 1 2 2 2 3

Assumptions:

At the occurrence of a shortage amount z, a part of the demand which was not satisfied is lost. The amount of lost sales is expressed by the function

$$h(z) = \mu z - aze^{-bz}$$

where μ is the mean demand rate, a and b are constant parameters. The amount $z - h(z)$ is backordered, and is satisfied after receiving the next order. The amount of demands in a unit time is random with a probability density function $f(x)$. The leadtime is also random.

Objective:

The loss of a unit of demand means a loss of profit p. The expected total cost including the profit loss is approximated by the cost function

$$G(R, Q) = \frac{T\mu U}{Q} + \frac{1}{2}QcIT + ITc(R - L\mu) + \frac{T\mu pE}{Q},$$

where E denotes the expected value of lost demands, I and U are the usual cost factors of inventory holding and ordering, the stock value of the item is c, the loss of profit from a "lost sale" is p, and R and Q are the decision variables.

Solution:

The expected value of lost demands is expressed by the convolutions of the distribution function. For gamma distributed demands, an explicit formula is derived which can be calculated by means of the tables of the X^2-distribution. An interative procedure is presented for determining the optimal value of R. Then the optimal value of Q can be calculated by substitution of the optimal R (into the first order optimality condition).

155. Inventory system for repairable items with random leadtime

Main codes:

1 1 1 1 0 1 2 2 1 3

Assumptions:

The item stored may break down. This failure can be either reparable or not. In the former case, it will be repaired in a random time. In the case of non-repairable failure, the item leaves the system. This is the output of the system which makes it necessary to place orders for new items. The decision maker knows the amount of items on stock, on repair and on order not yet delivered. The inventory is continuously reviewed; if it has decreased to, or below, the level r, an order is placed for an amount of nQ units, where n is the largest integer for which the inventory level does not increase above $r+Q$.

The occurrences of the two failure types (reparable and non-reparable failures) are geneıated by independent stochastic processes. In the paper, Poisson processes are considered.

Objective:

The expected total cost for a unit of time is

$$K(r, Q) = \frac{\lambda_n}{Q} A + H_1 CD(r, Q) + H_2 C \lambda_r T + \pi E(r, Q) + \pi' B(r, Q),$$

where the following notations are used:

λ_n, λ_r: expected number of non-reparable and reparable failures for a unit of time,

A: constant ordering cost,

C: unit purchasing cost,

$H_1 C$:   inventory holding cost per unit quantity and unit time,

$H_2 C$:   repairing cost per unit quantity and unit time,

T: expected repairing time for an item,

π: constant shortage cost by occurrence of shortage,

$E(r, Q)$: expected number of shortage occurrences per unit time,

π':   shortage cost per unit quantity and unit time,

$B(r, Q)$: the expected shortage per unit time,

$D(r, Q)$: the expected inventory per unit time.

Solution:

The expected shortage and inventory are derived as a function of r and Q, thus the cost function values can be calculated for all possible pairs of r and Q to select the optimal ones. No specific search procedure is suggested.

156. Determination of the cost-optimal service level

Main codes:

$$1\ 1\ 1\ 1\ 0\ 1\ 4\ 1\ 4\ 1\ 1\ 3$$

Assumptions:

The length of the order period is fixed. The demand of a period is random: it follows a normal distribution with standard deviation σ_b. The mean value of the random leadtime is k (not necessarily integer) periods. There may be a difference between the amount of order placed and received. This difference is a random variable with a normal distribution of zero mean and standard deviation σ_r. (Note that the expected values of the normal distributions involved do not play a role in the value of the optimal safety factor to be determined.)

246

Objective:

The safety stock factor is to be determined which corresponds to a minimal cost solution, when the usual unit cost factors c_1 and c_2 of inventory holding and shortage are considered. At safety stock $u\sigma_k$, the expected total cost can be expressed, as a function of the safety factor u, by

$$N(u) = c_1 u\sigma_k + c_2 \int_{u\sigma_k}^{\infty} \varphi(x)\, dx,$$

where $\varphi(x)$ denotes the standard normal density function and

$$\sigma_k = \sqrt{k(\sigma_b^2 + \sigma_r^2)}.$$

Solution:

The solution u_0 of the equation

$$\varphi(u_0) = \frac{c_1}{c_2} \sigma_k$$

yields the cost-optimal value of the safety factor. Thus the optimal safety stock is expressed by

$$K_0 = u_0 \sqrt{k(\sigma_b^2 + \sigma_r^2)}.$$

157. (s, q) model with Poisson demand and exponential leadtime

Main codes:

$$1\ 1\ 1\ 1\ 0\ 1\ 2\ 2\ 1\ 3$$

Assumptions:

The demand is generated by a Poisson process with a mean demand rate μ. There is a continuous reviewing of the inventory level: at the moment of reaching s, an order of amount q is placed. The leadtime has an exponential distribution with a mean value $1/\lambda$. The demand during the leadtime is proved to follow a negative binomial distribution.

Objective:

With the usual cost factors, the expected total cost per unit of time can be written as

$$L = c_1\left(\frac{q}{2} + s\right) + \frac{\mu}{q}\left(c_3 + c_2\frac{\mu}{\lambda}\left(\frac{\mu}{\lambda + \mu}\right)^s\right)$$

Solution:

An iterative procedure is suggested for solving the system of equations for the partial derivatives of the cost function (applying the first-order necessary optimality conditions). The approximate solution—which may serve as the starting point

17*

of the iteration—is the pair of values

$$q_0 = \sqrt{\frac{2\mu c_3}{c_1}},$$

$$s_0 = \frac{\ln\left[\frac{\mu c_2 \ln\left(1+\frac{\lambda}{\mu}\right)}{\lambda\sqrt{\frac{2c_3 \cdot c_1}{\mu}}}\right]}{\ln\left(1+\frac{\lambda}{\mu}\right)}.$$

158. Reorder-point lot-size model with random leadtime and deterministic demand

Main codes:

$$1\ 1\ 1\ 0\ 0\ 1\ 2\ 2\ 1\ 3$$

Assumptions:

There is a deterministic continuous demand process with known constant rate x. The leadtime is random with a known probability distribution function $G(r)$ on the interval $[0, \gamma]$, and its mean value equals M. There is a continuous inventory reviewing, and shortage is backordered. Inventory holding, shortage and ordering costs are considered with the usual dimensions and unit cost factors c_1, c_2 and c_3.

Objective:

The expected total cost for a unit of time,

$$K(s, q) = c_1\left(\frac{q}{2}+s-Mx\right)+\frac{c_1+c_2}{2q} \int\limits_{s/x}^{\gamma} (rx-s)^2\, dG(r)+c_3\frac{x}{q}$$

has to be minimized.

Solution:

The usual differentiation of the cost function yields

$$q_0 = \left[\frac{c_1+c_2}{c_1} \int\limits_{s_0/x}^{\gamma} (rx-s_0)^2\, dG(r)+\frac{2c_3}{c_1}x\right]^{1/2}$$

where s_0 is the solution of a rather complicated equation taking the next simpler form when supposing $q_0 = \gamma x$:

$$[c_1/(c_1+c_2)]\gamma x = \int\limits_{s_0/x}^{\gamma} (rx-s_0)\, dG(r).$$

248

159. *A special lot-size model*

Main codes:

$$1\ 1\ 1\ 0\ 0\ 1\ 4\ 7\ 0\ 0$$

Assumptions:

For the production of a given piece of equipment, special, expensive, components are necessary. The demand n is known, it is determined by the production plan. The delivery process of the components is also deterministic in time and amount. The only random factor is that a part of the components delivered may be defective. The number of faulty products becomes known only after quality control of a received lot. If the number of acceptable components is not sufficient, a new order has to be placed with an extra cost K. If there are surplus components, then they are transported back to the vendor to sell for a unit price v (which is smaller than the unit purchasing price c). A single period is considered, the surplus components cannot be used later.

Objective:

The expected total cost of the whole period by an order of n components is

$$f(n) = \min_{x \geq n} \left\{ c \left[x - \sum_{j=0}^{x} j p_x(j) \right] - v \sum_{j=0}^{x-n} (x-n-j) p_x(j) + \right.$$

$$\left. + \sum_{j=x-n+1}^{n} [K + f(n-x+j)] p_x(j) \right\}, \quad \text{for} \quad n = 1, 2, \ldots,$$

where $p_x(j)$ denotes the probability that, among x units, j are defective.

Solution:

The optimal n can be selected by a straightforward substitution of the $x \geq n$ values into the formula for $f(n)$. In the case of binomially distributed $p_x(j)$ values, the calculations are illustrated with a numerical example.

160. *Model with random leadtime and with a constrained number of periods with shortage*

Main codes:

$$1\ 1\ 1\ 1\ 0\ 1\ 2\ 2\ 1\ 3$$

Assumptions:

The expected demand of the whole planning period is A. The demand of a time unit has an expected value μ and a standard deviation σ. The leadtime is random with mean M and standard deviation V. The demand during the leadtime has an expected value $\mu_L = M\mu$ and variance $\sigma_L^2 = M\sigma^2 + V\mu^2$. The inverse of the probability distribution function of the standardized random variable $(L - \mu_L)/\sigma_L$ is denoted by $G(x)$. The expected number of periods with shortage may not exceed an upper bound N.

Objective:

The shortage cost is replaced by the reliability constraint $E(Y) \leqq N$, where Y is the random number of periods with shortage. The expected total cost of inventory holding and ordering (with unit cost factors H and K), equals

$$T(Q) = \frac{AK}{Q} + \frac{HQ}{2} + G\left(1 - \frac{NQ}{A}\right) H\sigma_L.$$

Solution:

The optimal order amount Q is the solution of the equation

$$-\frac{AK}{Q^2} + \frac{H}{2} - \frac{N}{A} H\sigma_L G'\left(1 - \frac{NQ}{A}\right) = 0.$$

Having determined the optimal Q (by iteration), the reorder point R can be computed from the reliability constraint. The sensitivity of the solution is analysed (varying the parameters M, V and σ/μ).

161. (s, q) *model for spare-parts*

Main codes:

$$1\ 1\ 1\ 1\ 0\ 1\ 2\ 2\ 1\ 3$$

Assumptions:

Spare-parts are demanded one-by-one in random instants according to a Poisson process. The orders are placed by a reorder point-lot size policy with integer decision variables. The leadtime is random with distribution function $F(x)$. The average time between two consecutive orderings is u. The amount of shortage is restricted, the reliability constraint may concern either the probability of shortage or the conditional expectation of the shortage under the condition that a shortage occurs in a given period.

Objective:

No explicit shortage cost is considered, it is replaced by the reliability constraint, which determines the value of the reorder point s. The optimal order amount q_0 has to be determined in such a way that the expected total cost of inventory holding and ordering per unit of time

$$f(q) = \frac{a + bq}{u} + \frac{c_1 h}{u}$$

takes its minimum in q_0. Here a is the constant cost of ordering, b is the unit purchasing price, c_1 is the unit cost factor of inventory holding and h is the average inventory per time unit.

Solution:

After determining s from the reliability constraint, then using this value, one can determine

$$q_0 = \sqrt{\frac{A - BD + CD^2}{C}} - D$$

is the value of q which yields the minimum of $f(q)$. The values of the parameters A, B, C and D are expressed in terms of the cost factors a, b, c_1, u and h. The optimal order amount is the integer part of $[q_0+1]$.

VII.1.2. Inventory Models Based on Queuing Theory

The models of this group express the inventory control problem by means of queuing theory. The following analogies are used: the maximal inventory on hand is considered as the total number of service channels, the actual physicaf stock as the number of free channels, the leadtime as the time of service, and the amount of shortage as the length of the queue.

162. *Poisson demand, arbitrary leadtime system, grouped deliveries*

1ain codes:

$$1\ 1\ 1\ 1\ 0\ 1\ 2\ 3\ 1\ 3$$

Assumptions:

Units are demanded at random instants generated by a Poisson process with a mean demand rate μ. The leadtime is random with a probability density function $f(x)$ and mean τ. The shortage is backordered. Three different shortage costs are considered:

— rejecting a demand with unit cost p_1 per unit demand,
— waiting of customers with unit cost p_2 per unit demand and unit time,
— loss of demand with unit cost p_3 per unit demand.

The maximal amount of shortage has an upper bound r. The constant ordering cost is g and the unit cost of inventory holding is s.

Objective:

The expected total cost is equal to

$$L(\hat{y}, q) = s \sum_{y=1}^{\hat{y}+q} yP_y - p_2 \sum_{y=-r}^{-1} yP_y + \frac{1}{\tau} g \sum_{y=-r}^{\hat{y}-q} P_y +$$

$$+ \mu \left[g P_{\hat{y}+1} + p_1 \sum_{y=-r}^{0} P_y + p_3 P_{-r} \right],$$

where P_y is the probability of an inventory y; furthermore, $\hat{y}$ denotes the reorder point and q is the lot size.

Solution:

Using the equivalence with the respective $M(G)/r$ type queuing model, on the basis of queuing theory the following solution procedure is suggested:
— determine the transition probabilities,
— formulate the differential equations which characterize the system,
— determine the stationary solution by formulating the initial conditions and integration coefficients,

251

— derive the state probabilities P_y,
— determine the expected cost function,
— minimize the cost function.

163. *Poisson demand, arbitrary leadtime system with lost sales*

Main codes:

$$1\ 1\ 1\ 1\ 0\ 1\ 2\ 2\ 0\ 3$$

Assumptions:

This model is the lost sales case version of the previous model, thus the time-dependent shortage cost p_2 is omitted; the other shortage cost factors p_1 and p_3 have the same meaning. A further assumption states that $q > \hat{y}$. All previous notations and assumptions remain unchanged.

Objective:

$$L(\hat{y}, q) = \frac{(sq - \mu p_0) \sum\limits_{z=0}^{\hat{y}-1} (\hat{y}-z) A_z + \mu[g + p_0(\mu\tau - \hat{y})] + \dfrac{sq(q+1)}{2}}{\mu\tau + (q-\hat{y}) \sum\limits_{z=0}^{\hat{y}-1} (\hat{y}-z) A_z}$$

where $p_0 = p_1 + p_3$ and A_z is the probability of a leadtime demand z.

Solution:

The theory of queuing models is applied to derive the explicit cost function, then necessary conditions are given for the optimal values of the parameters $\hat{y}$, q in the form of the inequalities

$$L(\hat{y}+1, q) - L(\hat{y}, q) \geq 0,$$

$$L(\hat{y}, q) - L(\hat{y}-1, q) \leq 0,$$

$$L(\hat{y}, q+1) - L(\hat{y}, q) \geq 0,$$

$$L(\hat{y}, q) - L(\hat{y}, q-1) \leq 0,$$

164. *Delivery based on inventory level*

Main codes:

$$1\ 1\ 1\ 1\ 0\ 1\ 2\ 2\ 1\ 3$$

Assumptions:

A reorder level—lot size policy is applied, where the reorder point is denoted by $\hat{y}$. The ordered quantities q may be delivered in several lots with a random leadtime which has mean τ. New transports are necessary, until the inventory reaches the level $\hat{y}+q$. Two types of ordering costs are considered: g is the fixed cost of an order and g_0 is the unit price of transportation. The demand is generated by a Poisson process.

Objective:

The expected total cost is equal to

$$L(\hat{y}, q) = s\left[\sum_{y=1}^{\hat{y}+q-1} y(P_y + P_y') + (\hat{y}+q) P_{\hat{y}+q}'\right] -$$

$$- p_2 \sum_{y=-r}^{-1} yP_y + \frac{g_0}{\tau} \sum_{y=-r}^{\hat{y}+q-1} P_y + \mu\left[gP_{\hat{y}+1} + p_1 \sum_{y=-r}^{0} P_y + p_3 P_{-r}\right],$$

with the same notations as used in the previous models (162, 163). The following additional symbols are used: P_y is the probability of having an inventory y when delivery occurs, and P_y' is the same probability when delivery does not occur.

Solution:

The steps of the method of solution are the same as described in connection with model 162.

165. Recurrent demand, exponential leadtime

Main codes:

$$1\ 1\ 1\ 1\ 0\ 1\ 2\ 2\ 1\ 3$$

Assumptions:

The time between two consecutive occurrences of unit demand is a random variable with an arbitrary given density $f(x)$. These random variables are independent and have the same distribution with mean θ. This type of demand is called recurrent demand (as the demands are generated by a recurrent stochastic process). The leadtime has an exponential distribution with parameter λ. This model is opposite to the previous ones, in which the leadtime has an arbitrary distribution and the interarrival time is exponentially distributed. The maximal amount of shortage is constrained by the constant value r, as in the previous models.

Objective:

The expected total cost is equal to

$$L(\hat{y}, q) = s\sum_{y=1}^{q+\hat{y}} yP_y + \frac{1}{\theta}(gP_{\hat{y}+1} + p_0 P_0),$$

where

$$p_0 = p_1 + p_3; \quad p_{\hat{y}+1} = C_{\hat{y}}\theta \quad \text{and} \quad P_0 = \frac{C_{\hat{y}} B_0^{\hat{y}}}{\lambda}$$

with

$$C_{\hat{y}} = \frac{1}{\theta\left(q + \frac{B_0^{\hat{y}+1}}{1-B_0}\right)}$$

253

finally, B_0 is the probability of the event that during the interarrival time of two consecutive demands no delivery occurs. (For other notations, see the previous models.)

Solution:

By using the results concerning the corresponding queuing model $(GI/M/r)$; the state probabilities P_y of the inventory level are determined. For the optimal decision parameters, only in the case $r=0$ and $q<\hat{y}<2q$ are some conditions given.

166. *Recurrent demand, delivery based on inventory level*

Main codes:

$$1 1 1 1 0 1 2 2 1 3$$

Assumptions:

This model is a modified version of Model 164 in the sense that the leadtime has an exponential distribution and the demand is recurrent, as is defined in the previous model.

Objective:

With the notations of the previous models, we have

$$L(\hat{y}, q) = s \left[\sum_{y=1}^{\hat{y}+q-1} y(P_y+P_y')+(\hat{y}+q) P_{\hat{y}+q}' \right] - p_2 \sum_{y=-r}^{-1} yP_y +$$

$$+ g_0 \lambda \sum_{y=-r}^{\hat{y}+q-1} P_y + \frac{1}{\theta} \left[gP_{\hat{y}+1} + p_1 \sum_{y=-r}^{0} P_y + p_3 P_{-r} \right].$$

Solution:

The state probabilities P_y and P_y' of the inventory level are determined using the results concerning the corresponding queuing model.

167. *Recurrent demand and lost sales*

Main codes:

$$1 1 1 1 0 1 2 2 2 3$$

Assumptions:

The leadtime has an exponential distribution and the demand is given by a special recurrent process: the so-called Palme process. There is a continuous reviewing, and demand is lost at shortage. The reorder point $\hat{y}$ is smaller than the amount q ordered by supposition.

Objective:

Using the notations of the previous models, we have

$$L(\hat{y}, q) = \frac{sq\left\{ \frac{q+2\hat{y}+1}{2} - \frac{1}{\lambda\theta} + B_0^{\hat{y}}\left[\frac{1}{\lambda\theta} + \frac{1}{2}\left(\frac{D_\theta}{\theta^2}-1\right) \right] \right\} + \frac{1}{\theta}\left(g + \frac{p_2 B_0^{\hat{y}}}{\lambda}\right)}{q + B_0^{\hat{y}}\left[\frac{1}{\lambda\theta} - 1 + \frac{1}{2}\left(\frac{D_\theta}{\theta^2}+1\right) \right]}$$

where B_0 is the probability of the event that during the interarrival time of two consecutive demands no delivery occurs. Further on, the interarrival time has mean value θ and standard deviation D_θ.

Solution:
The state probabilities of the inventory level are derived using the results of the respective queuing model.

168. Erlang-type demand, leadtime with an arbitrary distribution

Main codes:

$$1 1 1 1 0 1 2 2 1 3$$

Assumptions:
Units are demanded and the time between two consecutive demands has a gamma distribution with parameters k and λ. This stochastic process is the so-called Erlang process. (The interarrival time of demands can be considered as the sum of k independent, exponentially distributed random variables with the same parameter λ, i.e., k consecutive exponential interarrival times are to be considered before a demand occurs.) The leadtime is random with an arbitrary distribution.

Objective:
Using the same notations as in previous models, the total expected cost, as a function of reorder point and lot-size, can be expressed by

$$L(\hat{y}, q) = g\left[\mu P_{\hat{y}+1,k-1} + \frac{1}{\tau}\left(\sum_{y=-r+1}^{\hat{y}-q}\sum_{i=0}^{k-1} P_{y,i} + P_{-r,0}\right)\right] +$$

$$+\frac{p_1}{\theta}\left(\sum_{y=-r+1}^{0}\sum_{i=0}^{k-1} P_{y,i} + P_{-r,0}\right) - p_2\left(\sum_{y=-r+1}^{-1} y\sum_{i=0}^{k-1} P_{y,i} - rP_{-r,0}\right) +$$

$$+\frac{p_3}{\theta} P_{-r,0} + s\sum_{y=1}^{\hat{y}+q} y\sum_{i=0}^{k-1} P_{y,i},$$

where τ is the expected value of the leadtime, θ is the average interarrival time, $P_{y,i}$ is the probability of an inventory level y in transport phase i, and, finally, $\mu = k/\theta$.

Solution:
Only the state probabilities of the inventory level are derived using the analogous model $(E_k/G/r)$ of queuing theory.

169. Recurrent demand and leadtime with an Erlang distribution

Main codes:

$$1 1 1 1 0 1 2 2 0 3$$

Assumptions:
The main assumptions are the same as in the previous model, the difference being only that here the leadtime is described by a gamma distribution with

parameters k and λ, while the demand is generated by a recurrent process described by Model 166. Again, the lot-size q is supposed to be larger than the reorder point $\hat{y}$.

Objective:

The expected total cost, by using the notations of the previous models, can be expressed in the form

$$L(\hat{y}, q) = g\left[\frac{1}{\theta}P_{\hat{y}+1} + \lambda \sum_{y=-r}^{\hat{y}-q} P_{y, k-1}\right] + \frac{p_1}{\theta} \sum_{y=-r}^{0} \sum_{i=0}^{k-1} P_{y, i} -$$

$$- p_2 \sum_{y=-r}^{-1} y \sum_{i=0}^{k-1} P_{y, i} + \frac{p_3}{\theta} \sum_{i=0}^{k-1} P_{-r, i} + s \sum_{y=1}^{\hat{y}+q} y \sum_{i=0}^{k-1} P_{y, i}.$$

Solution:

The state probabilities $P_{y, i}$ are determined, using the solution of the respective queuing model $(GI/E_k/r)$.

170. Erlang distributed leadtime and delivery based on the inventory level

Main codes:

$$1\ 1\ 1\ 1\ 0\ 1\ 2\ 2\ 0\ 3$$

Assumptions:

All the assumptions are similar to those of the previous model. The only change is that the delivery is based on the inventory level as described in Model 164.

Objective:

The explicit cost function is not given by the author because of its complexity.

Solution:

The state probabilities are given on the basis of the results of the respective queuing model.

171. Poisson demand and exponential leadtime

Main codes:

$$1\ 1\ 1\ 1\ 0\ 1\ 2\ 2\ 2\ 3$$

Assumptions:

The demands occur in units, the random time interval between two consecutive demands has an exponential distribution with parameter μ. It means that the demand is generated by a Poisson process. The leadtime has also an exponential distribution with parameter λ. Such systems are called Poisson systems. There is a continuous reviewing, and it is supposed again that the lot-size q is greater than the reorder point $\hat{y}$. Demand is lost at shortage.

Objective:

With the unit cost factors of inventory holding s, the fixed cost of an ordering g and the fixed cost of a shortage p_0, the total expected cost is

$$L(\hat{y}, q) = \frac{\left(1+\frac{\lambda}{\mu}\right)^{\hat{y}}\left[sq\left(\frac{\lambda}{\mu}\frac{q+2\hat{y}+1}{2}-1\right)+g\lambda\right]+\mu p_0+sq}{1+\frac{\lambda}{\mu}q\left(1+\frac{\lambda}{\mu}\right)^{\hat{y}}}.$$

Solution:

The partial derivatives of $L(\hat{y}, q)$ are equal to zero at the optimal point (y_0, q_0). The system of two equations which is obtained can be solved by an iterative method using the starting point

$$q_0 = \sqrt{\frac{2\mu g}{s}}, \quad \hat{y}_0 = \frac{\ln\dfrac{p_0\mu^2\ln\left(1+\frac{\lambda}{\mu}\right)}{sq_0\lambda}}{\ln\left(1+\frac{\lambda}{\mu}\right)}.$$

172. Poisson system with constrained shortage

Main codes:

$$1\ 1\ 1\ 1\ 0\ 1\ 2\ 2\ 1\ 3$$

Assumptions:

Both the interarrival time and leadtime are exponentially distributed with parameters μ_0 and λ, respectively. Shortage is back-ordered; the maximal amount of shortage is constrained by the constant value r. There are two additional assumptions: $r<2q$ and $q>\hat{y}$, where q denotes the lot-size and $\hat{y}$ denotes the reorder point.

Objective:

With the notations of the previous models, the expected total cost of the system is

$$L(\hat{y}, q) = s\sum_{y=0}^{q+\hat{y}+1}(q+\hat{y}-y)P_y+\mu_0\,gP_{q-1}+p_1\mu_0\sum_{y=q+\hat{y}}^{r-1}(r-y)P_y+$$

$$+p_2\sum_{y=\hat{y}+q+1}^{r-1}(y-q-\hat{y})P_y.$$

Solution:

The state probabilities of the inventory level are derived on the basis of the corresponding $(M/M/r)$ queuing model.

257

173. *Poisson system without shortage constraint*

Main codes:

$$1\ 1\ 1\ 1\ 0\ 1\ 2\ 2\ 1\ 3$$

Assumptions:

Both the interarrival time and leadtime are exponentially distributed with respective parameters μ and λ, as in the previous model, but no shortage constraint is considered. By the continuous reviewing, when the inventory level decreases to the reorder point $\hat{y}$ an order is placed for an amount q, where $q > \hat{y}$ is assumed.

Objective:

With the notations of the previous models, we can write

$$L(\hat{y}, q) = s(q+\hat{y}) + p \sum_{y=q+\hat{y}}^{\infty} y P_y - s \sum_{y=0}^{q+\hat{y}-1} y P_y +$$

$$+ g \left(\lambda \sum_{y=2q}^{\infty} P_y + \mu P_{q-1} \right) + [p_0 \mu - (p+s)(\hat{y}+q)] \sum_{y=q+\hat{y}}^{\infty} P_y.$$

Solution:

A differential equation system is derived for the determination of the state probabilities, based on the corresponding $(M/M/\infty)$ type queuing model.

174. *The ZIP-system for reparable machines*

Main codes:

$$1\ 1\ 1\ 1\ 0\ 1\ 2\ 2\ 1\ 3$$

Assumptions:

The so-called ZIP system is a reserve of machines or spare-parts. If a machine breaks down during the production process, it is immediately replaced from the ZIP reserve, and then the repair starts. The number of working machines is checked continuously, it is filled up by orders placed for a unit at a time $(q=1)$. The inventory holding cost must be considered according to the total number of machines (including both good and out-of-order ones) in the ZIP system. When no working machine is available to replace the faulty one in the production, then a shortage cost occurs which is proportional to the time of shortage. The random failures of the machines produce the demand for other machines taken from the ZIP inventory. The demand is generated by a Poisson process. Both leadtime and interarrival time of demands are assumed to be exponentially distributed.

Objective:

The total expected cost for a reorder point $\hat{y}$ is

$$L(y) = s(\hat{y}+q) + p P_{\hat{y}+q-1} + g \mu P_{q-1},$$

where P_y is the probability of an inventory level y. s, p and g denote the usual cost factors of inventory holding, shortage and ordering (setup).

Solution:

The state probabilities of the system are determined.

175. *Continuous demand—continuous delivery Poisson system*

Main codes:
$$1\ 1\ 1\ 1\ 0\ 1\ 9\ 0\ 2\ -1$$

Assumptions:

The inventory system is considered similarly to a queuing system, where the order is delivered according to a Poisson process with a mean delivery rate μ. The inventory is a queue waiting for "service", i.e., for demand to remove it from inventory. The demand process is also described by a Poisson process with a mean demand rate λ. The amount of inventory is the queue waiting for demand. The process can be controlled only through the intensity of the delivery. The inventory holding and shortage costs are considered with the unit cost factors and dimensions:

$$C_n = \frac{[\$]}{[Q][T]}, \quad C_z = \frac{[\$]}{[T]}.$$

Objective:

The stationary solution of the state equations yields the following total expected cost function:

$$E(\mu) = C_n i\, \frac{P}{1-P} + C_z(1-P),$$

where $P = \dfrac{\lambda}{\mu}$ and i is the interest rate of capital invested in the inventory.

Solution:

The optimal P can be explicitly expressed

$$P_0 = 1 - \sqrt{C_n i/C_z}.$$

VII.2. Stochastic Demand, Stochastic Delivery, Order Level Models

A part of the models of this group has periodic reviewing, where at each reviewing an order is placed. The order level is S. The length of the reviewing period may be prescribed by external conditions, but it may be also a decision variable of the system. The delivery is random: either the leadtime is random or the delivery of an order is realized in several lots according to a random process.

A specific subgroup of (s, S) models is the $(S-1, S)$ type models, where an order is placed whenever a demand for a unit of item has occurred. The interarrival time of demands is a random variable and the leadtime is also random. The task is to determine the optimal value of the single decision parameter S.

VII.2.1. Order Period, Order Level (t, S) Type Models

176. *Order level system with random delivery process of interval-type*

Main codes:
$$1\ 1\ 1\ 1\ 0\ 1\ 1\ 1\ 1\ 1$$

Assumptions:

The demand of the interval $[0, s]$ is a random amount which has a probability density function $G(x, s)$. The demand is supposed to form a stationary stochastic process with independent increments. The mean demand rate is m. The length of an order period T is prescribed. The delivery of an order placed for an amount x is realized in several deliveries. The delivery is a random process described by the stochastic process η_t with the following properties:

$$\eta_\tau = 0, \quad \eta_{\tau+T} = x,$$

$$M(\eta_s) = (s-\tau)x, \quad \tau \leq s \leq \tau+T,$$

$$M(\eta_s \eta_t) - M(\eta_s) M(\eta_t) = \sigma^2 x^2(s-\tau)(\tau+T-t), \quad \tau \leq s \leq t \leq \tau+T$$

where the value of σ is defined by an empirical formula which involves the statistical data (based on earlier observations). Thus, the delivery of an order placed at $t=0$ randomly occurs on the interval $(\tau, \tau+T)$. The usual cost factors of inventory holding, shortage and ordering (denoted by IC, $\hat{p}$ and A) are considered.

Objective:

The expected total cost for a time unit is expressed explicitly when the demand process is a Wiener process, i.e. a stationary stochastic process with independent increments and normal distributed demand

$$G(x, s) = \Phi\left(\frac{x-ms}{\sigma\sqrt{s}}\right)$$

where Φ means the standard normal distribution function. In our case we have

$$H(S) = \frac{A}{T} + (IC+\hat{p})[V(T, S)+(S-m\tau-mT)U(T, S)] + \hat{p}(m\tau+mT-S),$$

where

$$U(T, S) = \frac{1}{T}\int_\tau^{\tau+T} \Phi\left(\frac{S-m\tau-mT}{U(S)}\right) dS;$$

$$V(T, S) = \frac{1}{T}\int_\tau^{\tau+T} U(S)\varphi\left(\frac{s-m\tau-mT}{U(S)}\right) dS,$$

and

$$U^2(S) = (\sigma^2 T + m^2 T^2)\left[(S-\tau)(\tau+T-S)\sigma^2 + \left(\frac{\tau+T-S}{T}\right)^2\right] + \sigma_S^2$$

(As earlier, φ denotes the density function of the standard normal distribution.)

Solution:

A numerical function minimization procedure has to be applied to the cost function given above.

177. Comparison of the cost functions of (t_p, S) models

Main codes:

$$1\ 1\ 1\ 1\ 0\ 1\ 1\ 1\ 1\ 1$$

Assumptions:

The cumulated demand during the time period $[t_1, t_2]$ is denoted by $\alpha(t_1, t_2)$. This is a random amount which depends only on the length of the period. The cumulated delivery until time s is also a random amount denoted by $\eta(s)$. This is also a stationary process, which is independent of $\alpha(0, s)$. The cumulated decrease of the inventory level in time interval $[0, s]$ is $\xi_S = S - (\eta(\tau+T) - \eta(s)) - \alpha(0, s)$ with a distribution function $H_s(x)$: here τ is the leadtime and T is the length of the order period. Let $H_s(x)$ be the probability distribution of ξ_S, then

$$H(x) = \frac{1}{T} \int_{\tau}^{\tau+T} H_s(x)\, ds, \quad h(x) = H'(x).$$

The usual cost factors of inventory holding IC and shortage $\hat{p}$ are considered. Furthermore,

$$\beta = \frac{\hat{p}}{IC + \hat{p}}.$$

Objective:

The expected total cost of an order period is equal to

$$K(S) = IC \left[S - \int_0^{\infty} x\, dH(x) \right] + (IC + \hat{p}) \int_S^{\infty} (x - S)\, dH(x).$$

The main purpose of the study is to compare the optimal costs for different stochastic processes ξ_s and different cost parameters β.

Solution:

For an arbitrary value of β, the inequality $K_2^*(\beta) \leq K_1^*(\beta)$ holds for the optimal values of the cost functions if $h_1(S_1^*(\beta)) \leq h_2(S_2^*(\beta))$. If $H(x)$ has a normal distribution, then the relation $\sigma_2 < \sigma_1$ for the standard deviations of $H_1(x)$ and $H_2(x)$ is a sufficient condition of the above inequalities.

178. A min—max inventory model with random leadtime

Main codes:

$$1\ 1\ 1\ 1\ 0\ 1\ 1\ 1\ 1\ 3$$

Assumptions:

A (t, S) model is considered for stochastic demand and stochastic leadtime which are independent random variables. An order placed earlier may be delivered later than an order which was placed later, due to the randomness of lead-time. The length of the order period is N units of time: N is also a control variable of the system, together with the order level S. The probability that the stock deficit is less than or equal to x given orders is denoted by $\varphi(x; N)$. The usual unit cost factors of inventory holding, shortage and ordering are considered and denoted by C_h, C_p and C_k. There is an additional unit cost C_w, which is the warehousing cost.

Objective:

The expected total cost for a unit of time is

$$C(N, S) = C_k/N + (C_h + C_w)S + C_p \int\limits_0^\infty x\, d_x \varphi(x; N) -$$

$$- (C_h + C_p) \int\limits_0^\infty \min(x, S)\, d_x \Phi(x; N).$$

Solution:

A special min-max solution procedure is suggested. Under fixed parameters of the demand and leadtime distributions, the parameters of the shortage probability distribution are determined. Thereafter, the distribution is selected which has the same parameter values and maximizes the expected cost for a unit of time. In the last step, the cost function corresponding to the above distribution is minimized with respect to N and S.

VII.2.2. Reorder-Point, Order Level (s, S) Type Models

179. *Poisson system with* $(S-1, S)$ *inventory policy*

Main codes:

$$1\ 1\ 1\ 1\ 0\ 1\ 6\ 2\ 2\ 3$$

Assumptions:

Units are demanded at random instants according to a Poisson process with mean rate λ. When a unit is sold, an order is placed for another unit. Thus the number of units on inventory plus those on order is a constant M at each time moment. This is the order level which represents the maximal level of the service. If the inventory level is zero, the new demands are lost for the system, until the next arrival (delivery) of an order. The leadtime is random with an exponential distribution, with parameter μ.

Objective:

The expected profit of the system is equal to

$$P_r(M) = gM - (C_i + g)\frac{ME_M(\lambda T) - \lambda TE_{M-1}(\lambda T)}{E_M(\lambda T)},$$

where g signifies the profit after selling a unit of the item and $C_i = C_n \cdot i$ is the unit cost of inventory holding, (C_n and i are the same as in Model 175), further on,

$$E_M(x) = \sum_{n=0}^{M} \frac{x^n}{n!} e^{-x}.$$

Solution:

The optimal value of M is the solution of the equation

$$\frac{C_i}{C_i+g} = \lambda T \frac{E_{M_o}(\lambda T)}{E_{M_o+1}(\lambda T)} - \frac{E_{M_o-1}(\lambda T)}{E_{M_o}(\lambda T)}.$$

180. $(S-1, S)$ inventory policy with arbitrary leadtime distribution

Main codes:

$$1\ 1\ 1\ 1\ 0\ 1\ 6\ 0\ 1\ 3$$

Assumptions:

The demand is generated by a Poisson process with parameter λ, as in the previous model. The random leadtime may have an arbitrary distribution: its expected value is denoted by τ. The order policy is the same as in the previous model. The demand is lost for the supplier in the case of shortage. The inventory holding and the shortage costs for a unit of item are denoted by H and L. No ordering cost is considered.

Objective:

The expected total cost for a given order level S is expressed by $C(S)=$ $=H[S-(1-p(S))\lambda\tau]+\lambda Lp(S)$, where $p(S)$ denotes the probability of the event that the leadtime demand is greater than S.

Solution:

The order level

$$S_0 = \lambda\tau + \alpha\sqrt{\lambda\tau}$$

with

$$\alpha = \left[2\ln \frac{\alpha\left(1+\dfrac{c_2}{c_1\tau}\right)}{\sqrt{2\pi}} \right]^{1/2}$$

provides an acceptable approximation of the optimal order level.

181. $(S-1, S)$ policy with restricted back-ordering time

Main codes:

$$1\ 1\ 1\ 1\ 0\ 1\ 6\ 2\ 9\ 3$$

Assumptions:

A Poisson system is considered: the demands are generated by a Poisson process, the leadtime is exponentially distributed. Their parameters are denoted by λ and μ, respectively. When a demand for a unit of item occurs, an order for another unit is immediately placed. In case of a shortage the demand is waiting a given time T. If during this time it is not satisfied, then the demand is lost to the supplier. Demands are satisfied according to the FIFO (first in—first out) service rule. When some demand is lost, the same amount of order is immediately revoked. The state of the system is described at time t by the function $R(t)$ which means the number of items ordered, but yet undelivered.

Objective:

The expected total cost for a unit of time is equal to $K(N)=C_lH+C_dD+C_l\pi_l$, where C_n and C_d denote the usual cost factors of inventory holding and shortage (both with dimension $[\$]/[Q]/[T]$.). The cost of a lost sale is denoted by C_l. H denotes the expected number of items in the inventory and D is the expected value of demand waiting for service in the case of stockout. Both expectations are considered at a random time. The average number of lost sales per unit time is denoted by π_l.

Solution:

Sufficient conditions are derived which make it possible to decide about the optimality of a parameter value N. Numerical examples are given.

182. $(S-1, S)$ *policy for demands generated by a compound Poisson process*

Main codes:

$$1 1 1 1 0 1 6 2 1 3$$

Assumptions:

Random discrete amounts are demanded at random instants which are generated by a Poisson process with parameter ϱ. The amount of demand j has the probability f_j. (The defined random demand process is a compound Poisson process.) The leadtime is also random with an arbitrary probability density function $\psi(t)$ having an expected value T.

The service level is defined in three different ways for the backorders case:

1. $R(S)$ is the probability that an item observed at a random point in time has no backorders.
2. $F(S)$ is the expected number of demands per time period for an item that can be met immediately from stock on hand.
3. $Q(S)$ is the expected number of units in routine resupply at a random point in time.

The service levels of types 2 and 3 are defined also for the lost sales case.

Objective:

The service level and the inventory level are weighted by the cost factors k_1 and k_2, where the first one characterizes the profit and the second one the costs. The net profit is the difference

$$H(S) = k_1 G(S) - k_2 cS,$$

which has to be maximized. Here $G(S)$ represents an arbitrary performance measurement and may be one of the functions R, F or Q (for the lost sales case only F or Q) and c denotes the unit purchase price.

264

Solution:

Algorithms are derived for the calculation of the service level R, F and Q (for arbitrary fixed parameter S) in the special case, when f_j has a geometric distribution:

$$f_j = (1-\varrho)\varrho^{j-1} \quad (0 \leqq \varrho < 1),$$

and the demand is backordered.

FORTRAN programs and tables are included to help the numerical calculations; optimization aspects are not studied.

183. *A special model: capacity optimization in a storage system*

Main codes:

$$1\ 1\ 1\ 1\ 0\ 1\ -1\ -1\ 7\ -1$$

Assumptions:

The system is analysed in a fixed time interval $[0, T]$. The input and output processes are signed by $X(t)$ and $Y(t)$, respectively. They are stochastic processes with known probability distribution. Thus the inventory level $Z(t)$ is also random. It can be expressed at time $t+1$ by

$$Z(t+1) = [\min\{Z(t)+X(t), K\} - Y(t)]^+,$$

where $X(t)$ and $Y(t)$ are the delivery and the demand between t and $t+1$, respectively, K is the storage capacity and $[x]^+ = \max\{0, x\}$. The expected value of $Z(t)$ is assumed to be independent of t, but this depends on the value of K.

Objective:

The purpose of the study is to determine the capacity K of the store which ensures a maximal expected net profit. Let $r(T, K)$ be the profit of a unit inventory per unit time for capacity K at time T. The interest rate is denoted by δ. Thus the present value of the total profit is

$$r_0(T, K) = \int_0^T e^{-\delta t} Z(t) r(t, K)\, dt.$$

The present value of the total operating cost is

$$c_0(T, K) = \int_0^T e^{-\delta t} Z(t) c(t, K)\, dt.$$

The building cost of capacity K is $c_f(K)$. Thus the present value of the net profit is

$$p_0(T, K) = r_0(T, K) - c_0(T, K) - c_f(K).$$

The objective function is the expected value of this expression.

Solution:

In the interesting cases in practice, the objective function has a single local optimum which can be calculated (e.g., from the partial derivatives of the objective function which are equal to zero at the minimum point). For exponential demand

265

and delivery distributions, the expected value of $\bar{Z}(K)$ is determined: here $\bar{Z}(K)$ denotes the expected value of $Z(t)$ which is supposed to be dependent only on the capacity K.

184. Special model for distribution of leadtime demand based on empirical data

Main codes:

1 1 1 1 7 3 7 2 7 3

Assumptions:

The probability that during the leadtime m customers arrive is denoted by $g(m)$. These customers require a total amount q with probability $P(q|m)$. The empirical sets of observations $d=[n_1, n_2, ..., n_m]$ are available where $1 \leq n \leq N$ is the demand of the j-th customer with probability $f(n_j)$.

Objective:

The purpose is to estimate the leadtime demand on the basis of the set of empirical data.

Solution:

The d vectors are partitioned in such a way that in a group
1. the sums of demands are the same,
2. the occurrence of a group has a constant probability.

The probabilities of the groups are estimated on the basis of the observations and on this basis the empirical distribution of the leadtime demand is derived. A FORTRAN program is included which is applicable for the above estimation procedure. This model can be used in combination with other models which provide a decision principle, given the knowledge of the above distribution.

VIII. Dynamic Reorder-Point Models

Dynamic models belonging to this family are those where the inventory policy is of (s, S) or (s, q) type.

Given the essence of the policy, the conditions concerning delivery have been taken into account in forming the subgroups. In the models of the first subgroup the ordered lot will be delivered without leadtime. For these models—in compliance with Bellman's dynamic optimization principle—a recursive relation can be formulated for the optimization of the cost function. The solution methods of the models, as a rule, differ from each other: besides the dynamic programming algorithm, the theory of Markov chains, and renewal theory or linear programming may be applied for their solution.

In the second subgroup, deliveries take place with a known leadtime. In some of the models, inventory review is periodic with a fixed length of the reviewing period. The ordered quantity arrives here after a leadtime of one or several periods. In the models of continuous inventory reviewing the ordered batch arrives also after a known leadtime. Due to the leadtime, joint consideration of several variables is necessary in the cost function: this makes optimization by means of dynamic programming very difficult and inefficient. Thus, finding ways for a quicker algorithmic solution is of special importance here; in some cases, an explicit solution is given for special demand distributions using an iterative procedure which quickly approaches the solution. In several models, the authors only verify the optimality of the (s, S) model (under various conditions), but they do not deal with the algorithmic solution.

Models of periodic inventory reviewing belong to the third subgroup, where the delivery of the ordered batch has a random character and the leadtime is a random variable. This fact makes the exact expression of the objective function of these systems especially complicated and causes difficulties in determining the optimal decision parameters. Thus, these models usually deal with the description of the system from a mathematical aspect, (i.e., with its stability and with the determination of the type of optimal ordering policy), but they do not give a solution algorithm.

We present below a detailed description of Model 185, which represents well the properties of the dynamic models. This model is a member of the family VIII.1.1.2.

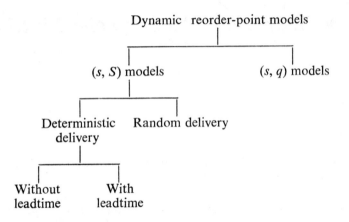

Dynamic reorder-point models

(s, S) models   (s, q) models

Deterministic   Random delivery
delivery

Without   With
leadtime   leadtime

185. Dynamic n period model with leadtime

Main codes:

$$1\ 1\ 0\ 1\ 1\ 1\ 5\ 1\ 1\ 1$$

Assumptions:

The demand arising in each period for a certain item in a store is considered as a random variable. Demand for period i are denoted by ξ_i, with a distribution function $F_i(x)$. Demands in various periods are independent of each other and their distributions may be different.

An order can be placed at the beginning of each period, and the entire quantity will arrive immediately (this will be generalized later on, when the occurrence of leadtime will be allowed).

By inventory level the sum of the volume on hand and on order is meant. The level of inventory before ordering is x_i. It will be raised up to level y_i by placing the order.

In the general model, the following expression may be given for the inventory level of the next period:

$$x_{i+1} = v_i(y_i, \xi_i) = \begin{cases} a(y_i - \xi_i) & \text{if } y_i \geq \xi_i, \\ b(\xi_i - y_i) & \text{if } y_i < \xi_i; \end{cases}$$

where $0 \leq a,\ b \leq 1$. This general equation may be specialized for several cases,

— if $a=1$ and $b=1$, then in the case of shortage the demand will wait (back-orders case);
— if $a=1$ and $b=0$, then in the case of shortage the demand will be lost;
— if $a=0$, the inventory left by the end of the period will be lost (e.g., perishable goods);
— if $0<b<1$, one part of the uncovered demand will wait, the other part will be lost;
— if $0<a<1$, only one part of the inventory left by the end of the period can be transferred to the next period, the other part $(1-a)$ cannot be used up in the next period.

268

Based on the observed inventory level x_i, the volume of order to be placed is $z_i = \bar{y}_i(x_i) - x_i$. The $\bar{y}_i$ inventory level is the function of x_i. The sequence of functions $\bar{y}_i(x_i)$ $(i=1, 2, ...)$ defines the ordering policy.

Objective:

Three types of costs will occur. The ordering cost may alter in each period, it equals

$$B_i(z) = K_i \delta(z) + c_i z,$$

where $\delta(0) = 0$ and $\delta(z) = 1$, if $z > 0$; K_i is the fixed ordering cost and c_i is the unit purchasing price in period i. The sum of inventory carrying and shortage costs is $l_i(y_i, \xi_i)$ in period i, and the expected value of l_i, assuming that $y_i = y$, is equal to

$$L_i(y) = \int_0^\infty l_i(y, \xi_i) \, dF_i(\xi_i).$$

The costs of period i can be weighted by a discount factor α_i $(0 \leq \alpha_i \leq 1)$, and the following symbol may be introduced:

$$\beta_i = \prod_{j=1}^{i-1} \alpha_j \quad \text{in the case of } i > 1, \text{ and } \beta_1 = 1.$$

If the initial inventory level is x_i, then the expected discounted total cost of the first n periods is equal to

$$C^n(x_1|\bar{y}) = E\left\{ \sum_{i=1}^n \beta_i[K_i\delta(y_i - x_i) + c_i(y_i - x_i) + l_i(y_i, \xi_i)] - \beta_{n+1}c_{n+1}x_{n+1} \right\}$$

in the case of a given ordering policy $\bar{Y} = (\bar{y}_1, ..., \bar{y}_i, ..., \bar{y}_n)$. Here and in the following, E denotes the expected value operator. This may be written also in the form

$$C^n(x_1|\bar{y}) = E\left\{ \sum_{i=1}^n \beta_i[K_i\delta(y_i - x_i) + c_i y_i + l_i(y_i, \xi_i) - \alpha_i c_{i+1} v_i(y_i, \xi_i)] \right\} - c_1 x_1,$$

where the term $c_1 x_1$ can be neglected, because it is not influenced by the decision policy, but depends only on the initial inventory level. The optimal ordering policy is where the above cost function will take its minimum. The cost function may be written in the following simple form:

$$C^n(x_1|\bar{y}) = \sum_{i=1}^n \beta_i[K_i\delta(y_i - x_i) + G_i(y_i)],$$

with the following notation:

$$G_i(y_i) = c_i y_i + L_i(y_i) - \alpha_i c_{i+1} \int_0^\infty v_i(y_i, \xi_i) \, dF_i(\xi_i).$$

If the quantity, ordered at the beginning of period i, arrives at the beginning of period $L+i$, the objective function will be modified in a suitable manner.

In the case of shortage, the demands occurring during the leadtime will be added to the unsatisfied demand:

$$u = \xi_i + \xi_{i+1} + \ldots + \xi_{i+L-1}.$$

The distribution function of u is denoted by $F_{i,i+L-1}(u)$.

The expected value of the inventory holding and the shortage costs in period $i+L$ in the case of $y_i = y$ is

$$L_i(y) = E\{l_i(y_i, \xi_{i+L}) | y_i = y\} =$$

$$= \int_0^\infty \int_0^\infty l_i(y_i - u, \xi_{i+L})\, dF_{i+L}(\xi_{i+L})\, dF_{i,i+L-1}(u),$$

depending on the demands of L periods. Hence, in the cost function $C^n(x_i | y)$ we can write

$$G_i(y_i) = c_i y_i + L_i(y_i) - \alpha_i c_{i+1} \int_0^\infty (y - \xi_i)\, dF_i(\xi_i).$$

In the case of shortage the expression of the cost function for the case $L > 0$ is known only for independent, identically distributed demands in each period. This will be described later in Model 199.

Solution:

The optimal ordering policy of an n period model is given by the vector of functions $Y^* = (\bar{y}_1^*, \ldots, \bar{y}_i^*, \ldots, \bar{y}_n^*)$, which minimizes the objective function

$$C^n(x_1 | \bar{Y}^*) = \min_Y C^n(x_1 | \bar{Y}).$$

The optimal quantity to be ordered in period i depends on the initial inventory x_i, i.e. $\bar{Y}_i^* = \bar{y}_i^*(x_i)$. To determine the optimal ordering policy, Bellman's dynamic programming principle can be applied; thus,

$$C^i(x) = \min_{y_i \geq x_i} \left[K_i \delta(y_i - x_i) + G_i(y_i) + \alpha_i E\{C^{i-1}[v_i(y_i, \xi_i)]\} \right]$$

for $i = 1, 2, \ldots, n-1$. This necessitates the solution of a functional equation system which implies very serious computational difficulties in the general case. Practical solutions are available for special cases, when the distribution of demands is identical in the different periods (the so-called stationary demand case). In the case of a discrete demand distribution the applicable general methods are based on two versions of dynamic programming: value iteration and decision iteration (in the case of continuous distribution discretization will be applied). These methods will be described below. There is a number of special methods which are more efficient, quicker and require less calculation than the general method. Those procedures which give approximations to the optimal solution are also very important in practice. All these will be described in connection with specialized models.

270

The value iteration method of dynamic programming may be started with the initial value $C^0(x) \equiv 0$. Starting from this,

$$C^1(x) = \min_{y \geq x} [K\delta(y-x) + G(x)]$$

will be calculated for each possible value x. In the course of iteration step k,

$$C^k(x) = \min_{y \geq x} \left[K\delta(y-x) + G(y) + \alpha E\{C^{k-1}[v(y, \xi)]\} \right]$$

will be determined for all possible values x. The iteration is to be continued, until the increase of the cost function is nearly the same for each examined value of x. (Note that this condition does not specify a well-defined stopping criterion.)

The decision iteration method of dynamic programming is based upon the fact that an objective function value corresponds to each decision rule $y=d(x)$ (ordering strategy) in the following form:

$$C(x) = K\delta[d(x)-x] + G[d(x)] + \alpha E\{C[v(d(x), \xi)]\},$$

which means a linear equation system for $C(x)$. On the other hand, a decision rule $y=d(x)$ can be given for each cost function, where

$$K\delta[d(x)-x] + G[d(x)] + \alpha E\{C[v(d(x), \xi)]\} =$$
$$= \min_{y \geq x} \{K\delta(y-x) + G(y) + \alpha E\{C[v(y, \xi)]\}\}.$$

The decision iteration algorithm is built up on the basis of these two systems of equations. In the first phase, the proper cost function will be determined for the given ordering strategy $y=d_0(x)$; in the second phase, minimization takes place in accordance with the aforesaid. In this way, a new $y=d_1(x)$ ordering strategy will be obtained, and the next step of the procedure continues with this in the two successive phases. An optimal policy will be gained if two successive steps give the same ordering strategy for each possible value of variable x.

The convergence of both procedures is proved under rather general conditions. The numerical efficiency of these procedures depends on the number of possible values of the inventory states x; different methods are known to decrease this number, especially for special model structures. This will be discussed later in connection with the analysed models.

Several authors have dealt with the determination of the type of the optimal $y=d(x)$ ordering strategy: this depends mainly on the value of the fixed ordering cost K.

In the case of $K=0$, the ordering cost will be proportional to the ordered quantity. If the distribution of demands $[F(x)]$ is the same in the different periods, then the optimal ordering strategy is of the (t_p, S) type with the convex one-period cost function $L(y)$ generally occurring in practice and expressing the sum of the inventory carrying and shortage costs. An order has to be placed in each period. In the case of immediate reception of the ordered batch, it is easy to determine the value of the optimal ordering level. In the case of shortages, the equation

$$C(1-\alpha) + L'(S) = 0$$

271

gives the optimal order level (case of unsatisfied demands), and in the case of lost sales the solution S of the equation

$$C+L'(S)-\alpha cF(S) = 0$$

is to be determined. The above equations may be written in the following simple form, with inventory holding c_1 and shortage cost factors c_2, and with linear cost function:

$$F(S) = \frac{c_3-c(1-\alpha)}{c_3+c_2} \quad \text{(in the case of unsatisfied demands)}$$

$$F(S) = \frac{c_3-c}{c_3+c_2-\alpha c} \quad \text{(in the case of lost demands).}$$

If the ordered batch does not arrive immediately, the ordering strategy is also (t_p, S), if the other assumptions are unchanged. The method of determination of the optimal S level will be discussed in connection with Models 185 and 199.

If the distribution of demands changes in each period, the optimal ordering policy is still (t_p, S) with a different order level S_n in the different periods if $K=0$ and $L_n(y)$ is convex.

If a fixed ordering cost arises $(K>0)$, then, generally speaking, the optimal ordering strategy is of the (s, S) type. A sufficient condition of its optimality is the convexity of the previously defined functions $G_i(y)$, which is valid in most important cases in practice. For the determination of the optimal parameters of the (s, S) models, several exact and approximate solution methods will be given later.

VIII.1. Dynamic (s, S) Models

VIII.1.1. Dynamic (s, S) Models with Deterministic Delivery

VIII.1.1.1. Delivery without Leadtime

186. Infinite horizon model with discounting

Main codes:
$$1\ 1\ 0\ 1\ 1\ 1\ 5\ 1\ 1\ 0$$

Assumptions:

Orders can be placed in each period. Let the initial inventory level be x, the order level $y=d(x)$, and the demand in each period ξ with distribution $F(\xi)$. The ordered quantity arrives immediately, demands arise also at the beginning of the period. The sum of inventory holding and shortage costs of one period is given by the function $L(y)$ depending on the order level. The cost of placing the order is $B(y-x)$ depending on the ordered quantity. The volumes of inventory and shortage are bounded.

Objective:

The expected value of the total cost of n periods, discounted by the factor α, equals

$$C^n(x) = \min_{y \geq x} \left\{ B(y-x) + L(y) + \alpha \int_0^\infty C^{n-1}(y-\xi)\, dF(\xi) \right\}.$$

This expression serves to determine also the minimal costs. Thus, we seek the optimal $y = d(x)$ ordering strategy which yields the minimal costs in the case of $n \to \infty$.

Solution:

The authors prove that under the given conditions the (s, S) strategy is optimal. Its determination is realized by using the decision iteration version of dynamic programming. (This means the iterative approximation of the optimal $d(x)$ strategy.) The convergence of the procedure is proved. The decision iteration was described with Model 185.

187. Random demand, immediate delivery—the dynamic case

Main codes:

$$1\ 1\ 0\ 1\ 1\ 1\ 5\ 1\ 1\ 0$$

Assumptions:

The demand of the time-interval T is a random variable, the density function of which is $f(x)$. Some initial inventory z is given at the beginning of the operation of the system (it may be negative), and an order will be placed in the initial moment which will raise the stock to y. We assume that the ordered batch is received prior to the first demand. i.e., an inventory level y is available by that time. If a superfluous inventory $y - x > 0$ remains by the end of the time interval T, the unit cost of this is denoted by s_T. For the unsatisfied demands, a penalty p_T is to be paid per unit. The inventory level will be reviewed in each period and an order will be placed according to the (s, S) policy. The purchasing cost is denoted by c.

Objective:

The minimal cost for one period is

$$L_T(z) = \min_{S > z} \left\{ c(S-z) + \int_0^S s_T(S-x)f(x)\, dx + \int_S^\infty p_T(x-S)f(x)\, dx \right\}$$

and that for n consecutive periods is expressed by

$$L_{nT}(z) = \min_{S > z} \left\{ L_T(S, z) + \alpha \left[L_{(n-1)T}^{(0)} \int_y^\infty f(x)\, dx \int_0^y L_{(n-1)T}(S-x)f(x)\, dx \right] \right\},$$

$L_{nT}(z)$ is the minimum of the $L_{nT}(S, z)$ function achieved with optimal S (S is the inventory established in the first period), and α is the discount factor.

Solution:

Bellman's dynamic programming optimum criterion may be applied. The author gives an iterative method to determine the functions L_{nT} for a Pólya-type distribution.

188. *An inventory model based on Markov chains*

Main codes:

$$1\ 1\ 0\ 1\ 1\ 1\ 5\ 1\ 0\ 0$$

Assumptions:

An item is measured in discrete units. There is one store, and the upper limit of its capacity is n units. There is no leadtime, the excess demand is lost but no shortage cost is calculated. Only ordering and storage costs are taken into consideration. The whole ordered quantity arrives at the beginning of the period. The volume of demands arising in the period is a random variable of a given distribution. The length of the periods is not prescribed, their sequence constitutes a Markov-process. The objective is the minimization of costs emerging during infinite time horizon.

Objective:

The optimization is formulated as a linear programming problem:

$$\max \sum_{i=1}^{n+m} v_i$$

subject to the constraints

$$v_i \le c_i^k + \sum_{j=1}^{n} q_{ij}^k v_j \quad \text{for} \quad i \le n,\ k \in L_i$$

$$v_i \le a_i + v_{n+i} \quad \text{for} \quad i \le m$$

$$v_{n+1} \le b_k + \sum_{j=1}^{n} q_j^k v_j \quad \text{for} \quad i \le m,\ k \in \varDelta_i$$

$$v_{n+1} \le v_{n+i+1} \quad \text{for} \quad i < m.$$

The dual problem with the dual variables x_i^k and x_i^t is the following:

$$\min \left\{ \sum_{i=1}^{n} \sum_{k \in L_i} c_i^k x_i^k + \sum_{i=1}^{m} \left[a_i x_i^t + \sum_{k \in \varDelta_i} b_n x_{n+i}^k \right] \right\}.$$

Subject to the constraints:

$$\sum_{x \in D_i} x_i^k - \sum_{j=1}^{n} \left[\sum_{k \in L_j} q_{ji}^k x_j^k + \sum_{k \in \varDelta_j} q_i^k x_{n+j}^k \right] = 1 \quad \text{for} \quad i \le n$$

$$\sum_{k \in D_i} x_i^k - x_{i-n}^t = 1 \quad \text{for} \quad i = n+1$$

$$\sum_{k \in D_i} x_i^k - x_{i-n}^t - x_{i-1}^t = 1 \quad \text{for} \quad n+1 < i \le n+m$$

$$x_i^k \ge 0 \quad \text{for all } i \text{ and } k$$

274

where D_i, L_i and Δ_i are decision sets defined in different ways;

$$q_{ij}^k = p_{ij}^k \int\limits_0^\infty e^{-\alpha x}\, dF_{ij}^k(x)$$

(in the state i we pass by decision k to state j), where:

p_{ij}^k: transition probability;

α: discount factor $(0<\alpha<1)$;

F_{ij}^k: the conditional, cumulative distribution function of the time of transition;

c, a and b are cost factors.

Solution:

The author proves that the optimal solution is of (s, S) type, and the optimal s is the fix point of the following operator A:

$$(Av)_i = \min_{k \in M_i} R(i, k, v),$$

where $R(i, k, v) = c_i^k + \sum_j q_{ij}^k v_j$;

v = n-component vector,

M_i = set of decisions.

Accordingly, the solution may be determined by some fix point searching algorithm.

189. Inventory model with transportation-efficient ordering

Main codes:

$$1\ 1\ 0\ 1\ 1\ 1\ 5\ 1\ 1\ 0$$

Assumptions:

Periodic inventory reviewing is applied. An order may be placed after each review (in each period) which arrives immediately. The demands of any two successive periods are pair-wise independent, and identically distributed, random variables.

The unsatisfied demands wait. Three types of costs will be considered: ordering, inventory holding and shortage costs. The sum of the latter two is characterized by a function $L(y)$ depending on the inventory level. It is assumed that the ordering cost of a unit z is the sum of a linear function cz and a jumping function $J(z)$, where a jump takes place in the $0, n, ..., 2n$ points; K is the size of the jump; c, M and K are known constants. Thus:

$$J(z) = K \cdot [z|M] \quad \text{and} \quad C(z) = cz + K \cdot [z|M].$$

For example, the capacity of the transportation vehicle is M; K is the hiring (inventory) cost and c is the unit procurement cost. The ordering policy may be characterized by the vector T_n, T_{n-1}, ..., T_1: if the inventory level at the beginning of period falls in the interval $[T_i - M_r,\ T_i + M - M_r]$ ($r \geq 1$ integer), an order of M_r is to be placed. No order is placed if the inventory level is higher than T_i (assume that $T_1 = -\infty$).

Objective:

$$f_n(x) = \inf_{y \geq x} \left\{ c(y-x) + J(y-x) + L(y) + \alpha \int_0^\infty f_{n-1}(y-\xi) \varphi(\xi) \, d\xi \right\}$$

where: $f_n(x)$: the cost function of period n,
  y: inventory level after a delivery,
  x: inventory level at the beginning of the period,
  $\varphi(\xi)$: density function of the probability distribution of demands,
  α: discount factor.

Solution:

The author verifies that the same T is optimal in each period; $T = \lim_{n \to \infty} T_n$, where T satisfies the following equation:

$$L(T+M) - L(T) = (\alpha - 1)(K + Mc).$$

190. Multi-class inventory model with inventory level function and optimal dynamic policy

Main codes:

$$1\ 1\ 0\ 1\ 1\ 1\ 5\ 1\ 1\ 0$$

Assumptions:

The model describes an inventory system, where two classes exist. The classification criterion: the "perfect" units belong to the first class, while the "worthless" units (perfect items may perish) get to the second class according to a stochastic process. The inventory reviewing is periodic, the ordered quantity is immediately at disposal. Only "perfect" items are ordered. The demand is (in each class and in each period) represented by the perished amount (getting into the lower class): it is a function of the initial stock in each class. The demand (deterioration of inventory) follows a binomial of uniform distribution; its value at the beginning of the inventory review period is denoted by $D(y)$.

Costs:

— ordering cost $c(Z)$, which is convex, increasing, $c(0) = 0$;
— inventory holding cost h, which is convex, increasing;
— shortage cost p, which is convex, increasing.

Inventory holding and shortage costs are defined at several special time-moments of the inventory review period, their sum is denoted by the function $L(y)$. A possible interpretation: for a working process a machines are needed. The machines break down randomly. If more than a machines are perfect, inventory holding costs arise; if less than a machines can operate, shortage cost has to be taken into account.

Objective:

$$C_n(x) = \min_{y \geq x} \left\{ c(y-x) + L(y) + \alpha E\left[C_{n-1}(y - D(y)) \right] \right\}, \quad x \geq 0,$$

where: $C_0(x) = 0$; α discount factor, $0 < \alpha < 1$.

276

Solution:

The optimal ordering policy may be characterized by the integer function $\bar{y}_n(x)$ and by the integer $\bar{x}_n$; these are parameters of the (s, S) ordering policy which alter in each period. If the inventory level $x < \bar{x}_n$, then it has to be replenished to $\bar{y}_n(x)$; if $x \geq \bar{x}_n$, no order has to be placed. The solution can be obtained by dynamic programming.

The author describes two special cases; when

— the ordering cost is linear, $c(x) = cx$, $c > 0$;
— the stocks left by the end of the last period can be sold at a price αc.

191. Discrete inventory model with arbitrary time and quantity distribution of demands

Main codes:

$$1\ 1\ 0\ 1\ 1\ 1\ 5\ 1\ 2\ 0$$

Assumptions:

The volume of demand is a positive integer random variable with a known probability distribution. Demand occurrences are generated according to independent, identical probability distributions. The maximal value of a single demand is given. The intervals between two successive demands are finite, discrete, positive numbers generated according to known independent, identically distributed probability distributions. The maximal value of the time between two subsequent demands is given. (Demand emerges always—if at all—at the end of the inventory review period.)

Orders are always placed depending on the inventory level subsequent to the occurrence of the actual demand. The ordered batch arrives without leadtime. Unsatisfied demands are lost for the system. Storage capacity constraints are also to be taken into consideration.

Objective:

The expected cost for a time unit is to be minimized. The exact expression of the objective function is not presented.

Solution:

The operation of the model takes place according to a Markovian renewal process. If i units are available in the store, this is qualified as state i. In this case, a number k will be selected as a decision from the set $i, i+1, \ldots, N$. This means: $k - i$ units are to be ordered. It can be proved by means of linear programming or iterative methods that an (s, S) policy is optimal.

192. Dynamic inventory model, unsatisfied demand is lost

Main codes:

$$1\ 1\ 0\ 1\ 1\ 1\ 5\ 1\ 2\ 0$$

Assumptions:

Unsatisfied demands are lost in each period. The ordered quantity arrives immediately. The demands arising in the periods are independent random variables with known continuous density functions. The ordering cost is linear: $c(y)=cy$, $c>0$. The inventory holding cost $h(y)$ is a continuous, convex, increasing function.

The cost $q(y)$ of unsatisfied demands y is a continuous, convex, increasing function. The discount factor is $0<\alpha<1$.

The shortage cost of the model is $p(y)=q(y)-\alpha \cdot cy$, i.e., if shortage occurs, then the discounted value of the quantity ordered during the period has to be added to the usual shortage cost.

Objective:

The goal is the minimization of the total discounted costs for n subsequent periods.

Solution:

The author shows the connection between this model and Model 292, thus the results and the solution procedure to be presented in the latter model may be applied to the present model, too.

193. Optimal pricing and ordering decision with random demands

Main codes:

$$1\ 1\ 0\ 1\ 1\ 1\ 5\ 1\ 1\ 0$$

Assumptions:

The system operates during N periods. Random demands arise in each period. The volume of demand depends on the price p of the item, which is a decision variable in each period. In the period n, with price p, the assumed distribution function of demand is $F_n(d|p)$. We assume that F_n is a stochastically decreasing function of p, i.e., $F_n(d|\hat{p})\geqq F_n(d|p)$ for arbitrary d in the case of $p>\hat{p}$.

The sets of the feasible inventory levels x, and prices p are given. A decision has to be made in each period for the price and the ordering (production) quantity. In this way, the optimal solution for each period and each inventory level is a two component vector of price and order quantity.

Let

— $L_n(x, p)$ denote the loss in period n with inventory level x and price p, the elements of which are: inventory holding cost, backordered shortage cost and profit;

— $K\delta(x-y)$ denote the ordering cost (the cost of replenishing the inventory from y to level x) where

$$\delta(x-y) = \begin{cases} 0, & \text{if } x \leqq y \\ 1, & \text{if } x > y; \end{cases}$$

— $n=1, 2, ..., N$ is the serial number of the periods, proceeding from the last one backwards.

Objective:

$G_n(y)$ is the expected minimal discounted cost if there are n periods left and the actual inventory level is y.

$$G_0 \equiv 0$$
$$G_n(y) = \min \{K\delta(x-y)+g_n(x, p)\} \quad (x \geqq y, \ p \in P).$$

278

$g_n(x, p)$ = the expected discounted cost with inventory level x and price p in period n, if the system operates optimally afterwards.

$$g_n(x, p) = L_n(x, p) + \int_0^\infty G_{n-1}(x-d)\, dF_n(d|p)$$

(α: discount factor).

Solution:

It is true for the majority of the examples examined by the author that the so-called (s, S, p) policy is optimal. According to this policy

1. there is a function $p_n(x)$ giving the optimal price for every inventory level x in a certain period;
2. if $x \geqq s_n$, there is no order; price is taken for $p_n(x)$;
3. if $x < s_n$, order will be placed to an inventory level S_n; price is taken for $p_n(S_n)$.

If the optimum for the (s, S, p) policy has already been established (by dynamic programming), then it can be seen by verifying certain conditions, whether the (s, S, p) policy is optimal or not. If the few feasible prices p are very different from each other, then the (s, S, p) policy is probably not optimal, whereas in the opposite case it is. It is hard to give easily verifiable, exact optimality criteria in advance.

194. Optimal policy in the case of inventory volume dependent demands and immediate delivery

Main codes:

1 1 0 1 1 1 5 1 1 0

Assumptions:

The assumptions are identical with those of Model 204, with the difference that there is no delay in delivery, and the demands depend on the stock at the beginning of the period.

Objective:

We seek an ordering policy which ensures the minimal expected discounted ordering cost for n periods while the probability of shortage should not exceed a prefixed bound.

Solution:

The authors express the shortage probability as a function of the sum of the inventory on hand and the ordered quantity, hence determining the ordering policy which ensures the lowest ordering cost for n subsequent periods.

195. Cost model of stochastic storing systems

Main codes:

1 1 1 1 1 1 5 1 1 0

Assumptions:

The input of the stochastic storing systems examined is a non-decreasing stochastic process, and their output mechanism works in intervals, in the form of emptying operations, leaving in the system an inventory of level m (m is not always zero). Emptying points are the replenishment points of the system. Emptying takes place if the inventory level in the system exceeds the level q. The author attaches a duration measure to the input process expressing the expected value of time the input process spends in a given set. Most of the one-item (s, S) inventory policies may be discussed as special cases of the stochastic storing system generalized by the author. In this way, the input process is in accordance with the demand arising in each time unit. There is no delivery delay (in the case of immediate emptying). The inventory level equals $I(t) = -v(t)$, where $v(t)$ is the net quantity in the system, s is equal to $(-m)$ and S equals $(-q)$.

Objective:

There is one fixed cost c in the system arising at emptying (ordering) and one time-proportional (inventory holding) cost $g(x)$ arising when the quantity inside the system is x. The total cost during an interval $(0, t)$ is to be minimized.

Solution:

The author seeks the optimal parameters according to the (s, S) policy, taking into account the regularities of the renewal processes.

196. (s, S) *model in the case of limited order quantities*

Main codes:

$$1\ 1\ 0\ 1\ 1\ 1\ 5\ 1\ 1\ 0$$

Assumptions:

The maximum ordering volume is R. The ordering policy is based on a periodic inventory review. The distribution of demands during a period is characterized by the density function $f(x)$. There is no leadtime. The ordering cost is proportional to the ordered quantity r; the usual fixed ordering cost is omitted. The total cost of inventory holding and shortage for one period is $L(y)$, where y is the initial inventory level of the period; L is a convex function.

Objective:

The optimal discounted total cost for n periods can be expressed as

$$C_n(x) = \min_{x \le y \le x+R} \left\{ (y-x) \cdot r + L(y) + \alpha \int_0^\infty C_{n-1}(y-\xi)f(\xi)\,d\xi \right\},$$

where $C_0 \equiv 0$, x is the inventory level before ordering, α is the discounting factor.

Solution:

The author proves that the optimal inventory policy satisfying the assumptions is of (s, S) type, but he does not deal with the determination of the optimal values s and S. (The methods of dynamic programming may be applied for determining the optimal s and S values.)

280

197. On the optimality of the generalized (s, S) policies

Main codes:

$$1\ 1\ 0\ 1\ 1\ 1\ 5\ 1\ 1\ 0$$

Assumptions:

A single-item, stochastic, dynamic inventory model is investigated, where stocks are periodically reviewed. The ordering cost function is a concave, monotonously increasing function. There is no leadtime of replenishment. Shortage is back-ordered. Demands of the periods are independent random variables with identical distribution (one-sided Pólya-distribution), the density function is denoted by $f(y)$. Costs arising after n periods are discounted by a discount factor α^n, $0 < \alpha < 1$. In a given period, the holding and shortage costs are calculated based on the volume of inventory at the end of the period, their expected value is indicated by a function $G(y)$, and the sum of the costs is denoted by a Pólya-type density function.

Objective:

The objective is the minimization of the expected total costs of the considered periods

$$K_n = \inf_{y \ge x} \{ C(y-x) + G(y) + \alpha K_{n-1} f(y) \},$$

according to the principle of dynamic programming.

Solution:

The author shows that in the case of the above assumptions, a modified (s, S) policy is the optimal solution (i.e. an ordering policy function, satisfying the two assumptions below):

a) $y(x) = x$, if $s \le x$;

b) $y(z) \ge y(x) \ge S \ge s$, if $z < x < s$.

198. Optimality of the generalized (s, S) policies in the case of a uniform demand distribution

Main codes:

$$1\ 1\ 0\ 1\ 1\ 1\ 5\ 1\ 1\ 0$$

Assumptions:

The assumptions are identical with those of the previous model; the only difference is that the demands in the periods follow a uniform distribution or the convolution of a finite number of uniform distributions.

Objective:

Identical with that of Model 197.

Solution:

The author shows the optimality of the modified (s, S) policy under these conditions, too.

281

199. *Dynamic n-period model with lost demands*

Main codes:
$$1\ 1\ 0\ 1\ 1\ 1\ 5\ 1\ 2\ 1$$

Assumptions:

The volume ξ of demands arising in each period is a random variable with a distribution function $F(\xi)$. The volume of orders placed in the period i is z_i, this quantity will arrive after an L-period leadtime in one lot. The expected value of the sum of inventory holding and shortage costs in L periods is $G(x, z_1, ..., z_{L-1}, z)$, where x is the initial inventory level and z_i is the volume of order placed in period i. The total cost of ordering and procurement is $c_3 + cz$.

Objective:

The expected discounted total cost of n periods are minimized:

$$C^n(x, z_1, ..., z_{L-1}) = \min_{z \geq 0} \{c_3 \delta(z) + cz + G(x, z_1, ..., z_{L-1}, z) +$$

$$+ \alpha C^{n-1}(z_1, ..., z_{L-1}, z) \int_x^\infty dF(\xi) + \alpha \int_0^x C^{n-1}(x + z_1 - \xi, z_2, ..., z_{L-1}, z) \, dF(\xi) \}$$

$$C^1(x, z_1, ..., z_{L-1}) = \min_{z \geq 0} \{c_3 \delta(z) + cz + G(x, z_1, ..., z_{L-1}, z)\}.$$

Solution:

The optimal ordering quantities $z_i, ..., z_n$ can be determined by dynamic programming; these values are shown to be in accordance with an (s, S) strategy. The values of s and S are given by the method of dynamic programming.

200. *Iterative solution of (s, S) models*

Main codes:
$$1\ 1\ 0\ 1\ 1\ 1\ 5\ 2\ 1\ 1$$

Assumptions:

The demand of a period is a discrete random variable, taking the value j with probability $\varphi(j)$. Inventory holding and shortage costs of one period are $g(j)$, if j is the level of inventory at the beginning of the period. The expected cost is $G(k) = \sum_{j=0}^\infty g(k-j)\varphi^L(j)$, where L is the leadtime; $G(k) > c_3 + G(S_0)$ where S_0 is the modus and k is a sufficiently large integer number, c_3 is the ordering cost. Let r be the smallest integer value for which $G(r) \leq c_3 + G(S_0)$.

Objective:

$$C(s, S) + \{G(S) + \sum_{k=0}^{S-s} G(S-k) m(k) + K\} / \{1 + M(S-s)\},$$

where

$$m(k) = \varphi(k) + \sum_{j=0}^{k} m(k-j)\varphi(j) \quad \text{and} \quad M(k) = \sum_{j=0}^{k} m(j).$$

Solution:

The optimal values of $s = \lim s_n$ and $S = \lim S_n$ $(r \leq s_n \leq S_n \leq R)$ can be determined by the following iteration method:

$$f_n(i) = \inf_{k>i} \left\{ c_3 \delta(k-1) + G(k) + \alpha_n \sum_{j=1}^{\infty} f_{n-1}(k-j)\varphi(j) \right\}, \quad f_0(i) = 0,$$

and S_n is the smallest integer for which $G_n(k) = L(k) + \alpha_n \sum_{j=0} f_{n-1}(k-j)\varphi(j)$
is minimal, while s_n is the smallest integer for which $G_n(s_n) \leq K + G_n(S_n)$ holds.
There are several assumptions to the series $0 \leq \alpha_n \leq 1$, the fulfilment of which
ensures convergence.

**201. Stochastic reorder-point, order-level system with inventory review
depending on a random delivery**

Main codes:
$$1\ 1\ 0\ 1\ 1\ 1\ 5\ 2\ 1\ 1$$

Assumptions:

Demands arise at random moments, and the time-intervals between two demand
moments are independent random variables having an identical distribution.
Furthermore, they are independent of the volume ordered. The demands are
independent random variables with density function $P(v)$. The inventory is
reviewed at each demand: in this way, the length of the inventory review periods
will be random. The orders are placed in such a way that the inventory position
becomes y (inventory on hand, plus on order, minus shortage). If the inventory
level has been reduced to level x by the demand, then $(y-x)$ is to be ordered.
The ordered quantity is constant, it is delivered with leadtime L. The possible
shortages are eliminated first. The density function of the demand volume is
$P_L(v_L)$ during the leadtime. The system operates during n inventory reviewing
periods. The model may be brought into agreement with the model of discrete
inventory review periods, where the length of the period is equal to the expected
length of the time interval between two demands.

The costs:

— ordering cost: $c_3 \delta(z) + cz$, where
   z: the ordered quantity,
   c_3: the fixed ordering cost,

$$\delta(z) = \begin{cases} 1 & (z > 0) \\ 0 & (z = 0) \end{cases}$$

   c: volume dependent ordering cost;

283

— c_1: inventory holding cost per unit amount and average inventory review period;
— c_2: shortage cost per unit amount and average inventory review period.

Objective:

We want to choose the inventory level y according to the following functional relation:

$$C_n(x) = \min_{y \ge x} \left\{ c_3 \delta(y-x) + c(y-x) + G(y) + \int_0^\infty C_{n-1}(y-v) P(v)\, dv \right\},$$

where $C_n(x)$ is the expected total cost according to the principle of dynamic programming, if the inventory position is x after the demand occurred and an optimal ordering policy is applied in the remaining n periods. (Sequence of the periods: $n, (n-1), \dots, 2, 1$.) Finally,

$$G(y) = c_1 \int_0^y (y-v_L) P_L(v_L)\, dv_L + P \int_0^\infty (v_L - y) P_L(v_L)\, dv_L$$

is the expected shortage cost and inventory holding cost during an inventory review period of average length.

Solution:

The solution is given by a discrete dynamic programming algorithm. It can be proved that if $c_3 = 0$, $n \to \infty$, then such an order level S is optimal, for which $G'(S) = 0$. If $c_3 > 0$, $n \to \infty$, then the (s, S) policy is optimal. (In this case, S is constant in time.)

202. Optimal inventory policies in the case of stochastic demands and linear ordering cost

Main codes:

$$1\ 1\ 0\ 1\ 1\ 1\ 5\ 1\ 1\ 1$$

Assumptions:

Unsatisfied demands are backordered in each period. Quantities ordered at the beginning of the i-th period will be delivered at the beginning of the $(i+L)$-th period, where L (leadtime) is a known non-negative integer. The demands $D_1, D_2, \dots$ arising in subsequent periods are independent, non-negative random variables with known continuous density functions $\varphi_1, \varphi_2, \dots$ The ordering cost $c(y)$ is linear; $c(y) = cy$, $c > 0$. The inventory holding cost $h(y)$ is a continuous, convex and increasing function of y. The shortage cost $p(y)$ is also a convex, continuous, increasing function. The author applies a discount factor α for the subsequent periods ($0 < \alpha < 1$). Furthermore, $\lim_{y \to \infty} p'(y) > c\alpha^{-L}$ is assumed.

284

Objective:

The goal is to minimize the expected discounted costs emerging in the periods
1, 2, ... and to determine the relevant ordering policies, respectively.

$$C(z, x_1, x_2,..., x_{L-1}; \varphi_1, \varphi_2, ...) = G(z, \varphi_1) +$$

$$+ \alpha G(z+x_1, \varphi_{1,2}) + \alpha^{L-1} G(z+x_1+...+x_{L-1}; \varphi_{1,L}) +$$

$$+ f(z+x_1+...+x_{L-1}; \varphi_1, \varphi_2, ...),$$

where

$$f(x, \varphi_1, \varphi_2, ...) = \min_{y \geq x} \{c(y-x) + \alpha^L G(y; \varphi_1, L+1) +$$

$$+ \alpha \int_0^\infty f(y-v; \varphi_2, \varphi_3, ...) \varphi_1(v) \, dv\};$$

$$G(y; \varphi) = \int_0^y h(y-v) \varphi(v) \, dv + \int_y^\infty p(v-y) \varphi(v) \, dv.$$

Solution:

The author shows that the optimal ordering follows an (s, S) policy, where
an order up to a "critical volume" is always placed. He does not deal with the
determination of the optimal s and S parameters; this may be accomplished
by dynamic programming.

203. Model of continuous inventory review

Main codes:

$$1\ 1\ 0\ 1\ 1\ 1\ 5\ 2\ 1\ 1$$

Assumptions:

The model considers a constant leadtime L; shortage is backordered. $Z(v)$
denotes the discounted expected cost for the future, if the system follows an optimal
policy and v is the initial inventory level. There is no shortage and inventory
holding cost, until the first ordered batch has not arrived. The fixed cost of ordering is c_3. In addition to the quantity and time-dependent inventory and shortage costs, a shortage cost h depending only on quantity emerges as well. The
demand process is stochastic (both as far as the interarrival times and its volume
are concerned). A demand of volume j will emerge with a probability $p(j)$.

Objective:

The discounted expected inventory holding and shortage costs for intervals
L and $L+t$: can be expressed by

$$f(v) = c_1 D(v) + hE(v) + c_2 B(v),$$

where:

$D(v)$ = expected inventory in the period $[L, L+t]$;
$B(v)$ = expected volume of shortage in the period $[L, L+t]$;
$E(v)$ = the expression involving both duration and volume of shortages.

Solution:

The functional to be considered equals

$$Z(v) = f(v) + \alpha \sum_{j=1}^{\infty} p(j) \{ \min_{y \geq 0} [c_3 \delta + cy + Z(v+y-j)] \},$$

where α is the expected value of the discount factor.

$\delta = \begin{cases} 0, & \text{if ordering takes place} \\ 1, & \text{if ordering does not take place} \end{cases}$

$c = $ procurement cost of unit item, y is the quantity ordered.

This functional differs from other previously presented dynamic programming problems, since y^* is the function of $v-j$, too, and not only that of v.

The authors show the optimality of the (s, S) policy. An explicit solution is given for the case when a Poisson process generates demands, and unit demands arise. An explicit solution is given for the general case and for the model of a stabilized state and periodic inventory review. The case of lost demands is analyzed, too.

204. Optimal policies corresponding to the shortage probability criterion for independent demands and arbitrary leadtime

Main codes:

1 1 0 1 1 1 5 1 1 1

Assumptions:

The inventory is reviewed periodically and a decision regarding ordering is made subsequently. The ordered batches arrive after a leadtime of length L. Demands emerging between two inventory reviewing points are stochastic. Unsatisfied demands are backordered. Demands arising in different periods are independent of each other and their distribution is not necessarily identical. The ordering cost is proportional to the quantity ordered.

Objective:

For an n-period inventory problem an ordering policy is sought which ensures for n periods the minimal expected discounted cost (with a constant discount factor) while the probability of shortage at the end of any period must not exceed a given value.

Solution:

First the authors determine the probability of shortage, then this probability will be expressed as a function of the inventory on hand and the quantity ordered. The ordering policy yielding minimal cost is determined on this basis.

286

205. *Inventory model in the case of independent and unknown demands*

Main codes:

1 1 0 1 1 1 5 1 1 1

Assumptions:

The inventory is periodically reviewed and a decision on ordering is made at the checkpoints. The ordered batches arrive with one period leadtime. The demand arising between two checkpoints has a binomial distribution with parameters x and p. x is the number of items at the beginning of the period, and p $(0<p<1)$ is the unknown parameter of the binomial distribution. Unsatisfied demands are backordered. Demands arising in the different periods are independent of each other. The ordering cost is proportional to the quantity ordered.

Objective:

It is identical with that of Models 194 and 204.

Solution:

A Bayesian procedure is developed for estimating the parameter p. By means of this estimation, the solution is reduced to the solution of Models 194 and 204.

206. *Inventory model for forecasting and dependent demands*

Main codes:

1 1 0 1 1 1 5 1 1 1

Assumptions:

The authors consider a single-item inventory model, in which demands arise in subsequent discrete periods and are not independent of each other. At deliveries, a constant leadtime $L>0$ is supposed. The demand distribution is F_n in the nth period, the ordering cost is denoted by $c_n(y)$, and the inventory holding and shortage cost is denoted by $G_n(y)$. The demand is forecasted for each period n by the estimation of the parameters of the distribution function F_n.

Objective:

The authors consider the holding, shortage and ordering cost (the sum of their expected value) during the forecasting time horizon as an objective function:

$$f_n(x) = \min_{y \geq x} \{c_n(y-x) - G_n(y) + E[f_{n+1}(y-F_n)]\},$$

and the (s, S) ordering strategy which minimizes $f_n(x)$.

Solution:

The solution of the problem may be accomplished by applying dynamic programming methods (either by decision or by value iteration). The authors illustrate the model by an example, where demand is represented by the failures of an operating system, while some of the failures can be repaired during time λ ($0 \leq \lambda \leq L$).

287

207. *Production control via inventory handling—an application of the Wiener filter theory*

Main codes:

$$1\ 1\ 0\ 1\ 1\ 1\ 5\ 2\ 1\ 1$$

Assumptions:

The author describes a continuous, linear, stochastic production-inventory model, in which there is one store and one item, and unsatisfied demands are backordered. The system may be described by the following equations:

1. $\dfrac{dI(t)}{dt} = v(t) - r(t),$

2. $v(t) = G_f\{n\},$

3. $n(t) = -G\{I\},$

where $I(t)$ is the inventory (state variable), $n(t)$ is the production volume (control variable), $r(t)$ is the demand, G_f is a time-independent linear operator partly characterizing the system (in the example shown by the author: the production delay), and G is a time-independent control operator. For example, the three equations can be summed as an

$$n(t) = - \int_{-\infty}^{t} G(t-t^1)\, I(t^1)\, dt^1$$

type linear control policy, where $G(t)$ is the "weight function" of G.

It is assumed that the demand is a stationary stochastic process with an expected value $E\{r(t)\}=0$ and auto-correlation function $R_{rr}=E\{r(t)r(t+\tau)\}$.

Objective:

The author applies a quadratic cost function:

$$Q = C_n E\{n^2\} + C_I E\{I^2\} + C_w E\{w^2\},$$

where the first two terms represent deviations (from given target values) of the production volume $n(t)$ and the inventory $I(t)$ as well as the cost consequences of these deviations.

In G_n denotes the differential operator, then $w(t)=dn(t)/dt$; thus, the last term in Q denotes the costs caused by the speed of change in the production rate.

Solution:

The system equation will be transformed by the Fourier-transformation first, then the author applies the Wiener–Hopf procedure in the course of solution.

288

VIII.1.2. Random Delivery

208. *Discrete dynamic ordering strategy*

Main codes:

$$1\ 1\ 1\ 1\ 1\ 1\ 5\ 1\ 2\ 3$$

Assumptions:

The capacity of the store is Q. Periodic ordering is possible. The demand of the nth period has a distribution $G(\bar{x})$. Unsatisfied demands are lost. The ordering strategy depends on the state of the inventory: there is either no ordering (d_0) or an order is placed to level y (d_y), $y \in [x, Q]$. Denote by p the probability of immediate delivery, $1-p$ is the probability that the delivery is delayed by one period. The cost $r(x, d)$ is the sum of the $r_1(x)$ inventory holding cost, $r_3(x)$ shortage and $r_2(x)$ ordering costs: all three are non-negative, bounded, monotonously non-decreasing functions of x, $r_3(0) = 0$. Furthermore,

$$\sum_{x=1}^{\infty} r_3(x) g(x) < \infty,$$

where $g(x)$ is the density function of demands for a period.

Objective:

A strategy minimizing the supremum of the expected cost is sought:

$$\varphi(x, d) = \lim_{n \to \infty} \sup \frac{1}{n+1} M_x^{\delta} \sum_{k=0}^{n} r(x_k, d_k),$$

where δ is the inventory control strategy, x is the initial stock and the expected value operator is denoted by M_x^{δ} referring to the total costs of n periods. The strategy δ^* is optimal, if

$$\varphi(x, \delta^*) = \inf_{\delta \in R} \varphi(x, \delta).$$

Solution:

If

$$p r_1(x) + p \sum_{y=x}^{\infty} r_3(y-x) g(y) - r_2(Q-x)$$

is a monotonously decreasing function of x, then the optimal strategy δ^* is of the (s, S) type, where $S = Q$. The determination of $x^* = s$ is not described.

209. *Brown's dynamic inventory system*

Main codes:

$$1\ 1\ 1\ 1\ 1\ 3\ 5\ 1\ 1\ 3$$

Assumptions:

This is a single-item control theoretical model described by stochastic differential equations. The author defines—after Brown—the concept of "maximally reasonable demand during the leadtime", consisting of the average demand arising during the leadtime plus the mean absolute deviation of demand (MAD) multiplied by a safety factor k. In each ordering period, this value is calculated and is defined as "normal ordering". The order to be placed is the difference between this value and the available stock. In this way, the probability of shortage occurrence can be controlled by modifying the safety factor. Demand and leadtime are random variables.

Objective:

There is no objective function, since the aim of the model is the state analysis of the system.

Solution:

A solution of the stochastic difference equations which describe the system and an examination of the system are presented.

VIII.2. Dynamic (s, q) Models

210. *A model with constrained order capacity*

Main codes:

$$1\ 1\ 0\ 1\ 1\ 1\ 2\ 1\ 1\ 0$$

Assumptions:

The differentiable density function of the demands of the inventory review period is denoted by $\varphi(\xi)$. Order quantities are constrained by a maximal volume R; this is chosen as the lot-size q according to the inventory policy. The objective is the determination of the reorder point. The sum of inventory holding and shortage costs is expressed by the function $L(y)$, where y is the inventory level after ordering; $L(y)$ is a convex function. Ordering has a fixed cost K and a volume-dependent cost r. There is no leadtime.

Objective:

The total cost of n periods is to be minimized:

$$C_n(x) = \min_{x}\left\{K\delta(y-x)+(y-x)\,r+L(y)+\alpha \int\limits_0^{\infty} C_{n-1}(y+\xi)\,\varphi(\xi)\,d\xi\right\};$$

$$C_0(x) \equiv 0,$$

where x is the inventory before ordering (the reorder point, the optimal value of which is sought), α is the discount factor,

$$\delta(z) = \begin{cases} 0 & \text{if } z = 0, \\ 1 & \text{if } z > 0. \end{cases}$$

Solution:

By applying the results of dynamic programming it can be proved that there is an order point s, for which the total cost of n periods takes its minimum, even in the case of $n \to \infty$. No method is given to determine the optimal level s.

290

211. *Optimal inventory policy for batch ordering*

Main codes:

1 1 0 1 1 1 2 1 1 1

Assumptions:

The model may operate during a finite or infinite number of periods. Demands in the single periods D_1, D_2, ... are all independent, identically distributed random variables with distribution function $\varphi(t)$. The inventory is reviewed at the beginning of the period, and if the inventory on stock and on order together do not reach a level k, then an order will be placed for such an integer multiple of a given unit Q (e.g. one truck load), that the inventory on hand plus on order will reach or exceed level k.

After λ periods the total ordered batch will arrive at the beginning of the next period. Demand arises—by supposition in one lot—at the beginning of the periods. Possible shortages are backordered.

The costs:

— ordering cost: $c \cdot z$ for z units (the cost arises at delivery);
— inventory holding cost and shortages cost $g(y, t)$: were y is the inventory after delivery; and t is the demand of the period.

With this notation, we can write

$$L(y) = \int\limits_0^\infty \int\limits_0^\infty g(y-u, v) \, d\varphi^\lambda(u) \, d\varphi(v)$$

the expected inventory holding and shortage cost in the $(i+\lambda)$th period, if the inventory level in the i-th period equals y, Φ^λ is the λ-fold convolution of Φ, and

$$G(y) = (1-\alpha)cy + L(y), \quad (0 < \alpha < 1 \text{ discount factor}).$$

Assume that $G(y) \to \infty$, if $|y| \to \infty$ and $-G(y)$ is unimodal. The decision maker will decide upon the ordering policy, i.e., the functions $Y = (Y_1, Y_2, ...)$ on the basis of the vector H_i of the "history" at the beginning of the i-th period (H_i contains the inventory levels before and after the orders of previous periods, and the demands). If x_i is the inventory level, then $Y_i(H_c) - x_i$ will be ordered (of course, this is the integer multiple of Q). The model is dynamic.

Objective:

The expected discounted cost of the $\lambda+1, ..., \lambda+n$ periods (if x_1 is the inventory level at the beginning of the first period and Y is the ordering policy) can be written as

$$f_n(x_1|Y) = \sum_{i=1}^n \alpha_{i-1} E\{G(y_i)\}.$$

The objective function in the finite case:

$$f_n(x_1|Y^*) = \min_y f_n(x_1|Y).$$

In the case of an infinite number of periods:

$$f(x_1|Y) = \lim_{n \to \infty} f_n(x_1|Y),$$

$$f(x_1|Y^*) = \min_Y f(x_1|Y).$$

If $\alpha = 1$, then:

$$a(x_1|Y) \equiv \lim_{n \to \infty} \frac{1}{n} f_n(x_1|Y); \quad a_1(x_1|Y^*) = \min_Y a(x_1|Y)$$

Solution:

It can be proven that the (k, Q) policy is optimal for both finite and infinite horizons (the same k in each period). Selection of k: Let $\bar{y}$ be the minimum of $G(y)$. In this case k is an arbitrary number, for which $k \leq \bar{y} \leq k+Q$ and $G(k) = = G(k+Q)$. It is a fundamental optimality condition of the (k, Q) policy that Q is a fixed, natural (e.g. transportation) unit lot. It can be verified that for the non-stationary case (distribution of demands as well as the cost function may alter from period to period) the generalized (k, Q) policy is optimal (the value of k will alter in each period), if $k_i - k_{i-1} \leq a_i$, where $a_i = \sup \{a|\varphi_i(a) = 0\}$.

212. Control theory model for gamma distributed production leadtime

Main codes:

$$1\ 1\ 1\ 0\ 1\ 3\ 2\ 2\ 1\ 3$$

Assumptions:

The production of a single item is continuously controlled, depending on the inventory level and the order volumes. The known demand is satisfied continuously from a store, shortage is backordered. The production time follows a gamma probability distribution.

Costs are not dealt with in the model, (only for verbal considerations, in connection with the modification of production and inventory levels and their volume).

Objective:

The stability of the system is investigated. The speed of eliminating "perturbations" (which means satisfying demand) and the possible minimization of the relevant inventory holding and production costs are described as the objective.

Solution:

By means of a Laplace transformation, the author examines the development of the decision operators (in the case of different demand processes), the changes of the modifying transformation characteristic to the system, and the stability of the system.

IX. Dynamic Ordering Time Models

The dynamic models belonging to this category are those in which the length of the order period is fixed. The subgroups below have been created according to the assumptions concerning the demands.

Models with a deterministic demand belong to the first subgroup. In this case, the volume of demand in each period is known (they may be different). An order is placed at the end of each period. The objective is to determine the optimal lot size to be ordered. Most models here are production-inventory models, namely, replenishment takes place by producing the appropriate amount. The ordering strategy, minimizing the total costs of a certain fixed number of periods, can be determined by means of dynamic programming. These models improve the efficiency of the traditional dynamic programming procedure by introducing various assumptions.

In the models belonging to the second subgroup, the demand of the period is not known at the time of placing an order; it is considered to be a random variable with an identical probability distribution in each period. An order is placed at the end of each period in order to replenish the inventory level up to a given order level. The objective is the determination of the optimal value of the order level, but—with some of the models—determination of order lot-size is a decision factor as well. The expected value of the total costs for a given finite (or infinite) horizon is to be minimized; these problems may be solved by a recursive relation based on dynamic programming theory.

In the models of the third subgroup, the demand is also considered to be a random variable with some distribution which may be different for different periods. The objective is the determination of the lot-size or the order level. The optimal value of this can generally be different in each period in accordance with the distribution of the demand in the different periods. The ordering policy, which minimizes the expected value of the total costs, is to be determined. For some of these models, only the type of the optimal ordering policy is determined, the specification of the optimal parameters may, in principle, be accomplished by applying dynamic programming. Nevertheless, this method requires so much calculation in practice that only a few periods can be taken into account simultaneously. Thus, simplified algorithms applied under special conditions, as well as procedures yielding approximate solutions are of great importance here.

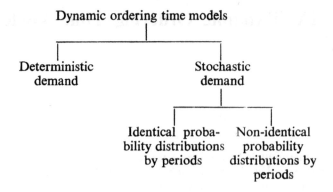

Dynamic ordering time models

Deterministic demand — Stochastic demand

Stochastic demand → Identical probability distributions by periods — Non-identical probability distributions by periods

IX.1. Dynamic Reorder-Time Models with Deterministic Demand

213. *Deterministic multiperiod production scheduling model with backlogging*

Main codes:

$$1\ 1\ 0\ 0\ 1\ 1\ 4\ -1\ 1\ 0$$

Assumptions:

The vector of demands $r=(r_1, r_2, ..., r_n)$ for n subsequent periods is known. The length of a period is fixed, the volume of demands may alter in each period. If shortage occurs, the demand will wait through α periods. If $\alpha=0$ then we have the lost demand case; if $\alpha>n$ there is a backorder case and, otherwise, demands remain through α periods. We seek the optimal production sizes for each period denoted by the vector $z=(q_1, q_2, ..., q_n)$. Instead of production, the model could be interpreted in terms of orders: in this case, the q_i, $i=1, 2, ..., n$, quantities mean the lot-size of ordering per periods.

The stock at the end of the i-th period can be expressed in the form:

$$I_i = \sum_{h=1}^{i} (g_h - r_h) \quad (i = 1, 2, ..., n).$$

If the value of I is negative, it indicates a waiting demand (we suppose that the initial stock is $I_0=0$). All demands are to be satisfied at most with α periods of leadtime, therefore we have

$$\sum_{h=1}^{i} g_h \geqq \sum_{h=1}^{i-\alpha} r_h$$

or equivalently,

$$I_i \geqq - \sum_{h=i-\alpha+1}^{i} r_h$$

to be fulfilled in the case of each $1 \leqq i \leqq n$, where $r_h=0$, if $h \leqq 0$. The assumption for the final stock of the planning horizon is $I_n=0$.

294

Objective:

A $P(z)$ production (ordering) cost belongs to the production (ordering) vector z in the i-th period.

The costs of inventory holding and shortage in the i-th period are expressed by the function:

$$H_i(z) = H_i[I_i(z)].$$

Assume that the function

$$F(z) = \sum_{i=1}^{n} [P_i(z) + H_i(z)]$$

expressing the total costs is concave in the intervals $(-\infty, 0)$ and $(0, \infty)$. The special form of the above cost function is not given, it may be selected as appropriate to the considered practical problem. Only the fulfilment of the above assumption must be ensured.

Solution:

A dynamic programming algorithm is applied which utilizes the special structure of the problem. The previously described constraints are linear, thus it is enough to deal only with the extremal points of the constraint set. This means that the following assumption can be given for the optimal production plan q_i $i=1, ..., n$:

$$q_i = \sum_{h=1+k_{i-1}}^{k_i} r_h,$$

where $0 = k_0 \leq k_1 \leq ... \leq k_n = n$ are integer numbers.

Let $F_i(k)$ be the minimal cost, if the production plan is optimal from the i-th to the n-th period, and the final stock of the $i-1$-th period is

$$I_i = \sum_{h=i+1}^{k} r_h$$

then, on the basis of the dynamic programming principle, the following recursive relation can be written:

$$F_i(k) = \begin{cases} P_i^*(K, k) + H_i^*(k) + F_{i+1}(k) & \text{if } k \geq i, \\ \min_{n \geq j \geq N} P_i^*(k_{ij}) + H_i^*(j) + F_{i+1}(j) & \text{if } k < i \end{cases}$$

where $N = \max\{k, i-\alpha\}$ and $\max\{0, i-1-\alpha\} \leq k \leq n$, furthermore,

$$H_i^*(k) = H_i \left(\sum_{h=i+1}^{k} r_h \right),$$

$$P_i^*(k, j) = P_i \left(\sum_{h=k+1}^{j} r_h \right)$$

are the forms of the cost functions expressed according to the new variable.

The values of k and j are integers for which the above equations have to be valid. On this basis, the possible optimal solution is to be sought on a rather limited set, increasing the efficiency of dynamic programming to a considerable extent.

According to the assumptions $I_n=0$ for the final stock, we have

$$F_n(k) = P_n^*(k, n) + H_n^*(n),$$

by which the value of $F_{n-1}(k)$ can be determined on the basis for each feasible integer k (considering also the constraints). This is summed up in a table, for each k, then the same is done in the case of $n-2$, $n-3$, etc., using the results of the previous step, and finally on the basis of the relation

$$F_1(0) = \min_{n \geq j \geq \max\{1-\alpha, 0\}} \{P_i^*(0, j) + H_i^*(j) + F_2(j)\},$$

the optimal value of $k_2, k_3, ..., k_n$ may obtained from the above mentioned table by means of the optimal value of $k_1=j$, progressing forward step-by-step. From this the optimal production (ordering) lot-size of the i-th period is

$$q_i = \sum_{h=1+k_{i-1}}^{k_i} r_h \quad (i = 1, 2, ..., n),$$

where $k_0=0$.

214. Non-stationary deterministic demand

Main codes:

$$1\,1\,0\ \ 0\,1\,1\,4\,0\,0\,0$$

Assumptions:

The demand is given as successive values of the total volume of consumption in the subsequent intervals of length T. These values are denoted by x_k $(k=1, 2, ..., n)$. The model does not allow shortages, thus only two cost types exist: s_k inventory holding and c_k ordering cost, both costs are volume-dependent. An order q_k is placed at the beginning of the k-th period. The ordered batch arrives immediately, without leadtime. The model is dynamic; this is ensured by taking into consideration the z_k inventory left over at the end of the periods.

Objective:

$$L_{nT} = \sum_{k=1}^{n} [C_k(q_k - z_k) + S_k(q_k - x_k)]$$

Solution:

Optimization takes place in accordance with Bellman's optimum principle, by which the decisions made in each interval are optimal for the remaining time intervals, independent of earlier decisions and the initial state, if the given sequence of decisions is optimal as a whole.

The author gives a solution for the case, when $c_k(u)$ and $s_k(u)$ are increasing and equal to 0 for $u=0$. He gives a block scheme for the calculation of the cost function values L_{nT} and the lot-sizes g_k.

296

215. Lot-size model with varying costs

Main codes:

$$1\ 1\ 0\ 0\ 1\ 1\ 4\ -1\ 0\ 2$$

Assumptions:

The model examines a finite number of periods. The demand rate is deterministic, shortage is not allowed. The unit procurement price (C) is constant, orders can be placed at n time moments. The leadtime is known, it may change in each period, but orders must not "cross" each other; i.e. an order placed at a later date will arrive later, too. The fixed cost of ordering A_j and the inventory holding cost factor I_jC may alter in each period. The order lots q_j are to be determined by minimizing the summed ordering and inventory carrying costs, for the whole planning horizon.

Objective:

$$K = \sum_{j=1}^{n} [A_j\delta_j + I_jCy_{j+1}],$$

where

$$\delta_j = \begin{cases} 0 & \text{if} \quad q_j = 0, \\ 1 & \text{if} \quad q_j > 0, \end{cases}$$

and

$$y_{j+1} = y_j + q_j - x_j,$$

y_j: initial stock of the j-th period,
x_j: demand of the j-th period.

Solution:

The solution can be found by the method of dynamic programming on the basis of the following function:

$$Z_k(\xi) = \min_{q_1 \cdots q_n} \sum_{j=1}^{k} (A_j\delta_j + I_jCy_{j+1}) \quad k = 1, ..., n$$

where

$$y_{k+1} = \xi$$

for all possible values of the state variable ξ.

Since both the initial and the final stock are fixed, the solution can be computed in both directions, starting with $k=1$ or $k=n$. The authors give some other assumptions which simplify the calculation and also include a table to aid calculations.

216. Capacity expansion with inventory

Main codes:

$$1\ 1\ 0\ 0\ 1\ 1\ 4\ -1\ 0\ 0$$

Assumptions:

T successive periods are examined with a known demand r at the end of each period. The demands may be satisfied by production and from the inventory.

Production takes place at the beginning of each period. Production capacity can be increased but cannot be decreased; production itself must not decrease in any of the periods.

Objective:

The optimal production and capacity value are to be determined in each period in such a way that the discounted total costs of production P_i, inventory holding H_i, capacity increase K_i and unutilized capacity C_i should be maximized. The cost function is assumed to be monotonously non-decreasing, concave.

Solution:

It is a dynamic programming algorithm with two variables. The number of feasible decisions (possible values) can be decreased in different ways, which improves considerably the efficiency of the dynamic programming.

217. Elementary dynamic inventory model

Main codes:

$$1 1 0 0 1 1 4 - 1 0 0$$

Assumptions:

The demand D_t for each period t is known at the beginning of the period. Replenishment of the inventory is ensured (inflow process) by the production x_t. The quantity of stock at the end of the t-th period is i_t. The cost of the t-th period depends on the production volume and the final stock. The variables are discrete, the final stock of the last period is 0. The cost functions in each period are non-linear. The cost function is given in the form $C_n(x, j)$, where x is the production and j is the final stock in the period, when there are n time periods still ahead.

Objective:

$$F_n(i) = \min_x \{(x, i+x-d_n) + F_{n+1}(i+x-d_n)\}, \quad n = 1, 2, ..., N.$$

where:

$F_n(i)$ is the optimal policy cost with an initial stock i and n further periods,
$\quad d_n$ is the demand of the period, when n periods still lie ahead.

Solution:

The problem may be solved by means of dynamic programming. The author illustrates the method with a numerical example. The calculation of the objective function follows the technique of progressing backwards in time. The author shows that the problem can be solved by a forward technique too.

218. Deterministic demand of changing intensity

Main codes:

$$1 1 0 0 1 1 4 - 1 0 0$$

Assumptions:

A demand z arises continuously with a periodically different, but known intensity in each period. The ordered quantity arrives immediately. If the delivery date of the ordered batch is the beginning of the periods, then the order points and lot-sizes will be obtained by the usual dynamic programming algorithm. If an order can be placed at any time, then the results obtained by dynamic programming can be improved by considering non-integer values of the time units for which a purchase order may be given. The leadtime is zero and shortages are not allowed.

Objective:

Ordering and holding cost is considered. The objective is to determine an improvement of the dynamic programming solution which will maximize the net decrease in inventory while the number of orders remains unchanged. This can be achieved by considering non-integer multiples of the basic time period. For two orders, the decrease in inventory is

$$Z(x) = d_1 x (T_1 - x) - D_2 x,$$

where T_1 is the time period covered by the first order, d_1 is the rate of demand nearing the end of T_1, D_2 is the total amount procured in the second order, and the first order covers the demand during $T_1 - x$.

Solution:

For two orders the optimal value of x is expressed by

$$x_0 = \min\left[\max\left(\frac{d_1 T_1 - D_2}{2 d_1}, 0\right), t_1\right],$$

where t_1 is the time in periods for which the rate of demand is constant (d_1). For more than two orders an iterative procedure is given based on the two-orders solution.

219. Deterministic lot-size model, the number of orders is limited

Main codes:

$$1\ 1\ 0\ 0\ 0\ 1\ 2\ -1\ 0\ 0$$

Assumptions:

The annual demand is given by the rates λ_j ($j=1, ..., n$) in n successive periods. A total of h orders can only be placed during the n periods. In the j-th year Q_j lots are ordered, for these amounts the inequality

$$\sum_{j=1}^{n} \frac{\lambda_j}{Q_j} \leq h$$

has to be fulfilled.

There is no ordering cost, the specific cost of inventory carrying is constant, and the cost factor for the average inventory is $I_j C_j$; ($j=1, ..., n$) $/volume. Whether the number of orders is integer or not is ignored. Shortage is not allowed.

Objective:

$$K = \sum_{j=1}^{n} I_j C_j \frac{Q_j}{2} \rightarrow \min_{Q}, \quad j = 1, ..., n.$$

Solution:

In this model, the constraint is always active, thus the following Lagrange function is to be minimized:

$$j = \sum_{j=1}^{n} I_j C_j \frac{Q_j}{2} + \eta \left(\sum_{j=1}^{n} \frac{\lambda_j}{Q_j} - h \right),$$

where $\eta > 0$ is the Lagrange multiplier.

The optimal Q'_js and η can be calculated from the derivatives of the function according to Q_j and η. Here η^*, the optimal value of η, can be interpreted as the cost of placing an order.

220. Planning and forecast horizons for the dynamic lot-size model in the case of constant production costs

Main codes:

1 1 0 0 1 1 0 1 1 0

Assumptions:

A single item is produced in period i in the amount x_i. Production and satisfaction of demands take place in each period. Constant K_i and variable c_i costs arise in the course of production. They may alter from period to period. The production cost has the form $K_i \delta(x_i) + c_i x_i$. The shortage cost is concave and non-increasing, and the inventory carrying cost is a concave and non-decreasing function of the stock on hand at the end of the i-th period. It is denoted by $W_i(y_i)$ and designates the holding cost if the end-of-period inventory $y_i \geq 0$, and the shortage cost if $y < 0$.

Objective:

The optimal production plan $(x_1, x_2, ..., x_t)$ is achieved by minimization of the production holding costs during t production and demand covering periods. It requires the minimization of

$$\sum_{i=1}^{t} K_i \delta(x_i) + W_i(y_i), \quad \text{where} \quad y_i = y_{i-1} + x_i - d_i$$

with a known demand d and an initial stock y_0.

Solution:

The problem may be considered as a concave programming problem, but, in this case, an optimal decision can be made in the first period only if demands and costs for the whole t-period horizon are known, which is an assumption valid often only for a short period.

Based on the information up to a $t_1(<t)$ period, the authors find conditions under which the optimal quantities produced in the 1, 2, ..., t_1-1 period also remain optimal for t_1+j-period problems, in the case of arbitrary $j>0$. If these

300

conditions are fulfilled, then t_1 is called the forecast horizon and (t_1-1) is called the planning horizon concerning any problem of (t_1+j)-period length. The authors developed an algorithm for the calculation of the optimal policies.

221. Planning and forecast time horizons for the dynamic lot-size model in the case of varying production costs

Main codes:

$$1 1 0 0 1 1 0 1 1 0$$

Assumptions:

The following differences are to be taken into account in comparison with the previous model:
1. the production cost is a function of x_i, the quantity produced at the beginning of period i

$$p_i(x_i) = K_i \delta(x_i) + c_i x_i, \quad i = 1, 2, \dots;$$

2. the inventory holding and shortage costs are linear

$$w_i(y_i) = h_i \max(y_i, 0) - b_i \min(y_i, 0),$$

where y_i is the inventory on hand ($y_i < 0$ in case of shortage)

$$\delta_i(x) = \begin{cases} 0 & \text{if } x = 0 \\ 1 & \text{if } x > 0 \end{cases}$$

finally, h_i and b_i are unit inventory holding and shortage costs which may be different in each period.

Objective:

It is identical with the objective function of the preceding model, and has the form

$$\min_{x_i} \sum_{i=1}^{t} p_i(x_i) + W_i(y_i) \quad \text{with} \quad y_i = y_{i-1} + x_i d_i,$$

with a known demand d_i and initial stock y_0.

Solution:

This is based on the forward-proceeding algorithm, elaborated by the author, which can be applied in the case of both the previous model and this model, respectively.

222. Planning horizons for a stochastic leadtime deterministic demand inventory model

Main codes:

$$1 1 1 0 1 1 9 -1 1 3$$

Assumptions:

The known D_i volume of demand arises at P not necessarily equidistant moments $(\eta_1, \eta_2, \dots, \eta_i, \dots, \eta_p)$. Each of the single demands is a "special demand", therefore they can be satisfied only by a corresponding production order. This is important, because the orders arrive after a stochastic leadtime of length L (with density function $g(L)$, $L_{min} \leq L \leq L_{max}$). This way, goods may be received for a later order earlier than for an earlier one; nevertheless, earlier demands cannot be satisfied from these possible deliveries.

The costs:

c_1 inventory holding cost,
c_2 shortage cost per one unit of item and unit time,
K fixed production cost.

Shortage is backordered, demands wait.

It can be verified that the optimal policy is always the same, as if some successive demands are drawn together into certain groups, and a production process is carried out for these. Logically this grouping is the first task, but, as far as the calculations are concerned, it is the second task. The other task is to determine the starting points of the production process. (Actually this is the time of placing the order; the production period is the leadtime.)

Objective:

$F(j)$ is the optimal expected cost if demand arises at j different time moments.

$$V(l, T, m) = F(l) + K + EIC(T, l+1, m),$$

where $EIC(T, l, m)$ is the expected inventory cost of the joint production for $D_i, \ldots, D_m$ demands:

$$EIC(T, l, m) = \sum_{i=l}^{m} D_i \left\{ c_1 \int_{L_{min}}^{\eta_i - T} (\eta_i - T - L) g(L) \, dL + \right.$$

$$\left. + c_2 \int_{\eta_i - T}^{L_{max}} (T + L - \eta_i) g(L) \, dL \right\}$$

and

$$F_j = \min_{0 \leq i \leq j-1} \left[\min_{T^*(i) < T \leq T^*(j)} \{ V(i, T, j) \} \right],$$

where for the optimal value $T^*(l, m)$ the following equation holds

$$\sum_{i=1}^{m} D_i G[\eta_i - T^*(l, m)] = \frac{c_2}{c_1 + c_2} \sum_{i=1}^{m} D_i$$

(G denotes the distribution function of the leadtime).

Solution:

1. Determination of $F(j)$ by dynamic programming.
2. T^* can be determined by some numerical technique (e.g. by Newton–Raphson iteration).

IX.2. Dynamic Reorder-Point Models with Stochastic Demand

IX.2.1. Random Demands with Identical Distribution in Each Period

223. Stationary demand n-period model for lost sales case

Main codes:

$$1\ 1\ 0\ 1\ 1\ 1\ 1\ 1\ 2\ 0$$

Assumptions:

The quantities to be ordered are to be determined for n subsequent periods. The demand arising in each period is a random variable with a distribution function $F(\xi)$ in all periods (i.e. the demand is stationary). The initial inventory level x is to be increased up to a level $y=x+z$ by an order z at the beginning of the period. There is no leadtime. Ordering cost is cz, the cost of inventory holding and shortage is $L(x+z)$ with the assumption that this is a convex function. If a demand cannot be satisfied immediately, it is lost.

Objective:

The total expected cost discounted by a factor α for n periods is:

$$C^n(x) = \min_{z \geq 0} \left\{ cz + L(x+z) + \alpha C^{n-1}(0) \int\limits_{x+z}^{\infty} dF(x) + \right.$$
$$\left. + \alpha \int\limits_{0}^{x+z} C^{n-1}(x+z-\xi)\, dF(x) \right\},$$

where C^{n-1} means the cost for $n-1$ periods.

Solution:

It can be shown that the optimal strategy is ordering up to an $\bar{x}_n$ level in a period n. The order level alters in each period, and $\bar{x}_1 \leq \bar{x}_2 \leq ... \leq \bar{x}_n$ form a convergent sequence which tends to $\bar{x}$. The author proves this by means of dynamic programming but does not give a numerical procedure to determine $\bar{x}_n$. The value of $\bar{x}$ can be obtained from the equation

$$c + L'(\bar{x}) - \alpha c F(\bar{x}) = 0.$$

224. Stationary demand model of n-periods for backordered demands

Main codes:

$$1\ 1\ 0\ 1\ 1\ 1\ 1\ 1\ 1\ 0$$

Assumptions:

All conditions of the previous Model 223 hold, except that shortage is back-ordered.

Objective:

The total cost discounted by the factor α for n periods is minimized:

$$C^n(x) = \min_{z \geq 0} \left\{ cz + L(x+z) + \alpha \int_0^\infty C^{n-1}(x+z-\xi) \, dF(\xi) \right\},$$

where C^{n-1} denotes the minimal total cost for $n-1$ periods.

Solution:

The authors prove, by means of dynamic programming, that the optimal policy is again ordering up to the level $\bar{x}_n$ in the n-th period, where $\bar{x}_1 \leq \bar{x}_2 \leq \ldots \leq \bar{x}_n$ constitute a convergent sequence with limit point $\bar{x}$. They do not give a numerical procedure to determine the order levels $\bar{x}_n$. The value of x can be determined by the equation

$$C(1-\alpha) + L'(x) = 0,$$

where $C = \lim_{n \to \infty} C^n$.

225. Random demands with varying distributions by periods

Main codes:

$$1\ 1\ 0\ 1\ 1\ 1\ 1\ 1\ 2\ 0$$

Assumptions:

An order can be placed in each period. The known density function of the demand of the i-th period is $\varphi_i(\xi)$, and this may alter in each period. The ordered lots arrive immediately and demands also emerge in one batch at the beginning of the respective period. The inventory holding and shortage cost $h(x)$ and $p(x)$ are convex functions of the inventory level x, $L(x, \varphi_i)$ denotes the expected value of the sum of $h(x)$ and $p(x)$. It is supposed that $h'(0) + p'(0) - \alpha c < 0$, where the ordering cost factor is quantity-dependent. There is no fixed order cost. Discounting takes place by applying a discount factor $0 < \alpha < 1$.

Objective:

The minimum of the expected discounted total costs determined for a number of periods, according to the principle of dynamic programming, can be expressed by

$$C(x, \varphi_1, \varphi_2, \ldots) = \min_{z \geq 0} \left\{ cz + L(x+z, \varphi_1) + \alpha C(0, \varphi_2, \varphi_3, \ldots) * \right.$$

$$\left. * \int_{x+z}^\infty \varphi_1(\xi) \, d\xi + \alpha \int_0^{x+z} c(x+z-\xi, \varphi_2, \varphi_3, \ldots) \, \varphi_1(\xi) \, d\xi \right\},$$

where x is the inventory level, and z is the ordered quantity.

Solution:

It is shown by means of dynamic programming that the optimal strategy is ordering up to an S_i level with $w = S_i$ altering from period to period. The value of this is given by the minimum of the convex function:

$$C(w, \varphi_1, \varphi_2, \ldots) = cw + L(w, \varphi_1) + \alpha C(0, \varphi_2, \varphi_3, \ldots) *$$

$$* \int_w^\infty \varphi_1(\xi)\, d\xi + \alpha \int_0^w C(w-\xi, \varphi_2, \varphi_3, \ldots)\, \varphi_1(\xi)\, d\xi.$$

The author does not give a specified algorithm to determine this minimum.

226. Dynamic model, the parameter of demand distribution is unknown

Main codes:

$$1\ 1\ 0\ 1\ 1\ 1\ 1\ 1\ 1\ 0$$

Assumption:

The demand of each period follows a gamma distribution with density function:

$$F(\xi|w) = \frac{w(w\xi)^{a-1}e^{-w}}{\Gamma(a)},$$

where a is a known constant; w is an unknown parameter with density function
$$f(w) = \frac{\lambda(\lambda w)^{b-1}e^{-\lambda w}}{\Gamma(a)}$$
(λ and b are known constants). N past periods were observed, the demands in which were $\xi_1, \ldots, \xi_N$. The parameter w is estimated on the basis of the statistics $s = \sum_{i=1}^{N} \xi_i/N$. Demands arise at the beginning of each period. There is no leadtime. The final stock is x, and the ordered quantity is y. The ordering cost c is quantity-dependent. There is no fixed ordering cost. The expected value of the inventory carrying and shortage cost in one period is $L(y, N)$, which is a convex function of y.

Objective:

The minimum of the expected value of total cost, discounted by the factor α, can be written according to the dynamic programming principle in the following form:

$$C(x|s, N) = (s+\lambda)c\left[\frac{x}{s+\lambda}, N\right],$$

where the following recursive formula is valid for $C(x, N)$:

$$C(x, N) = \min_{y \geq x}\left\{c(y-x) + L(y, N) + \alpha \int_0^\infty C\left[\frac{y-t}{1+t}\right](1+t)\varphi(t, N)\, dt\right\}.$$

Solution:

A recursive procedure based on the dynamic programming principle is given for the determination of the optimal cost function and the quantity to be ordered. It is proved that the optimal strategy is ordering up to a level S.

227. Periodic inventory review model with backordering

Main codes:

1 1 0 1 1 1 4 1 1 1

Assumptions:

The model investigates the steady state of a system, where the average rate of the demand process r is stochastic, but its distribution function does not change in time. The system operates for an infinitely long time. Shortage is allowed, and backordered. Costs: inventory review cost J, ordering cost C_3, unit price C (constant, independent of the quantity ordered), the inventory carrying charge is I, the shortage cost $\pi + c_2 t$. T is the length of one period. The leadtime is constant. The demand is independent in each period, and the probability of x units of demand in a period is $p(x, T)$. The present values of the future costs are the same at each inventory review point, i.e., there is no cost discounting. The variables are discrete. The quantity ordered in each period is q.

$$f(\xi+q, T) = ICD_T(\xi+q, \omega) + \pi E_T(\xi+q, T) + c_2 B_T(\xi+q, T),$$

with the notations:

ICD_T: expected value of inventory holding cost,
$\pi E_T + c_2 B_T$: expected value of shortage cost

from the delivery up to the end of the period, where both are proportional to quantity and time.

Solution:

The solution can be determined by applying dynamic programming:

$$Z(\xi, T) = \min_q \left[J + A\delta + Cq + f(\xi+q, T) + a \sum_{z=0}^{\infty} P(x, T) Z(\xi+q-x, T) \right],$$

where

$$\delta = \begin{cases} 0 & \text{if } q = 0, \\ 1 & \text{if } q > 0. \end{cases}$$

Remark: a functional appears in the solution (it contains an unknown function). The authors denote $Z(\xi, T)$ the basic functional for periodic inventory review models. In the case of continuous variables, an integral replaces the sum in $Z(\xi, T)$, while the discrete distribution is replaced by a continuous density function.

228. Periodic inventory review model with lost sales

Main codes:

1 1 0 1 1 1 4 1 2 1

Assumptions:

The decision for ordering is made at inventory reviewing at equal intervals. A finite planning horizon consisting of T periods is considered. The ordered

306

batch arrives after L periods of leadtime. Demand for short quantity is lost. The following process is considered for each period t:

— a decision for ordering is made,
— delivery of the ordered batch arrives (order of the $(t-L)$-th period),
— the consumer's demand arises; $p_t(x)$ denotes the probability that the demand of the t-th period is x ($x \geq 0$ integer variable),

The ordering cost is $C_t(q)$ (ordering of quantity q in the t-th period), inventory holding and shortage cost in the t-th period is $H_t(k, x)$, where

$$H_t(k, x) = \begin{cases} k_t(k-x), & \text{if} \quad (k-x) \geq 0 \\ \pi_t(x-k), & \text{if} \quad (k-x) < 0 \end{cases} \quad t = 1, 2, \ldots, T.$$

Objective:

The following recursive relation will be satisfied by the ordering strategy which minimizes the expected value of costs for the t-th period:

$$f_t(j) = \min_{y \geq j} \left\{ c_t(y-j) + L_t(y) + \alpha \sum_{x=0}^{\infty} f_{t+1}(y-x) p_t(x) \right\},$$

where

$$f_{T+1}(j) \equiv 0 \quad t = 1, 2, \ldots, T;$$

$$L_t(y) = \begin{cases} \sum\limits_{x=0}^{y} h_{t+\lambda}(y-x) p_{t,t+\lambda}(x) + \sum\limits_{x>y} \pi_{t+\lambda}(x-y) p_{t,t+\lambda}(x) & \text{if} \quad y \geq 0, \\ \sum\limits_{x=0}^{\infty} \pi_{t+\lambda}(x-y) p_{t,t+\lambda}(x) & \text{if} \quad y < 0; \end{cases}$$

Here the following notations are used: $j_t = j$ is the inventory level before decision making and $y_t = y$ is that of after decision making. Thus the ordered quantity $q_t = q$ is $q_t = y_t - j_t$ and $p_{t,t+1}$ is the probability that the total demand of the $t, t+1, \ldots, t+d$ periods is x (i.e. $x = x_t + \ldots + x_{t+d}$ will be fulfilled).

Solution:

The quantity q_t to be ordered (an integer number) may be calculated by means of dynamic programming for every possible inventory level j (integer) in the backward order $t = T, T-1, \ldots, 1$.

229. Infinite horizon production-inventory model

Main codes:

1 1 0 1 1 1 4 0 1 0

Assumptions:

The demand for the item in the i-th period is D_i. Demands in each period are independent, identically distributed, non-negative integer random variables with expectation μ ($0 < \mu < \infty$).

The initial stock of the i-th period is x_i, the quantity to be produced is q_i. Unsatisfied demands wait.

At the beginning of the i-th period $H_i=(q_{n+1}, y_n, \ldots, y_{i+1}, x_n, \ldots, x_i, D_n, \ldots, D_{i+1})$ describes the operation of the production-inventory system between the dates n and i $(n<i)$. The decision variable is $y_i=x_i+q_i$.

The inventory holding and shortage cost of the i-th period is $g(y_i, D_i)$, the expected value of this is $L(y)=Eg(y, D)$ and is assumed to be convex. The production cost per item is c, production smoothing costs $p \geqq 0$ in the case of increasing production, and $\gamma \geqq 0$ if production decreases.

Objective:

The expected discounted production, smoothing and inventory holding cost:

$$E\left\{ \sum_{i=1}^{n} \alpha^{n-1}[c(y_i-x_i)+g(y_i, D_i)+d(q_i-q_{i+1})+e(q_i-q_{i+1})]-c\alpha^n x_0 \right\},$$

where

$$d = \frac{(\gamma+p)}{2} \quad \text{and} \quad e = \frac{(\gamma-p)}{2}.$$

Solution:

The model determines the optimal value of y by dynamic programming. The state equation of the problem is:

$$f_i(x, q) = \min_{y \geqq x} \{d|y-x-q|+J_i(x, y)\},$$

where $J_i(x, y)=g(y)+\alpha E\{f_{i-1}(y-D, y-x)\}$. The optimal solution has the following form:

$$y = \begin{cases} y_1(x) & \text{if } x < y_1(x)-q, \\ x+q & \text{if } y_1(x)-q \leq x < y_2(x)-q, \\ y_2(x) & \text{if } y_2(x)-q \leq x \leq y_2(x) \\ x & \text{if } y_2(x) < x; \end{cases}$$

where

$$y_j(x) = \sup \{y|J^{(2)}(x, y) \leq (-1)^j d\}, x \in R \cdot J^{(2)}(x, y) = J(x, y)-J(x, y-1).$$

230. An application of a servo-mechanism to an inventory system with leadtime

Main codes:

$$1\ 1\ 0\ 1\ 1\ 3\ 4\ 1\ 1\ 1$$

Assumptions:

Ordering (demand) periods of equal length are defined where the t-th period lasts from $(t-1)$ to t. Inventory is reviewed and order is placed at the end of each period. A quantity q_t ordered at time t arrives in one batch after L periods at the end of the $(t+L)$-th period. X_t is the random demand of the t-th period. I_t denotes the inventory level at the end of the t-th period that is in the store as surplus or shortage compared to a prescribed safety stock. Our objective is to keep I_t as close as possible to the safety stock in time. Shortage is backordered.

308

The following relation is valid:

$$I_t = I_{t-1} + q_{t-L-1} - x_t \quad (t = 1, 2, 3, \dots).$$

It is assumed that X_t is generated by a stochastic process where demands are mutually independent with the common variance a^2 and given expectation function m_t $(t=1, 2, \dots)$.

The set of the possible ordering policies is limited to the family of linear decision rules corresponding to the above assumptions. The lot to be ordered at time t is expressed by

$$q_t = \sum_{j=0}^{t} A_j X_{t-j} + \sum_{j=0}^{t} B_j I_{t-j} \quad (t = 1, 2, 3, \dots),$$

where A_j, B_j are constants to be determined.

Objective:

To assure a minimal deviation from the prescribed safety stock, i.e., the control of the processes in such a way that the control variable (I_t) deviates from the standard value (0) to the least possible extent. This can be done in the model by selecting q_t (i.e. A_j and B_j, respectively).

Solution:

The Z transformation known from control theory is applied. The forecasting of X_t is also required for the ordering rule. It can be proved that exponential smoothing yields a prediction, corresponding to the assumptions, in the following form:

$$S_0 = X_0$$

$$S_t = \alpha X_t + (1-\alpha) S_{t-1} \quad (t = 1, 2, 3, \dots),$$

where α $(0 < \alpha < 1)$ is the smoothing coefficient.

It is assumed that the expected value of demand is constant in time, $m_i = a$. In this case, S_t is an a_t estimation of a and

$$q_t = \alpha(L+1)(X_t - a_{t-1}) + X_t.$$

231. Critical number ordering policies for perishable items

Main codes:

$$1\ 1\ 0\ 1\ 1\ 1\ 1\ 1\ 1\ 0$$

Assumptions:

The volume of demand is D_n in the n-th period $(n=1, 2, \dots)$. The D_n's are independent, identically distributed, non-negative random variables, their distribution function is continuous, strictly monotonously increasing, with continuous density. Demands arise in one batch after delivery. At the beginning of every period an order raises the inventory up to the level S. Inventory is reviewed at the beginning of each period.

The item is perishable: it may be used up during m periods, afterwards it has to be sorted out at cost v. The age of every unit of item is registered in the inventory, and though the same utility is attached to them (independent of age), demands are satisfied according to the FIFO system (always from the oldest stock available). The new stock arrives always from a fresh source assumed to be inexhaustible, at unit procurement cost C. The order of events in a period is: ordering, arising of demand, decreasing of shelftime and sorting out. Shortage is backordered.

The costs:

h = inventory holding cost;

p = shortage cost (h and p depends in each period on the volume of inventory and shortage, respectively);

c = procurement cost per unit item;

v = sorting out cost.

Objective:

This can be given in a closed form only in the case of $m=2$, but the author does not even present this explicitly. The objective is to minimize the expected average cost for one period.

Solution:

The Markov-chain model of the complete outflow in one period (demand+ +sorting out), as a stochastic process, has a constant probability distribution; the expected cost may, in principle, be written as a function of this and the known distribution of D_n. Nevertheless, this needs the solution of an integral equation which can be given only for the case $m=2$. In this case, the optimal S^* can be calculated. For the case of $m=3$, the author calculated the objective function values as a function of S for several discrete distributions, by successive approximation.

232. High-order approximations for ordering in the case of perishable items

Main codes:

$$1\ 1\ 0\ 1\ 1\ 1\ 1\ 1\ 1\ 0$$

Assumptions:

The duration of life of inventories in the system is n periods; after this time, stocks become obsolete. Demands emerging in subsequent periods are random variables independent of each other, with an identical distribution function F and density function f. $x=(x_{m-1}, x_{m-2}, ..., x_1)$ denotes the volume of inventories having a different span of life at the beginning of each period, where x_j is the available quantity which will be obsolete after j periods. The decision variable (y) is the quantity of the ordered new item. FIFO policy is applied, i.e., always the oldest item is used up first. Shortage is backordered.

Objective:

Convex storage and shortage costs (L) are present in the model with order costs (c), proportional to the ordered quantity, and sorting out costs (v), proportional to the stock having become obsolete (v arises at the end of the periods).

$$c_n(x) = \min_{y \geq 0} \left\{ cy + L(x+y) + v \int_0^y G_m(u, x)\, du + \alpha \int_0^\infty c_{n-1}(y, x, t) f(t)\, dt \right\},$$

where

$$x = \sum_{j=1}^{m-} x_j,$$

and $0 \leq \alpha \leq 1$ is the discount factor.

Solution:

The direct calculation of the optimal policy would mean the solution of a dynamic programming problem for an $(m-1)$-dimension state variable. If $m > 2$, then both the calculation of the optimal policy and its implementation are rather difficult. The author developed an heuristic estimation, which requires the solution of a problem of reduced dimension.

IX.2.2. Stochastic Demands of Different Distribution by Periods

233. Dynamic model with demands of different distribution and arbitrary number of periods

Main codes:

$$1 \ 1 \ 0 \ 1 \ 1 \ 1 \ 4 \ 1 \ 1 \ 1$$

Assumptions:

The demand of the i-th period is X_i with a density function f_i. The final stock is z, the ordered quantity q is delivered in one batch after L (integer number) periods of leadtime. Demands arise at the beginning of the period in one batch. The functions $h(z)$ and $p(z)$ give the inventory carrying and the shortage costs, the expected value of their sum is $L(q+z, f_i)$ in the i-th period. Assume that the functions $h(z)$ and $p(z)$ are convex, and $L'(0, f_i)+c<0$, where c means the procurement cost which is quantity-dependent. There is no fixed ordering cost.

Objective:

The minimal expected total cost, discounted by the factor α for an arbitrary number of periods, may be obtained according to the dynamic programming principle, applying the following recursive formula:

$$C(z, q_1, q_2, ..., q_{L-1}, f_1, f_2, ...) =$$

$$= \min_{q \geq 0} \left\{ cq + L(z, f_1) + \alpha \int_0^\infty C(z+q_1-x, q_2, ..., q_{L-1}, q, f_i, ...) f_1(x) \, dx \right\}.$$

The optimal $q_L = q$ order volume and the optimal final stock z at the end of the L-th period can be obtained from the above relation.

Solution:

The optimal size of the batch to be ordered may alter in each period. It can be determined by means of dynamic programming.

21*

234. *Dynamic lot-size model with a stochastic demand process and a fixed time horizon*

Main codes:

$$1\ 1\ 0\ 1\ 1\ 1\ 4\ 1\ 1\ 2$$

Assumptions:

We consider a finite number of periods. The demand is stochastic, and shortage is backordered. The leadtime L_j is constant, altering in each period but orders do not cross each other: batches ordered earlier will arrive earlier, too. The lot ordered even in the n-th period will arrive within the examined horizon. The initial stock of the first period is given. In an arbitrary interval T_j the demand has a Poisson distribution with parameter λ_j. Cost factors in the j-th period: order placing cost is A_j; the price of ordering q_j quantity is $C_j(q_j)$, this function may be non-linear. The inventory holding cost of a unit is $I_j\hat{C}_j$, where $\hat{C}_j$ is the average unit procurement cost. π_j is the shortage cost proportional only to quantity, whereas $\hat{\pi}_j$ is the shortage cost proportional to quantity and time. The stock left over at the end of the horizon will be sold, V is the sales price of one unit.

Objective:

$$K = \sum_{d_j \geq 0} \left[\prod_{j=0}^{n} p(d_j;\ \sigma_j T_j) \right] \left[\sum_{j=1}^{n} A_j \delta_j + C_j(q_j) \right] +$$

$$+ I_j \hat{C}_j \left[y_j + q_j - \frac{\lambda_j}{2} (T_j + L_{j+1} + L_j) \right] + \pi_j E(y_j + q_j, T_j + L_{j+1}, L_j) +$$

$$+ [(T_j + L_{j+1} - L_j)\hat{\pi}_j + I_j \hat{C}_j] B(y_j + q_j, T_j + L_{j+1}, L_j) - G(y_n + q_n)],$$

where $y_{j+1} = y_j + q_j - d_j$ is the initial stock in period $j+1$ (d_j is the expected volume of demand);

$$\delta_j = \begin{cases} 0 & \text{if } q_j = 0, \\ 1 & \text{if } q_j > 0; \end{cases}$$

$E(y_j + q_j, T_j + L_{j+1}, L_j)$ is the expected value of shortage in the j-th period; $G(y_n + q_n)$ is the expected profit from selling one lot; $(T_j + L_{j+1} - L_j)B(y_j + q_j, T_j + L_{j+1}, L_j)$ is the expected length of the shortage in the j-th period.

Solution:

The determination of $q_{j0}(y_j)$ is accomplished by the method of backward dynamic programming. The authors give the recursion formulae required for the calculation and analyze the possibilities of simplifying the procedure. Similar formulae can be given also for the case when the leadtime is random.

235. Dynamic lot-size model with a stochastic demand process and a variable horizon

Main codes:

1 1 0 1 1 1 4 1 1 2

Assumption:

The time horizon is a discrete random variable with a known discrete distribution function $p_i i = 1, ..., m$. The cumulated average demand as well as the dates when orders can be placed depend on the actual value of this variable. Other assumptions of the model are identical with those of the previous model depending on the length of the horizon.

Objective:

$$\min_{q_1} \left\{ \sum_{i=1}^{m} p_i w_i(y_1 + q_1) \right\},$$

where y_1 is the initial stock, $w_i(y_1 + q_1)$ is the expected total cost, if q_1 units are ordered in the "first" period and an optimal policy is followed afterwards.

Solution:

The solution can be obtained by means of dynamic programming (in a similar manner to the previous model). A solution has to be provided for m different cases, where m is the number of elements of the time horizon.

The authors also analyze the possibilities of forecasting the rate of future demands (by time series analysis, project analysis, economic forecast).

236. Lot-size system in the case of a periodically varying random demand

Main codes:

1 1 0 1 1 1 4 1 2 0

Assumptions:

The system operates through an infinite time horizon. A seasonal demand of random size arises for the given item, consequently the changes of the inventory can be described by a periodic, but non-stationary Markov process of discrete time, the transition probabilities of which periodically alter in time.

The paper presents the model by a numerical example. Three types of inventory state are possible: there are 0, 1 or 2 items on hand. Demand arises at the beginning of each period of fixed length, but only after placing the order (if any) and after the delivery of the ordered batch; its volume may be 0, 1 or 2. Two periods make one cycle, the probability of demand for 0, 1, 2 items is different in these two periods. Demands at shortage are lost. The decision variable is the quantity to be ordered at the beginning of the period, depending on the stock on hand as well as on the period. The total ordered batch arrives without leadtime. Therefore $0 \leq q \leq 2 - i$, where i is the number of items on hand, and q is the ordered quantity. (No more than two item units can be on hand in the store.)

The cost factors also vary in the two periods of the cycle:

— r_k is the unit profit in the k-th period ($k=1, 2$);
— $c_k(q)-q$ is the unit ordering cost in the k-th period (consisting of the one part proportional to the ordered quantity and one constant part);
— h is the unit inventory carrying cost in one period;
— shortage cost;
— β discount factor.

Objective:

The author does not give an explicit expression for the objective function, but seeks the ordering policy which results in a maximal expected profit.

Solution:

1. The author provides the transition probabilities $p_{ij}^q(k)$ as a function of q, k, i, j (i, j: states).

2. The $a_i^q(k)$ direct profits are calculated on the basis of the cost factors.

3. Applying a decision iteration method, the optimal matrix A is determined, which contains in the possible inventory states the ordered quantities in each period. As we have no knowledge of the optimal policy in advance, first the decision matrix is chosen maximizing the direct profit. Then the matrix is improved by carrying over the consequences of this on the "screen" of discounting until (nearly) the same matrix A is obtained in the last two steps.

237. Dynamic, stochastic order level system of two ordering periods

Main codes:

$$1\ 1\ 0\ 1\ 1\ 1\ 1\ 1\ 1\ 0$$

Assumptions:

The operation of the system extends through two ordering periods of prescribed length t_p. Only two decisions have to be made: these concern the order level of the two periods. Stochastic demand arises in the first period with a density function $g(y)$ and, in the second period, with a density function $f(x)$ in one batch at the beginning of the respective ordering periods. The ordered quantity arrives without leadtime in one batch at the beginning of the ordering period, before demand arises. S' is the order level of the first period, S is that of the second period. If $S'>S$, and does not go below S after the emergence of y, then the initial inventory level in the second period will not be S but $S'-y$. Also this possibility is taken into account during optimizing; thus S' and S influence each other mutually. Therefore, the model may be considered to be dynamic. The possible shortages are eliminated first. (The independence of $f(x)$ and $g(y)$ is not explicitly specified among the assumptions.)

The model counts with inventory holding cost C_1 and shortage cost C_2; these are proportional to quantity and time.

314

Objective:

$$C(S', S) = C_1 \int_0^{S'} (S' - y) g(y) \, dy + C_2 \int_{S'}^{\infty} (y - S') g(y) \, dy +$$

$$+ \int_0^{S'-S} \left[C_1 \int_0^{S'-y} (S' - y - x) f(x) \, dx + C_2 \int_{S'-y}^{\infty} (x - S' + y) f(x) \right] g(y) \, dy +$$

$$+ \int_{S'-S}^{\infty} \left[C_1 \int_0^S (S - x) f(x) \, dx + C_2 \int_0^{\infty} (x - S) f(x) \, dx \right] g(y) \, dy.$$

Solution:

Elementary implicit conditions can be obtained for the minimum of the objective function by differentiation of $C(S', S)$ with respect to S' and S. This system of equations may be solved by some numerical algorithm.

238. Dynamic lot-size system of three ordering periods

Main codes:

$$1\ 1\ 0\ 1\ 1\ 1\ 4\ 0\ 2\ 1.$$

Assumptions:

The operation of the system extends through three ordering periods of prescribed length t_p. Only three decisions have to be made for the q'', q' and q orderings of the three periods. The lots arrive after a constant leadtime $L = t_p$; thus, orders have to be placed always by a period earlier (i.e. the first one at a time t_p before the starting date of the system's operation). Demands arise in discrete units, and are stochastic, with distributions $R(z)$, $Q(y)$, $P(x)$ in the successive ordering periods. Demand emerges immediately after the delivery of the ordered quantity, at the beginning of the ordering period, in one batch. The above probability distributions are independent. The demand for shortage items is lost. An inventory holding cost exists in the third period only and is proportional to the number of the items left over.

The system is considered to be dynamic, since the values of q'', q' and q mutually influence each other at optimization, because the initial inventory level of the second and the third period do not depend only on the volume of the ordered batch but also on the beginning of the previous periods, as well as on the previous q values adjusted to them.

Objective:

The author does not give the cost function explicitly; the objective is the minimization of the expected total cost.

Solution:

The author recommends dynamic programming to solve a concrete numeric problem, he does not give a general method of solution.

315

X. Multi-Item Models

This group contains a relatively large number of different models. Three main classes of models can be distinguished. The first contains the models in which the different considered items are connected by some ordering constraint. In the second class are listed the models with some inventory constraint (capital, storage, shortage, etc.), while the third class contains the models which consider production and inventory together.

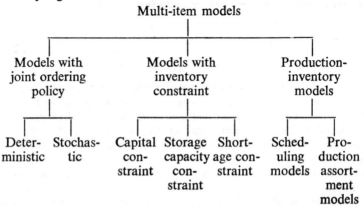

Multi-item models

Models with joint ordering policy	Models with inventory constraint	Production-inventory models
Deter-ministic Stochas-tic	Capital con-straint Storage capacity con-straint Short-age con-straint	Sched-uling models Pro-duction assort-ment models

X.1. Multi-Item Models with a Joint Ordering Policy

The models of this class have the specific property that the different items considered may be (or must be) jointly ordered and this fact influences the ordering cost. This class of models can be subdivided into groups of deterministic and stochastic models, where the stochastic nature of the models is due to the random fluctuation of demand (except one of the models). In the deterministic subgroup of models there are different specifications of the joint ordering cost that characterize the models. In the stochastic models, the main characteristic property is the ordering policy and the other one is the type of the joint ordering cost considered. Both groups contain some special models.

316

X.1.1. Multi-Item Deterministic Models with Joint Ordering

239. *Multi-item lot-size system*

Main codes:

$$8\ 1\ 0\ 0\ 0\ 1\ 3\ -1\ 0\ 0$$

Assumptions:

The system considers N different items. The constant ordering cost is denoted by c_3: it does not depend on the number of items ordered jointly nor on the amounts of order q_i ($i=1, 2, ..., N$). Let r_i denote the mean demand rate and c_{1i} the unit cost of inventory holding for the i-th item. If the length of the order period is t_i, then $q_i = r_i t_i$ must hold since no shortage is allowed.

If only the above type of ordering cost is considered, then the optimal length of the ordering periods is the same value t for all of the items. This t coincides with the minimum of ordering periods (say t_k), calculated for the items separately from each other:

$$t = t_k \leq t_i \quad (i = 1, 2, ..., N).$$

Suppose indirectly that for the j-th item $t_j > t$: this policy results in a mean inventory level I_{ij}. If the length of the ordering period is decreased also for this item to $t_j = t$, then the mean inventory level and, consequently, the inventory holding cost decreases. At the same time, no surplus cost appears at the ordering, since the ordering cost c_3 is assumed to be independent of the number of items ordered jointly. (This argumentation shows that the orders are to be placed for each item at the same time.)

Objective:

The total cost of inventory holding and ordering of N items for a unit of time is for a fixed t

$$C(t) = \sum_{i=1}^{N} \frac{c_{1i} r_i t}{2} + \frac{c_3}{t}.$$

Solution:

The derivative of the objective function is zero at the optimal value of t. The solution of the equation yields for the optimal length of the order period

$$t_0 = \sqrt{\frac{2c_3}{\sum\limits_{i=1}^{N} c_{1i} r_i}}$$

and for the optimal lot sizes $q_{i0} = r_i t_0$. In the case of $N=1$, the formula is the same as in the classical EOQ-model (Model 1).

240. Lot-size system, the ordering cost depends on the number of jointly ordered items

Main codes:
$$8\ 1\ 0\ 0\ 0\ 1\ 3\ -1\ 0\ 0$$

Assumptions:

This is a modified version of the previous model in the sense that the ordering cost equals $c_3 = c_{31} + c_{32} \cdot n$, where n is the number of items ordered at the same time, while c_{31} and c_{32} are constant values. The unit cost of inventory holding for item i is denoted by c_{1i}. The unit cost factors have the dimensions

$$[c_{1i}] = \frac{[\$]}{[Q][T]} \quad [c_{31}] = [\$]$$

$$[c_{32}] = \frac{[\$]}{[\text{number of items ordered}]}$$

The minimal time t between two subsequent orderings may be prescribed or subject to decision. The length of the ordering period t_j for item j is an integer multiple of t. This integer k_j means that the item j is ordered at each k_j-th ordering. Let r_j be the mean demand rate for the j-th item.

Objective:

The total cost for N items for a time unit

$$C(t, k_j) = \frac{t}{2} \sum_{j=2}^{N} k_j r_j c_{1j} + \frac{1}{t} \left(c_{31} + c_{32} \sum_{j=1}^{N} \frac{1}{k_j} \right)$$

has to be minimized.

Solution:

If t has a prescribed value, then k_j is considered as a continuous variable and the optimal value is calculated from the derivative of the objective function with respect to k_j. The integer value nearest to the continuous optimum is considered as the (approximate) discrete optimum.

If t is subject to control, then the optimal value of t and the minimal cost belonging to it can easily be expressed as a function of k_j ($j=1, 2, ..., N$). For fixed values of k_j, the optimum results. If the values of k_j are also to be optimized, then an iterative procedure is suggested based on the above two steps. (The initial values of the iteration are presented by the authors.)

241. Optimal ordering periods for joint and individual ordering costs

Main codes:
$$8\ 1\ 0\ 0\ 0\ 1\ 4\ -1\ 0\ 1$$

Assumptions:

In the model, n different items are considered which may be ordered jointly or individually. There is a fixed ordering cost S. At each ordering of item j,

318

an extra cost S_j appears. The annual number of orderings is N, for item j it is N_j, where $N/N_j=k_j$ is an integer. The demand rate is a known constant for each item. No shortage is allowed.

Objective:

The values of N and k_j $(j=1, 2, ..., N)$ which minimize the total cost of inventory holding and ordering for a unit of time, are to be determined:

$$C(N) = N\left(S+ \sum_{i=1}^{N} S_j/k_j\right)+\frac{1}{2N} \sum_{i=1}^{n} h_j D_j k_j$$

Solution:

For given values of k_j $(j=1, 2, ..., N)$, the optimal N and lot-sizes q_j are expressed, then an iterative procedure is derived for the calculation of the optimal values of k_j starting from $k_j=1$ $(j=1, 2, ..., N)$.

242. Optimal length of the ordering periods for N items with partial deliveries

Main codes:
$$8\ 1\ 0\ 0\ 0\ 1\ 9\ -1\ 0\ 0$$

Assumptions:

The ordering cost is given in the form

$$g_n = (1+\gamma n)g,$$

where n denotes the number of jointly ordered items, g and γ are constant cost factors. The length of the ordering period for item i is denoted by $k_i(T)$, where T the minimal time interval between two orderings. No shortage is allowed. A constant known demand rate is assumed for each item. The carrying cost is $\mu_i s_i$, where s_i is the value of the i-th item and μ_i is the rate which determines the carrying costs.

Objective:

The total cost of inventory holding and ordering for a unit time as a function of T and $k_i(T)$ is expressed as follows:

$$L(T) = \frac{1}{2} \sum_{i=1}^{N} \mu_i s_i k_i(T)+\frac{g}{T}\left\{\gamma \sum_{i=1}^{N} \frac{1}{k_i(T)}+1\right\}.$$

Solution:

A numerical iteration is suggested.

243. Optimal rate of return for N items

Main codes:
$$8\ 1\ 0\ 0\ 0\ 1\ 2\ -1\ 0\ 0$$

Assumptions:

A deterministic demand rate r_i is considered for N items. The constant ordering cost c_3 is independent of the number of ordered items. The unit cost of inventory holding for item i is c_{1i}. The lot-size q_i is ordered at each time for item i. The value of profit (without taking the ordering cost into account) is denoted by P. The capital invested (without inventory holding cost) is denoted by C.

Objective:

The rate of return as the quotient of profit and capital invested is

$$f(q_1, \ldots, q_N) = \frac{P - \sum\limits_{i=1}^{N} \dfrac{c_3 r_i}{q_i}}{C + \sum\limits_{i=1}^{N} \dfrac{q_i}{2}\left(c_{1i} + \dfrac{c_3}{q_i}\right)}$$

and has to be maximized.

Solution:

The optimal lot size for item i is

$$q_i = \sqrt{\frac{2c_3 r_i}{f^* c_{1i}}},$$

where the optimal rate of return f^* is the solution of the equation

$$\left(C + \frac{Nc_3}{2}\right)f^* + \sum_{i=1}^{N} \sqrt{2c_3 r_i c_{1i}} \cdot \sqrt{f^*} - P = 0.$$

X.1.2. Stochastic Multi-Item Models with Joint Ordering

244. *Periodic order-level system for N items with random demand*

Main codes:

8 1 0 1 0 1 1 0 1 0

Assumptions:

N different items are considered with a periodic ordering policy; for each item we have the same prescribed length T of a period. The demand in a period is a random variable with a probability density function $f_i(x)$ for the i-th item. There is an immediate delivery, thus the initial stock of each period equals the reorder level S_i. The purchasing price is c_{3i} for the i-th item. No constant ordering cost is considered. The inventory holding cost is proportional to the inventory level z_i at the end of the period: the unit cost factor is denoted by c_{1i}. The shortage cost is proportional to the expected value of the maximal shortage with a unit cost factor c_{2i}.

320

Objectives:

The expected total cost of a period is

$$L(S_1, ..., S_N) = \sum_{i=1}^{N} \left[c_{3i}(S_i - z_i) + c_{1i} \int_0^{S_i} (S_i - x) \cdot f_i(x)\, dx \right] +$$

$$+ \max_{1 \le i \le N} c_{2i} \int_{S_i}^{\infty} (x - S_i) f_i(x)\, dx.$$

Solution:

An iterative procedure based on the Newton method is suggested for the minimization of the objective function.

245. *Periodic order-level system with quantity-dependent costs*

Main codes:
$$8\ 1\ 0\ 1\ 0\ 1\ 1\ 0\ 1\ 0$$

Assumptions:

The same assumptions are considered as in the previous model. The only difference is that the inventory holding and the shortage costs are proportional to the amount of the total inventory and total shortage of a period, respectively.

Objective:

The expected total cost of a period:

$$L(S_1, ..., S_N) = \sum_{i=1}^{N} \left\{ c_{3i} S_i + c_{1i} \left[\int_0^{S_i} \left(S_i - \frac{x}{2} \right) f_i(x)\, dx + \frac{S_i^2}{2} \int_{S_i}^{\infty} \frac{1}{x} f_i(x)\, dx \right] \right\} +$$

$$+ \frac{1}{2} \max_{1 \le i \le N} c_{2i} \int_{S_i}^{\infty} \frac{(x - S_i)^2}{x} f_i(x)\, dx$$

has to be minimized.

Solution:

A system of equations is derived for the optimal solution which can be solved by an iterative procedure.

246. *Periodic order-level system with shortage cost proportional to the probability of shortage*

Main codes:
$$8\ 1\ 0\ 1\ 0\ 1\ 1\ 0\ 1\ 0$$

Assumptions:

There are the same assumptions as in Model 243. The only difference is that the inventory holding cost is proportional to the amount of the surplus stock at the end of the period and the shortage cost is proportional to the probability of shortage.

Objective:

The expected total cost of a period equals

$$L(S_1, ..., S_N) = \sum_{i=1}^{N} \left[c_{3i}(S_i - z_i) + c_{1i} \int_0^{S_i} (S_i - x) f_i(x) \, dx \right] + \max_{1 \le i \le N} c_{2i} \int_{S_i}^{\infty} f_i(x) \, dx.$$

Solution:

The optimal order levels S_i ($i=1, 2, ..., N$) are defined by a system of equations which can be solved using an iterative procedure.

247. A multi-item (t, S) policy with random demand

Main codes:

$$8\ 1\ 0\ 1\ 0\ 1\ 0\ 0\ 1\ 0$$

Assumptions:

The same assumptions are used as in Model 243. The difference is that the length of the order period T is also subject to control.

Objective:

Here the expected total cost per unit time has to be minimized, since the cost of a period depends on the decision variable T. The ordering cost can be expressed in the form $g(1+\mu N)$.

$$L(S_1, ..., S_N, T) = \frac{1}{T} \left\{ g(1+\mu N) + \sum_{i=1}^{N} \left[c_{3i}(S_i - z_i) + c_{1i} \int_0^{S_i} (S_i - x) f_i(x) \, dx \right] + \right.$$

$$\left. + \max_{1 \le i \le N} c_{2i} \int_{S_i}^{\infty} (x - S_i) f_i(x) \, dx \right\}$$

Solution:

The multivariety objective function can be minimized by, for example, the gradient method, where an approximate method is used for the calculation of the optimal T.

248. (s, q) policy with joint shortage cost

Main codes:

$$8\ 1\ 0\ 1\ 0\ 1\ 2\ 1\ 1\ 1$$

Assumptions:

The ordering rule is the following: if the inventory level of an item decreases to its own reorder point s_i, then an order is placed for all the items. The average quantity of order is $q_i = r_i k_i T$, where r_i is the mean demand rate and $k_i T$ is the average length of the order period for the i-th item. The demand during the lead-time is a random amount with a probability density function $f_i(x)$. The shortage cost is paid jointly, according to the largest amount of stockout during the order period. There is a periodic reviewing with the period length T. The unit cost of inventory holding is c_{1i}, the ordering cost of n items is $c_n = c_3(1+\gamma_n)$ and the shortage cost factor is p_{τ_i}, concerning the leadtime τ_i.

Objective:

The total expected cost for a unit of time is

$$L = \frac{T}{2} \sum_{i=1}^{N} r_i c_{1i} k_i + \frac{c_3}{T}\left(\gamma_i \sum_{i=1}^{N} \frac{1}{k_i} + 1 \right) + \sum_{i=1}^{N} c_{1i} s_i +$$

$$+ \frac{P_{\tau 1}}{T} \int_{s_i}^{\infty} (s_i - x) f_i(x)\, dx,$$

where the subscripts of the items are chosen in such a way that the following relation holds:

$$P_{\tau_1} \int_{s_1}^{\infty} (s_1 - x) f_1(x)\, dx = \max_{1 \le i \le N} P_{\tau_i} \int_{s_i}^{\infty} (s_i - x) f_i(x)\, dx.$$

Solution:

An iterative procedure is suggested, by which for a fixed value of s_1, the cost optimal T is calculated using the derivative of the cost function subject to T. Similarly, for fixed T, the optimal s_1 is the solution of the equation

$$\sum_{i=1}^{N} \frac{c_{1i} P_{\tau_1} \int_{s_1}^{\infty} f_1(x)\, dx}{P_{\tau_i} \int_{s_i}^{\infty} f_i(x)\, dx} - \frac{P_{\tau_i}}{T} \int_{s_1}^{\infty} f_1(x)\, dx = 0.$$

If the values of s_1 and T are approximated with a prescribed accuracy (applying an iterative scheme for their determination), then the optimal values of s_i ($i = 2, 3, ..., N$) are obtained solving the following system of equations:

$$\sum_{i=1}^{N} \frac{c_{1i}}{P_{\tau_i} \int_{s_i}^{\infty} \left(1 - \frac{s_i}{x} \right) f_i(x)\, dx} = \frac{1}{T}$$

$$P_{\tau_1} \int_{s_1}^{\infty} \frac{(x - s_1)^2}{x} f_1(x)\, dx = P_{\tau_i} \int_{s_i}^{\infty} \frac{(x - s_i)^2}{x} f_i(x)\, dx \quad (i = 2, 3, ..., N)$$

249. (s, q) **policy with independent shortage costs**

Main codes:

$$8\ 1\ 0\ 1\ 0\ 1\ 2\ 1\ 1\ 1$$

Assumptions:

The same assumptions are valid as in the previous model. The only difference is that the shortage costs of the different items are independent of each other.

Objective:
With the notations of the previous model we have

$$L = \frac{T}{2} \sum_{i=1}^{N} r_i c_{1i} k_i + \frac{c_3}{T} \left(\gamma_i \sum_{i=1}^{N} \frac{1}{k_i} + 1 \right) + \sum_{i=1}^{N} c_{1i} s_i +$$

$$+ \frac{1}{T} \frac{p_{\tau_i}}{k_i} \int_{s_i}^{\infty} (s_i - x) f_i(x) \, dx.$$

Solution:
A similar iterative procedure is suggested as for the previous model.

250. Optimization of ordering frequency and number of deliveries

Main codes:

$$8\ 1\ 0\ 1\ 0\ 1\ 0\ 1\ 0\ 1$$

Assumptions:
A number of items are jointly ordered with an order period length T which is subject to control. The delivery of an order occurs after a leadtime L in N lots. The value of N may be different for different items and can be chosen for an arbitrary integer. The input of the order happens at equidistant time instants and the size of the lots delivered are also equal. At a single delivery a cost c_4 is to be calculated. The unit cost of inventory holding is c_1, the constant cost of an ordering is c_3. The demand of a period is random. It is continuous with a mean demand rate r. The maximal rate of demand $k \cdot r$ is known. No shortage is allowed.

Objective:
The expected total cost of inventory holding, ordering and delivery is

$$C(T, N) = c_1(k-1) \cdot r(L+T) + \frac{c_3}{T} + \frac{c_1 rT}{2N} + c_4 \frac{N}{T}.$$

Solution:
The optimal value of the length of the order period is equal to

$$T_0 = \sqrt{\frac{c_3}{c_1(k-1)r}}.$$

For all items the reorder level is $S_0 = kr(L+T_0)$ and the number of deliveries is the integer part of the expression

$$N_0 = T_0 \sqrt{\frac{c_1 r}{2c_4}}.$$

251. Multi-item model with joint ordering discounts

Main codes:

$$8\ 1\ 0\ 1\ 0\ 1\ 5\ 2\ 1\ 1$$

Assumptions:
The inventory of N items is reviewed continuously. Each ordering has a fixed cost g_0 which is independent of the number of ordered items. At the ordering of

324

the i-th item an additional cost g_i has to be considered. Hence, if on one occasion, a larger number of different items are ordered, then the relative ordering cost is lower. A generalized (s, S) inventory policy to the lower reorder point $y_{1,i}$ (minimal level of inventory) than an order is placed for all the items which have z_i inventory level below their upper reorder point $y_{2,i}$ ($y_{1,i} \leq y_{2,i}$). The order level is denoted by Y_i. Thus the amount of order for item j is

$$q_j = \begin{cases} Y_j - z_j & \text{if } z_j < Y_{2,j}, \\ 0 & \text{if } z_j \geq Y_{2,j}. \end{cases}$$

Objective:

The cost function

$$L_1 = c_1\left(\frac{q}{2} + y_2 - rL\right) + \frac{r}{q}[g + \pi(y_2)]$$

is minimized for q and y_2: this yields the values of q_i and $y_{2,i}$ ($i=1, ..., N$). Here r is the mean demand rate and L is the leadtime; the unit cost of inventory holding is c_1, $g=g_i$ is the replenishment cost; the expected shortage cost is denoted by $\pi(y_2)$.

The optimal value of the vector y_1 with components $y_{1,i}$ is determined by minimizing the cost function

$$L_2 = c_1\left(\frac{q}{2} + y_1 - rL\right) + \frac{r}{q}[G_1 + \pi(y_1)],$$

where $G_i = g_0 + g_i$. This minimization can be carried out independently for each item ($i=1, ..., N$).

Solution:

An iterative algorithm is suggested for the minimization of L_1, while the minimum of L_2 is the solution of the equation

$$-\frac{c_1}{2r}(Y - y_1)^2 = \pi'(y_1)(Y - y_1) + G + \pi(y_1),$$

which can be determined by numerical techniques.

252. Stochastic order-level system for two items

Main codes:

$$8\ 1\ 0\ 1\ 0\ 1\ 6\ 2\ 0\ 0$$

Assumptions:

Two items are considered which have the same properties with respect to demand and cost. The demand is generated by a Poisson process for both items with the same parameter λ. The demand processes are independent for the two items. If for one of the items the inventory level decreases to 0, then an order is placed for both items: the order level is the same integer value S for both items. There is an immediate delivery. The inventory reviewing is continuous. The following cost factors are considered

c: purchasing cost of a unit of item,
rc: inventory holding cost factor,
A: constant ordering cost of one item,
mA: constant ordering cost of two items $(1 \leq m \leq 2)$.

Objective:

The expected total cost for a unit of time $g_J = \dfrac{cr\, S + \dfrac{\lambda m A}{S}}{1 - \left(\dfrac{2S}{S}\right) 2^{-2S}}$ has to be minimized.

Solution:

The cost function can be approximated by $g_J \approx \dfrac{cr\, S + \dfrac{\lambda m A}{S}}{1 - \dfrac{1}{\sqrt{\pi S}}}$, which can easily be minimized. Further on the probability of a surplus stock k before ordering and its expected value is expressed; the expected length of an ordering period is also derived. Finally, the optimal two-item policy is compared to the independent ordering policy for the two items considered.

253. (s, S) policy for two items

Main codes:

8 1 0 1 0 1 5 1 1 0

Assumptions:

A periodic review system is considered. The demand for two items during a review period is described by the joint probability distribution function $F(x_1, x_2)$. The demand of the different periods is independent of each other. The two items are jointly ordered with a constant ordering cost $K_{1,2}$. There is an immediate delivery. The expected value of the inventory holding and shortage cost for a unit time is denoted by $L(u, r)$.

Objective:

The expected total discounted cost (with $D_i = S_i - s_i$) equals

$$a = \dfrac{L(S_1, S_2) + \sum\limits_{u=0}^{D_1} \sum\limits_{r=0}^{D_2} L(S_1 - u, S_2 - r)\, m(u, r) + K_{1,2}}{1 + M(D_1, D_2)},$$

where $M(u, r)$ denotes the renewal function and $m(u, r)$ is its derivative.

Solution:

Recursive formulas are derived for the renewal function and its derivative and for $L(u, r)$. The minimization of the objective function can be accomplished by some nonlinear optimization procedure. No specific solution procedure is suggested.

254. Single-period model for two items

Main codes:

$$8\ 1\ 0\ 1\ 0\ 1\ 5\ 1\ 1\ 0$$

Assumptions:

If two items are jointly ordered, then the ordering cost is $K_{1,2}$; if only one item is ordered, then the cost is either K_1 or K_2. The following inequalities hold:

$$\max(K_1, K_2) \leqq K_{1,2} \leqq K_1 + K_2.$$

Only a single period is considered with a random demand which is characterized by the joint density function $f(u, r)$. The unit purchasing prices are denoted by c_1 and c_2, the expected inventory holding and shortage cost $L(y_1, y_2)$ is a function of the inventory level after ordering. There is an instantaneous delivery. The inventory levels before ordering are x_1 and x_2.

Objective:

The expected total cost of the period considered equals

$$C(y_1, y_2) = K(y_1 - x_1, y_2 - x_2) - c_1 x_1 - c_2 x_2 + G(y_1, y_2)$$

with the notation

$$G(y_1, y_2) = c_1 y_1 + c_2 y_2 + L(y_1, y_2).$$

Solution:

The domain of the possible inventory levels (x_1, x_2) is subdivided into four subdomains defined by 5 equations. (This is a generalization of the reorder-point system.) If $(x_1, x_2) \in B_0$, then no order is placed in B_1 or B_2 only for one of the items, while in $B_{1,2}$ an order is placed for both items.

255. Dynamic multi-item model

Main codes:

$$8\ 1\ 0\ 1\ 1\ 1\ 1\ 1\ 1\ 0$$

Assumptions:

There are n different items which have m classes of demands, and m is not necessarily equal to n. It is characterized in period i by the random vector $D_i = (D_{i,1}, ..., D_{i,m})$ which has the m-dimensional joint distribution function $\Phi_i(D_i)$. The demands in different periods are independent, and may have different distribution functions. The vector of inventory levels before ordering is denoted by x_i and after ordering by y_i ($i = 1, 2, ...$). The leadtime is zero, the vector of order levels is denoted by Y_i.

Objective:

The expected total discounted cost for $1, 2, ..., N$ periods (with discount factors β_i) can be expressed as

$$f_N(x_1|y) = E\left\{ \sum_{i=1}^{N} \beta_i [c_i(y_i - x_i) + g_i(y_i, D_i)] - \beta_{N+1} c_{N+1} x_{N+1} \right\},$$

where c_i denotes the unit cost of purchasing. The function g_i expresses the expected cost of inventory holding and of shortage (this function is not specified). It is assumed that any stock left over after N periods can be discarded with a return of c_{N+1}.

Solution:

The dynamic multi-period problem is reduced to the solution of single-period problems: $y_i = \min_y G_i(y)$ is the optimal value in period i, where

$$G_i(y) = \int_{D_i} W_i(y, t) \, d\Phi_i(t)$$

with the single-period total cost $W_i(y, t)$.

256. *Dynamic multi-item model with leadtime*

Main codes:

$$8\ 1\ 0\ 1\ 1\ 1\ 1\ 1\ 1\ 1$$

Assumptions:

This model is a generalized version of the previous model in the sense that the leadtime may be positive. It is deterministic and has the same length L for all items. The demand is separate for all the items, i.e., $n=m$ with the notation of the previous model.

Objective:

The expected total cost of the periods $L+1, ..., L+N$ is

$$f_N(x_1|Y) = \sum_{i=1}^{N} \beta_i EG_i(y_i) - \left[c_1 x_1 - \beta_{N+1} c_{N+1} \sum_{i=N+L}^{N+L} E(D_i) \right],$$

where

$$G_i(y) = \int_{D_i+L} W_i(y, t) \, d\Phi_{i+L}(t).$$

No constant ordering cost is considered, the vector of purchasing prices in period i is denoted by c. The expected total cost of inventory holding, shortage and purchasing is expressed in period i by the function $W_i(y, t)$ (that is not specified).

Solution:

The components of the cost function part in brackets are independent of the vector of order levels Y. Based on this fact, the minimization can be reduced by suitable substitutions to the solution of the zero leadtime problem described by the previous model.

257. *Multi-item model based on stochastic control*

Main codes:

$$8\ 1\ 0\ 1\ 1\ 1\ 4\ 1\ 2\ 4$$

Assumptions:

The demand process for n different items is described in a period $[0, T]$ by the stochastic differential equation $d\varrho(t) = \mu(t) \, dt + \delta(t) \, db(t)$, where $\mu(t)$ represents

328

the deterministic part and $b(t)$ is the joint normal distribution with covariance matrix $\delta(t)$. The delivery is characterized by a deterministic scalar-vector function $v(t)$. For a given initial stock vector x, the decision variables are the time $\vartheta_{x,t}$ and amount $\varDelta_{x,t}$ of the orders. The inventory level at time t is denoted by $y_x(t)$. This results in an inventory holding or shortage cost $f(y_x(t), t)$, which may depend also on time t. The cost of ordering (or producing) $k(t, \varDelta)$ may also depend on the time and amount.

Objective:

The order policy $V_{x,t}$ denotes the system of time and amount of orders $(\vartheta^1_{x,t}, \varDelta^1_{x,t}, \vartheta^2_{x,t}, \varDelta^2_{x,t}, \ldots)$. The expected total cost in the interval $[0, T]$ belonging to this policy is

$$J(V_{x,t}) = E\left\{\sum_i \exp\left[-\alpha(\vartheta^i_{x,t} - \varDelta^i_{x,t})\right] k(\vartheta^i_{x,t}, \varDelta^i_{x,t}) + \right.$$

$$\left. + \int_0^T \exp(-\alpha s) f(y_{x,t}(s), s)\, ds\right\},$$

where α is the discount factor.

Solution:

The dynamic programming principle of Bellman is used to reduce the determination of the optimal policy to the solution of a system of differential equations. The numerical solution can be performed by the method of finite differences.

X.2. Models with Inventory Constraint

The common property of the models in this group is that they all have some constraints concerning the inventory. The most common is that the capital invested in the inventories is bounded, thus the different items are connected by a joint capital constraint. The solution of these types of models can be derived usually by the Lagrange-multiplier method. The various models differ from each other by the deterministic or random nature of demand or delivery, by the different assumptions concerning ordering possibilities and by the different types of cost factors considered. Some specific models are listed here which contain investigations concerning the joint inventory level.

Another subgroup of models deals with the storage capacity constraint. The structure of these models is very similar to the models with capital constraint. In two of the models, the probability of shortage is constrained under stochastic demand or stochastic delivery conditions.

X.2.1. Models with Capital Constraint

258. *Deterministic lot-size model with constrained capital invested in an order amount*

Main codes:
$$8\ 1\ 0\ 0\ 0\ 1\ 2\ \ -1\ 0\ 0$$

Assumptions:

There is a deterministic demand with rate r_j for item j $(j=1, ..., n)$. The constant ordering cost is A_j. The unit cost of inventory holding is $I_j C_j$ with dimension [$]/[amount]. No shortage is allowed. The purchasing price is C_j. The capital invested at one occasion of ordering is bounded by the capital D:

$$\sum_{j=1}^{n} C_j q_j \leqq D.$$

Objective:

The total cost of ordering and inventory holding,

$$K(q_1, ..., q_n) = \sum_{j=1}^{n} \left[\frac{r_j}{q_j} A_j + I_j C_j \frac{q_j}{2} \right]$$

has to be minimized under the capital constraint.

Solution:

The unconstrained minimum of the cost function can be calculated by setting the partial derivatives equal to zero. If it satisfies the capacity constraint (i.e. the constraint is not active), then the optimum has been achieved. In the opposite case the constraint is satisfied in the form of an equation for the optimal solution, thus the Lagrange multiplier method can be applied. The optimal solution is the unconstrained minimum of the function (with the Lagrange multiplier θ):

$$J = \sum_{j=1}^{n} \left[\frac{r_j}{q} A_j + I_j C_j \frac{q_j}{2} \right] + \theta \left[\sum_{j=1}^{n} C_j q_j - D \right],$$

which can be calculated by setting its partial derivatives equal to zero and solving the resulting system of equations.

259. *Lot-size model with capital constraint and restricted number of annual orders*

Main codes:
$$8\ 1\ 0\ 0\ 0\ 1\ 2\ \ -1\ 0\ 0$$

Assumptions:

Similar to the previous model: here the capital invested in the order quantities is bounded and, besides this constraint, the number of annual orders is also restricted:

$$\sum_{j=1}^{n} \frac{r_j}{q_j} \leqq h.$$

330

The number of orders need not be an integer for a unit of time. Because of the second restriction, no explicit ordering cost has to be considered.

Objective:

The inventory holding cost, $K = \sum_{j=1}^{n} I_j C_j \frac{q_j}{2}$, has to be minimized under the above two constraints.

Solution:

The Lagrange function (with Lagrange multipliers θ and ϑ)

$$ J = \sum_{j=1}^{n} I_j C_j \frac{q_j}{2} + \theta \left(\sum_{j=1}^{n} \frac{r_j}{q_j} - h \right) + \vartheta \left(\sum_{j=1}^{n} C_j q_j - D \right) $$

has to be minimized. This can be calculated from the partial derivatives of J with respect to q_j, θ and ϑ; the numerical optimization, e.g., by the Newton method is efficient.

260. Multi-item model with capital constraint and distributed orders

Main codes:

8 1 0 0 0 1 4 1 0 0

Assumptions:

A deterministic multi-item lot-size model is considered, where the total value of the stock on hand may not be larger than the capital constraint X. The maximal value of stocks can be estimated by

$$ I_T = k \cdot \sum_{j=1}^{J} q_j C_{nj}, $$

where C_{nj} is the unit purchasing price and q_j is the lot size ordered for item j. The constant k ($0 \le k \le 1$) expresses the distribution of orders. If all the items are ordered together, then $k = 1$. In the case of different ordering times for different items, the mean capital invested in stocks is approximately half of the maximal capital invested, thus $k \approx 1/2$. The model is similar to Model 259, except with respect to the estimation of the parameter k.

S_j denotes the demand of the j-th item. C_{nj} is the purchasing price and C_{0j} is the ordering cost of the j-th item. i is the inventory carrying cost factor (%).

Objective:

The inventory holding and ordering cost is minimized under the above capital constraint.

$$ E_T = \sum_{i=1}^{J} \left(\frac{C_{0j} S_j}{q_j} - \frac{q_j}{2} C_{nj} i \right) + z \left(x - k \sum_{i=1}^{J} q_j C_{nj} \right), $$

where z is the Lagrange multiplier.

331

Solutions:

By using the Lagrange multiplier method, the optimal lot size is

$$q_j = \sqrt{\frac{2C_{0j}r_j}{C_{nj}(i-2kz)}}\ .$$

261. Continuous delivery and capital constraint

Main codes:

$$8\ 1\ 0\ 0\ 0\ 1\ 4\ -1\ 0\ 0$$

Assumptions:

A capital limit X is considered similarly as in the previous model. The main difference is that the delivery is continuous with a known finite rate P_{dj} for item j at day d. The order period has the same fixed length of N days for all items. The mean demand rate is r_j, the demand of a day is S_{dj}.

Objective:

The optimal solution given the capital constraint is determined from the Lagrange function

$$E_T = \sum_{j=1}^{J}\left[C_{0j}N + \frac{r_j}{2N^2}C_{nj}i\left(1-\frac{S_{dj}}{P_{dj}}\right)\right] +$$

$$+ Z\left[X - k\sum_{j=1}^{J}\frac{r_j}{N}C_{nj}\left(1-\frac{S_{dj}}{P_{dj}}\right)\right],$$

with the notations of the previous model.

Solution:

$$N^* = \sqrt{\frac{\sum\limits_{j=1}^{J}r_jC_{nj}\left(1-\dfrac{S_{dj}}{P_{dj}}\right)(i-2kZ)}{2\sum\limits_{j=1}^{J}C_{0j}}}$$

and $q_j^* = r_j/N^*$.

262. Order level for multi-item system with capital constraint

Main codes:

$$8\ 1\ 0\ 1\ 0\ 1\ 1\ 1\ 2\ 7$$

Assumptions:

N different items are considered which are demanded in each period in a random amount according to a joint normal distribution with independent components: the means and deviations are M_k and D_k $(k=1, ..., N)$. The capital used for filling up the inventories is limited. The upper limit is E. The unit purchase price is c_k. The order level is expressed in the form of $M_k + \lambda_k D_k$, where the optimal value of the safety factor λ_k has to be found. The unit cost of inventory holding and shortage are denoted by a_k and b_k for the k-th item.

332

Objective:

The expected total cost of inventory holding and shortage in a period,

$$K = \sum_{k=1}^{N} D_k \left[a_k \int_{\lambda_k}^{\infty} (1-\Phi(u))\,du + b_k \int_{\infty}^{\lambda_k} \Phi(u)\,du \right]$$

is to be minimized under the capital constraint

$$\sum_{k=1}^{N} c_k (M_k + \lambda_k D_k - r_k) \leq E,$$

where r_k denotes the initial stock of item k and Φ is the standard normal distribution function.

Solution:

The constrained minimization problem is solved by the Lagrange multiplier method. We obtain that $\lambda_k = \Phi^{-1}\left(\dfrac{a_k - \mu c_k}{a_k + b_k}\right)$, where μ is the solution of the equation

$$\sum_{k=1}^{N} c_k D_k \Phi^{-1}\left(\frac{a_k - \mu c_k}{a_k + b_k}\right) = E - \sum_{k=1}^{N} c_k (M_k - r_k)$$

and Φ^{-1} denotes the inverse of Φ.

263. Distribution of a restricted amount of capital for safety stocks of several items

Main codes:

8 1 0 1 0 1 4 0 1 0

Assumptions:

The total value of safety stock for n items has an upper bound. The mean demand rate of item j is r_j. The deviation from the mean rate is a random variable with density function $p_j(t)$ and standard deviation δ_j. The length of the order cycle is the same for all items, $c = r_j/q_j$. The amount of safety stock is $k_j \delta_j$, where k_j is to be determined by the model.

Objective:

The expected value of the total loss of profit has to be minimized:

$$C(k_1, \ldots, k_n) = \sum_{j=1}^{n} \delta_j v_j \frac{r_j}{q_j} \int_{k_j}^{\infty} (t - k_j)\, p_j(t)\, dt,$$

where v_j is the value of the items.

Solution:

The objective function is minimized under the safety stock restriction. The optimal values of k_j are determined by the Lagrange multiplier method as the solution of the equation.

$$c\lambda_0 = \int\limits_{k_j}^{\infty} p(t)\,dt,$$

where λ_0 is the optimal value of the Lagrange multiplier.

264. Distribution of safety stock at (s, q) policy

Main codes:

8 1 0 1 0 1 2 2 1 0

Assumptions:

The same assumptions are valid as in the previous model. The only difference is that the length of the order period is not prescribed, thus an (s, q) policy is applied instead of a (t_p, q) policy.

Objective:

The expected value of the profit lost (due to shortage) has to be minimized under the constraint that the total value of safety stock for n items is limited from above.

Solution:

The solution procedure based on the Lagrange multiplier method is similar to that of the previous model.

265. Cost-optimal distributions of a limited value of safety stock

Main codes:

8 1 0 1 0 1 2 2 1 0

Assumptions:

This is a version of the previous models, where the purchasing cost is taken also into consideration for an (s, q) policy.

Objective:

The expected total cost of ordering and shortage is minimized for the n items:

$$C(k_j, q_j) = \sum_{j=1}^{n} c_j \frac{r_j}{q_j} + \sum_{j=1}^{n} v_j \frac{r_j}{q_j} \delta_j \int\limits_{k_j}^{\infty} (t - k_j) p(t)\,dt$$

under a constraint on the total value of the average stock on hand. (The notations of previous models are used.)

Solution:

Applying the Lagrange multiplier method, for each j a system of two equations is derived for the optimal values of k_j and q_j.

334

266. Minimization of the value of safety stocks under a prescribed service level

Main codes:

$$8\ 1\ 1\ 0\ 0\ 1\ 1\ 1\ 1\ 4$$

Assumptions:

There is a continuous demand in the order period $[0, T]$ with a known constant rate c for item i $(i=1, ..., k)$. The delivery of an order for any of the items is described in the following way. The order is delivered in n equal lots. The minimal time between two consecutive deliveries is a known constant γ, $\gamma < \dfrac{T}{n}$. The moments of deliveries are random points of the interval $[0, T]$ according to the following model. The interval $(0, T-n\gamma)$ is subdivided by n independent random points which are uniformly distributed on this interval. Let $u_1 < u_2 < ... < u_n$ be a realization of these random points, then the corresponding realization of the j-th delivery moment is $t_j = j\gamma + u_j$. The safety stock at the beginning of the period is denoted by M_i for the i-th item. Continuous supply is provided if $c_i t \leq b_i(t) + M_i$ holds for all $0 \leq t \leq T$, where $b_i(t)$ denotes the cumulative amount delivered until time t according to the above model of delivery process.

Objective:

The capital $\sum\limits_{i=1}^{k} d_i M_i$ invested in safety stock has to be minimized under a prescribed service level, which is expressed by the probability of the continuous supply:

$$\min \sum_{i=1}^{k} d_i M_i$$

$$\prod_{i=1}^{k} P\left(\sup_{0 \leq t \leq T} \{c_i t - b_i(t)\} \geq M_i \right) \leq p.$$

Solution:

The probability of the continuous supply is approximated by a simple expression, which enables the calculation of the optimal safety stock.

267. Multi-item reliability-type model for inhomogeneous deliveries

Main codes:

$$8\ 1\ 1\ 0\ 0\ 1\ 1\ 1\ 7\ 4$$

Assumptions:

There is a continuous demand with a known constant rate c_i for $i=1, ..., k$ different items. The length of the order period T is fixed in advance. The delivery of an order for an item is described in the following way. The order is delivered in n lots at random instants of the order period $[0, T]$. There is a minimal time interval $\gamma < \dfrac{T}{n}$ between two consecutive deliveries. The interval $[0, T-n\gamma]$ is subdivided by n independent random points which are uniformly distributed on

335

this interval. These points are in increasing order: $t_1 < t_2 < ... < t_n$. Thus, by the model the subsequent delivery times are $\gamma + t_{j1}$, $2\gamma + t_{j2}$, ..., $n\gamma + t_{jn}$. By statistical methods, this model of a delivery process can be fitted to the observations of earlier deliveries also in the case when no time-homogeneity is assumed. The amount of lots delivered can be modelled in a similar way. (That is to say, the total amount ordered can be arbitrarily subdivided among the delivery instants by random points, according to the statistical data of earlier observations. The delivery processes for the different items are supposed to be independent of each other.)

Objective:

The value of the safety stock $\sum_{i=1}^{k} d_i M_i$ is minimized under the reliability constraint that the probability of the continuous supply of the k items considered has to be at least $1 - \varepsilon$.

Solution:

The cost minimization under a probabilistic constraint is reduced to the solution of a nonlinear programming problem, where the function values must be calculated by stochastic simulation. It is solved by the SUMT method for which the global convergence is proved.

268. *The estimation of aggregated inventory characteristics*

Main codes:

$$8\ 1\ 0\ 1\ 0\ 3\ 2\ 2\ 1\ 0$$

Assumptions:

The order amount of M items is determined by the Wilson formula, the reorder level is proportional to a fixed power of the mean demand. The demand is random and has lognormal distribution for each item.

Objective:

The purpose is to estimate the aggregated average level of the safety stock, cycle stock, order frequency and rate of return for M items.

Solution:

The above aggregated inventory characteristics are expressed by the moments of the lognormal probability distribution, then they are represented as analytical functions of the expected value and mode. Thus the estimation of the aggregated characteristics can be achieved on the base of a sample of demand or usage.

269. *Lower limit for the joint inventory value of* n *items*

Main codes:

$$8\ 1\ 1\ 1\ 0\ 1\ 2\ 2\ 1\ 7$$

Assumptions:

Continuous random demand is considered for n items. The leadtime may also be a random variable. The demand during the leadtime is characterized by its density function $f_i(x)$ for item i. The value of the joint inventory level must exceed the minimum level T. Thus for the order levels R_i, the inequality

$$\sum_{i=1}^{n} v_i R_i \geq T$$

has to be valid, where v_i denotes the unit value of product i.

Objective:

The expected total cost of inventory holding, shortage and ordering with cost factors c_i, u_i and k_i is expressed in the form

$$\sum_{i=1}^{n} \left\{ c_i \left[\frac{q_i}{2} + \int_0^{R_i} (R_i - x_i) f_i(x_i)\, dx_i \right] + \frac{s_i k_i}{q_i} + \frac{s_i u_i}{q_i} \int_{R_i}^{\infty} (x_i - R_i) f_i(x_i)\, dx_i \right\}.$$

Solution:

The cost function is minimized under the inventory value constraints using the Lagrange multiplier technique.

X.2.2. Models with Storage Capacity Constraints

270. *Deterministic lot-size model with constrained storage capacity*

Main codes:
$$8\ 1\ 0\ 0\ 0\ 1\ 2\ \ -1\ 0\ 0$$

Assumptions:

There is a deterministic demand with rate r_j for item j, $j = 1, ..., n$. The constant ordering cost is A_j, the unit cost of inventory holding is $I_j C_j$. The inventory holding cost is proportional to the amount of inventory on hand. No shortage is allowed. A unit of item j needs a storage place f_j. The storage capacity of the store is limited: this is expressed by the constraint

$$\sum_{j=1}^{n} f_j q_j \leq f$$

for order amounts q_j.

Objective:

The cost of inventory holding and ordering is

$$K = \sum_{j=1}^{n} \left(\frac{r_j}{q_j} A_j + I_j C_j \frac{q_j}{2} \right).$$

Solution:

The cost function is minimized with respect to q_j $j = 1, ..., n$ under the storage capacity constraint applying the Lagrange method.

271. *The optimal length of order period by limited storage capacity*

Main codes:

$$8\ 1\ 0\ 0\ 0\ 1\ 4\ \ -1\ 0\ 0$$

Assumptions:

The assumptions are the same as in the previous model. Here, instead of the amount of orders, the optimal length of the order period has to be determined. All the items are ordered together.

Objective:

Similar to those of the previous models, only the decision variable has to be replaced; here we have $t = q_i/r_i$.

Solution:

Fon the optimal length of the joint ordering period we have

$$t = \sqrt{\frac{2 \sum\limits_{j=1}^{n} A_j}{\sum\limits_{j=1}^{n} I_j C_j r_j}},$$

if this satisfies the storage capacity constraint; otherwise t is selected as the solution of the equation

$$f = \frac{t}{2}\left[\sum_{j=1}^{n} f_i r_j + \frac{\sum\limits_{j=1}^{n} f_j^2 r_j^2}{\sum\limits_{j=1}^{n} f_j r_j}\right].$$

272. *Order level for N items and limited storage capacity*

Main codes:

$$8\ 1\ 0\ 0\ 0\ 1\ 1\ \ -1\ 1\ 0$$

Assumptions:

The same assumptions are valid as in the previous model, but a (t_p, S) policy is used.

Objective:

The cost of inventory holding and shortage with cost factors c_{1i} and c_{2i} is

$$K(S_1, \ldots, S_N) = \sum_{i=1}^{N} \frac{c_{1i} S_i^2 + c_{2i}(q_i - S_i)^2}{2q_i},$$

where q_i is the amount of order which increases the inventory level to S_i.

Solution:

The cost function is minimized under the storage capacity constraint by the Lagrange multiplier method.

273. Order level for random demand and limited storage capacity

Main codes:

$$8\ 1\ 0\ 1\ 0\ 1\ 1\ 0\ 1\ 0$$

Assumptions:

A similar multi-item (t_p, S) policy is considered as in the previous model under limited storage capacity, but here the demand of a period is a random variable with a probability density function $f_{i,t}$ for item i.

Objective:

The expected inventory holding and shortage cost of a period can be expressed as

$$L(S_1, ..., S_N) = \sum_{i=1}^{N} c_{1i} \int_0^{S_i} (S_i - x) f_{i,t}(x)\, dx + \frac{c_{2i}}{t} \int_{S_i}^{\infty} (x - S_i) f_{i,t}(x)\, dx.$$

Solution:

Based on the Lagrange multiplier method a system of equations is derived for the optimal order levels S_i ($i=1, ..., N$).

274. Changes in the aggregate inventory level of a multi-item lot-size system

Main codes:

$$9\ 1\ 0\ 0\ 0\ 3\ 4\ -1\ 0\ 0$$

Assumptions:

The inventory policy of N items is connected by the limited joint storage capacity. The demand has a known constant rate r_i for each item. The length of the order period τ_i is fixed. No shortage is allowed. Any replenishment is the integer multiple of constant reorder volume q_i. The reorder cycle $\tau_i = q_i/d_i$ is a known constant.

Objective:

The aggregate inventory level at time t is expressed by

$$S(t, z_1, z_2, ..., z_N) = \sum_{i=1}^{N} z_i r_i + \sum_{i=1}^{N} \left[\frac{t + \tau_i - z_i}{\tau_i} \right] q_i - t \sum_{i=1}^{N} r_i,$$

where z_i is the time of ordering for item i and $[a]$ is the largest integer smaller than or equal to a. The purpose of the model is to smooth the changes in the aggregate inventory level that enables it to have a lower storage capacity.

Solution:

Only the case $\tau_i = \tau$ $(i=1, ..., N)$ is considered, i.e., the same length of order for each item is supposed. A critical storage capacity c^* is derived which enables a given ordering policy without overflow:

$$c^* = \frac{\tau}{2}\left(D_N + \frac{D_N^2}{D_N}\right)$$

with

$$D_N = \sum_{i=1}^{N} d_i \quad \text{and} \quad D_N^2 = \sum_{i=1}^{N} d_i^2.$$

275. A single-period model: The flyaway-kit problem

Main codes:

8 1 0 1 0 1 4 0 2 7

Assumptions:

A single period is considered with stochastic demand. The probability of demand x for item i is $p_i(x)$, where x is an integer. The initial stock of the period which is subject to control is denoted by h_i. The items are connected by the joint storage capacity constraint

$$\sum_{i=1}^{N} v_i h_i \leq V,$$

where v_i is the storage place necessary for a unit of the i-th item.

Objective:

The total shortage cost is minimized:

$$K(h_1, ..., h_N) = \sum_{i=1}^{N} \pi_i \left[\sum_{x=h_i}^{\infty} (x - h_i) p_i(x) \right]$$

with unit shortage costs π_i.

Solution:

If the values h_i are large enough, then the discrete problem is approximated by a continuous one and is solved by the Lagrange multiplier method, which results in an iterative procedure for the optimal values of the initial stock. For small values of the initial stock $(h_i = 0, 1$ or $2)$ another iterative procedure is suggested based on a discrete enumeration technique.

X.2.3. Models with Shortage Constraint

276. *Multi-item stochastic model with constrained probability of shortage*

Main codes:

$$8\ 1\ 0\ 1\ 0\ 1\ 1\ 0\ 7\ 0$$

Assumptions:

A period with length T is considered. The item i is ordered periodically with the length of an order period $k_i T$ $(0 < k_i \le T)$. The random demand of this period is characterized by the probability density function $f_{i,k_iT}(x)$. The order level of item i is denoted by S_i. The leadtime is zero. The shortage cost factor is not known, therefore the probability that no shortage occurs (i.e. the probability of continuous supply for all of the N items) is constrained by the service level Q:

$$\prod_{i=1}^{N} \int_0^{S_i} f_{i,k_i,T}(x)\, dx \ge Q.$$

Objective:

The inventory holding cost of the items

$$L = k_i T \sum_{i=1}^{N} c_{1i} \int_0^{S_i} (S_i - x) f_{i,k_iT}\, dx(x)$$

has to be minimized under the above constraint.

Solution:

A system of equations is given for the optimal order levels based on the Lagrange multiplier method.

277. *Cost minimization under random deliveries and expected shortage constraint*

Main codes:

$$8\ 1\ 1\ 0\ 0\ 1\ 1\ 1\ 1\ 4$$

Assumptions:

There is a continuous demand with a known constant rate for t different items. The length of the order period T is fixed. The delivery of an order is described similarly to Model 266. The order is delivered in n lots. The delivery moments $x_{j_1}^*, x_{j_2}^*, \ldots, x_{j_m}^*$ and the delivery lots $y_{k1}^*, y_{k2}^*, \ldots, y_{kn}^*$ may be random. Their modelling is based on choosing certain elements $j_1 < j_2 < \ldots < j_n$, and $k_1 < k_2 < \ldots < k_n$ of a random sample according to the statistical data of earlier observations. The delivery process need not be homogenous in time. The decision variables are the initial stocks $M^{(j)}$ for the t different items $(j=1, \ldots, t)$.

Objective:

The expected inventory holding and shortage cost

$$\sum_{j=1}^{t} d^{(j)} M^{(j)} + \sum_{j=1}^{t} \sum_{i=1}^{n} E(\beta_i^{(j)})$$

is minimized in a period under the constraint that the conditional expectation of the shortage may not exceed a prescribed level in the periods when shortage occurs:

$$E(\varrho_i^{(j)} - M \,|\, \varrho_i^{(j)} - M^{(j)} > 0) \leqq g_i^{(j)} \quad \text{for } i = 1, ..., n; \; j = 1, ..., t$$

where $\varrho_i^{(j)}$ is a function expressed by means of the random delivery moments $x_{j_i}^*$ and delivery lots $y_{k_i}^*$. The constants $g_i^{(j)}$ are prescribed bound.

Solution:

An optimal safety stock plan for the k different items is derived.

The cost minimization under the constrained expected shortage is reduced to the solution of a nonlinear programming problem, where the function values are to be calculated by some simulation technique. It is solved similarly to Model 266, by the SUMT method for which the convergence is proved.

X.3. Production-Inventory Models

X.3.1. Scheduling Models

The models of this group deal with the joint investigation (usually with the joint optimization) of production and inventory. The two considered subgroups contain models for scheduling and models for production assortment. The former can be classified further according to the property that the inventory is located on the side of input or output.

278. *The lot release times of a multi-item production system*

Main codes:

$$8\ 1\ 0\ 0\ 0\ 1\ 4 \quad -1\ 0\ 1$$

Assumptions:

There is a constant demand with known rates r_i $i=1, ..., n$ for n items. The items are produced by the same machine (only one type of item) at one time in production lots. The intensity of production is p_i, the setup time by changing the production to item i is t_i. The setup cost is s_i, the inventory holding cost is h_i for the item i. The optimal cycle lengths T_i of the lot releases are to be determined (T_i is the time between two consecutive lot releases for the same item). T_i is assumed to be an integer multiple of a basic cycle length T: $T_i = k_i T$, furthermore, we have the constraint:

$$\sum_{i=1}^{n} \left(\frac{k_i T r_i}{p_i} + t_i \right) \leq T.$$

Objective:

The total cost per unit time is

$$F(K, T) = \sum_{i=1}^{n} \left(\frac{s_i}{k_i T} + h_i \frac{(p_i - r_i) r_i k_i T}{2 p_i} \right).$$

Solution:

The dynamic programming principle is used. Accordingly, if the optimal plan is prepared for periods $i=1, ..., m-1$ which use up a time τ, then the minimal cost for the remaining periods $i=m, m+1, ..., n$ is equal to

$$F_m(T-\tau, T) = \min_k \sum_{i=m}^{n} f_i(k_i, T),$$

with the constraint

$$\sum_{i=m}^{n} \left(\frac{k_i r_i}{p_i} + \frac{t_i}{T} \right) \le \frac{T-\tau}{T}.$$

As $T_{n+1}(T-\tau, T)=0$, thus starting with $m=n$ and with the discretization, $\tau=\Delta T, 2\Delta T, ..., T$, the backward iteration can be continued until $m=1$.

279. Deterministic production-inventory system for two items and a single machine

Main codes:

8 1 0 0 0 3 4 2 0 0

Assumptions:

A machine produces two types of items in time-sharing. The setup time from the production of one type of item to the other is S. The production rate is a known constant p and the demand rate D is also known $(D<p/2)$. No shortage is allowed. The state of the system is described by the inventory levels (I_1, I_2) expressed in hours.

Objective:

No control action is investigated, the changes in the state of the system during a production-inventory cycle are described. The initial state of the system which ensures its stability is to be determined.

Solution:

The steps of the solution are illustrated graphically. The critical initial stock of the system is

$$x_p = \frac{S}{1-2D/p},$$

which provides stability. If $x<x_p$, then the subsequent production-inventory cycles have a decreasing sequence of initial stocks which leads to shortage. If $x>x_p$, then the sequence of initial stocks tends to infinity. Thus the production must be changed for the other item when the stock level is decreased to x_p.

280. Stochastic production-inventory system for two items and a single machine

Main codes:

8 1 0 0 0 1 7 2 2 9

Assumptions:

The basic assumptions are the same as in the previous model, with the difference that the amount produced is a random variable with a normal distribution. It is characterized by four parameters. These are the expected amount of production, the time of production, the standard deviation of production, and the expected production intensity factor. The last parameter takes into consideration extensive possibilities, such as overtime or increase of work forces. Shortage is permitted, but is lost in the system.

The concept of desired stock level is defined. If the closing stock of a production period exceeds this level, idle hours are included. There are possibilities also to increase the production intensity by external sources. The following cost factors are considered: setup cost, production and material cost of an item, inventory holding cost, shortage cost, and cost of including external sources (with fixed and variable cost factors). The aggregated characteristics of the system, the lot sizes, the utilization of production capacity, and the economic production costs are also considered.

Objective:

The expected total cost of production and inventory is expressed as a function of the control parameters (namely, as a function of lot-size and production intensity).

Solution:

The cost consequences of different changes in the parameter values are analyzed using the Markovian property of the state probabilities. The internal and external sources of increasing production intensity are also examined.

281. *A production scheduling problem with batch processing*

Main codes:

$$8\ 1\ 0\ 0\ 1\ 1\ 9\ 1\ 1\ 0$$

Assumptions:

N different products are produced in parallel to M similar facilities during H periods.

x_{ik} denotes the number of facilities used to produce product i in period k.

$\bar{I}_{ik}$ denotes the inventory of product i at the end of period k and $\bar{B}_{ik}$ is the back-orders of product i at the end of period k.

The cost factors: f_{ik} is the inventory holding cost and $\bar{b}_{ik}$ the backordering cost for product i at the end of period k,

c_{ik} is the cost of producing product i in period k.

A production schedule is to be determined which minimizes the production, inventory holding and shortage cost for the production horizon.

$$\min \sum_{i=1}^{N} \sum_{k=1}^{H} c_{ik} x_{ik} + \sum_{i=1}^{N} \sum_{k=1}^{H} f_{ik} \bar{I}_{ik} + \sum_{i=1}^{N} \sum_{k=1}^{H} \bar{b}_{ik} \bar{B}_{ik}.$$

Solutions:

The scheduling problem is formulated as a mixed linear integer programming problem. It is transformed into a pure integer programming problem which can be solved by efficient standard algorithms of minimum-cost network flow problems.

344

282. Multi-item supply problem of a production equipment by fixed ordering cost

Main codes:

$$8\ 1\ 0\ 1\ 0\ 1\ 2\ 2\ 2\ 1$$

Assumptions:

For an equipment n different items are necessary in a production period. This demand is random with an expected value d_i for a unit of time. The demand of the fixed leadtime is also random with expected value L_i and density function $f_i(x_i)$ for item i $(i=1, ..., n)$. The demand cumulated during a production period is satisfied at the beginning of the next period. Shortage may occur, and, in this case, the unsatisfied part of demand is lost. The items are independent.

If the inventory level decreases to the reorder point P_i, then an order of amount q_i is placed. During a production period only one order may be placed. The cost factors are the following: the production cost of a unit of item i is c_i, the inventory holding cost factor is r, the shortage cost factor is m. The constant ordering cost is S, the purchasing unit cost of the items is s. For N items the total cost of ordering is $A=S+Ns$.

Objective:

The expected total cost for a unit of time

$$C = c_i r \left(P_i - L_i + \frac{q_i}{2} \right) + \frac{d_i A}{q_i} + \frac{mc_i d_i}{q_i} \int\limits_{x_i = P_i}^{\infty} (x_i - P_i) f_i(x_i)\, dx_i$$

has to be minimized.

Solution:

From $\dfrac{\partial C}{\partial q_i}=0$ we obtain that the optimal amounts of order are

$$q_i^2 = \frac{2A d_i}{c_i r} + \frac{2m d_i}{r} \int\limits_{x_i = {}_i P}^{\infty} (x_i - P_i) f_i(x_i)\, dx_i$$

and from $\dfrac{\partial C}{\partial P_i}=0$ it follows that P_i is the solution of

$$\int\limits_{x_i = P_i}^{\infty} f_i(x_i)\, dx_i = \frac{r q_i}{m d_i},$$

while $P_i=0$, if

$$\int\limits_{x_i = P_i}^{\infty} f_i(x_i)\, dx_i < \frac{r q_i}{m d_i}.$$

If the integral cannot be explicitly calculated, an iterative solution method is suggested.

283. *Multi-item inventory problem of a production equipment with joint ordering cost*

Main codes:

$$8\ 1\ 1\ 1\ 0\ 1\ 2\ 1\ 2\ 1$$

Assumptions:

It is a modified version of the previous model. The cost structure is different and the ordering process has also a more complicated form. Two different input processes are examined, in both cases the considered time interval $[0,\ T]$ consists of $1/T$ number of scheduling periods.

1. The times of ordering are generated by a Poisson process with parameter

$$\lambda = \sum_{i=1}^{n} (d_i/q_i).$$

In this case, the total cost of ordering for a unit of time can be expressed in the form $A = s + S[1 - \exp(-\lambda T)]/\lambda T$, where S means the constant cost of an ordering and s is the unit purchasing price.

2. If the ordering actions of the n items are independent, then

$$A = s + \frac{S}{\lambda T}\left[1 - \sum_{i=1}^{n}(1 - Td_i/q_i)\right],$$

where $\prod\limits_{i=1}^{n}(1 - Td_i/q_i)$ is the probability of no production jobs (idle time), whatever is run in some given scheduling period.

Objective:

The same expression as in the previous model, except the value of A given above.

Solution:

Setting the partial derivatives of the objective function (with respect to P_i and q_i) to zero, a system of $2n+1$ nonlinear equations is to be solved. This may be rather difficult in general, therefore an iterative procedure is suggested which consists of the following steps:

1. Let $A = S$;
2. Assuming a constant ordering cost calculate P_i and q_i according to the previous model;
3. Using these values of q_i, A is determined; if it is sufficiently close to the previous value of A, then P_i and q_i are accepted, otherwise Step 2 has to be repeated with the new value of A.

(The convergence of the procedure is proved in the paper.)

346

284. Economic packing frequency for jointly replenished items

Main codes:
$$8\ 1\ 0\ 0\ 0\ 1\quad -1\quad -1\ 0\quad -1$$

Assumptions:

There are m different items in the system. The annual demand for item j is Q_j. It has a constant, known demand rate. The whole demand is to be satisfied by N production lots for each item. The lots are packed. Item 1, which is the most intensively demanded, is packed after each production lot, i.e., N times per year. The other items are packed T_j times, where $T_j \leq N$, $K_j = N/T_j$ with an integer K_j. The value of N and K_j $(j=2, ..., n)$ are subject to control. Each production lot has a fixed production cost S and each packing a fixed package cost S_i. The annual inventory holding cost of an item is denoted by h.

Objective:

The total annual cost

$$C(N, K_j) = N\left[S + S_1 + \sum_{j=2}^{m} \frac{S_j}{K_j}\right] + \frac{h}{2N}\left[Q_1 + \sum_{j=2}^{m} Q_j K_j\right]$$

has to be minimized.

Solution:

The method is based on the dynamic programming principle. The optimal cost of the m items, if the first item is packed N times, is denoted by $f_m(N)$. N can be expressed by the package and inventory holding cost of the first item $P_1(N)$ and by the optimal cost of the other $m-1$ items $f_{m-1}(N)$ in the form

$$f_m(N) = \min[P_1(N) + f_{m-1}(N)].$$

Here

$$P_1(N) = S_1 N + \frac{hQ_1}{2N}$$

and

$$f_{m-1}(N) = \min \sum_{j=2}^{m}\left(S_j T_j + \frac{hQ_j}{2T_j}\right) = \min \sum_{j=2}^{m}\left(\frac{S_j N}{K_j} + \frac{hQ_j K_j}{2N}\right),$$

where the K_j-s can be found separately (for fixed N). For the calculation of the approximately optimal N, a numerical procedure is given.

X.3.2. Models for the Optimization of Production Assortment

285. Production assortment problem with nonlinear cost functions

Main codes:
$$8\ 1\ 0\ 0\ 0\ 1\ 4\quad -1\ 0\ 0$$

Assumptions:

In the system n different quality or size of a product is demanded with constant rate d_j $(j=1, ..., n)$. All demands are to be satisfied. The demand for a given quality can be satisfied also by another quality or size which is superior. The initial stock is zero. The ordering and inventory holding cost for an order of amount $x_i \geqq 0$ from quality group i is

$$c_i(x_i) = \begin{cases} \pi_i + \sqrt{2s_i h_i x_i} \\ 0, \end{cases}$$

where π_i is the constant ordering cost, s_i is the purchasing price and h_i is the unit holding cost.

The cost of satisfying an amount y_{ij} of demand i with a superior quality j is the replacement cost c_{ij}/y_{ij} which may have four different structures involving additive and proportional, half additive and proportional, additive and fixed, half additive and fixed cost factors.

Objective:

The nonnegative values of x_i and y_{ij} are to be determined which minimize the total cost

$$\sum_{i=1}^{n} c_i(x_i) + \sum_{i=1}^{n} \sum_{j=1}^{n} c_{ij}(y_{ij})$$

under the constraints

$$\sum_{i=1}^{j} y_{ij} = d_j \quad j = 1, ..., n$$

$$\sum_{j=1}^{n} y_{ij} = x_i \quad i = 1, ..., n.$$

Solution:

In principle, the technique of dynamic programming which could be applied is not effective for practical calculations. If the cost function of replacement is additive and proportional, then the optimal policy is segmented. It means that the qualities can be subdivided in groups. In all of the groups, the best quality supplies all the demands of the group. Under weaker conditions the optimal policy is quasi-segmented. It means that the best quality supplies almost all of the demands of the group, only some qualities are in the group with self-supply. These specific policies can be more effectively calculated than is the general case.

286. Production-inventory system with resource constraints

Main codes:

$$8 1 0 0 1 1 4 \ -1 0 0$$

Assumptions:

The system is considered during T periods. In each period t, a deterministic demand d_{it} appears for N different items. Demand must be satisfied by inventory and production within the period, no shortage is allowed. The production plan x_{it} of N items has to be determined under the constraint that each period t has finite resources R_{kt} from the k-th type of resource. The number of resources is

348

K. The inventory level I_{it} at the end of period t from item i is expressed as $I_{it}=$ $=I_{i,t-1}+x_{it}-d_{it}$ $(i=1, ..., N; \quad t=1, ..., T)$. The setup cost of product-type i is denoted by s_i, and the production cost of a unit of item is v_i.

Objective:

The total cost of T periods is

$$\sum_{i=1}^{N} \sum_{t=1}^{T} [s_i \delta(x_{it})+v_i x_{it}+h_i I_{it}]$$

with

$$\delta(x_{it}) = \begin{cases} 1 & \text{if } x_{it} > 0 \\ 0 & \text{if } x_{it} = 0. \end{cases}$$

Solution:

The cost function has to be minimized under the constrained capacity of resources expressed in the form

$$\sum_{i=1}^{N} [r_{ik}^{\delta} \delta(x_{it})+r_{ik}^{*} x_{it}] \leq R_{kt} \quad k = 1, ..., K; \quad t = 1, ..., T,$$

where r_{ik}^{δ} is the capacity absorption for one set-up of product i on resource k and r_{ik}^{*} is the per-unit capacity absorption of product i on resource k.

A lower bound for the optimal cost can be derived using linear programming. Other model-variants are also described, where the capacity constraint is not a constant but is also a decision variable. The capacity can be increased by overtimes, new workers, etc., but these options also imply different cost factors and new constraints.

287. *Assortment problem for size-combinations by deterministic demand*

Main codes:

$$8\ 1\ 0\ 0\ 0\ 1\ 4\ -1\ 0\ 0$$

Assumptions:

In system K different sizes of an item are demanded in a known quantity. In general it is not economic to produce and store all of the sizes. The demand for the missing sizes can be met from a larger size by cutting the surplus amount, where the loss of material has cost consequences.

If the number of stored sizes R is presribed, the optimal sizes can be determined, which provides for a minimal loss of cutting. By increasing R, these costs decrease; however, the cost of production and inventory holding increase.

Objective:

The minimal loss of material for a given R can be expressed by

$$A_{R+1}(I, K) = \min_{I+1 \leq J \leq K-R+1} [A_x(I, J)+A_R(J, K)],$$

349

where I is the smallest and J is the largest of the sizes considered, and $A_x(I, J)$ expresses the minimal loss of material in the interval of sizes (I, J), if x different sizes are stored.

Solution:

The expression of the minimal loss given above allows the reduction of computations to a moderate amount. Thus, by calculating the total costs for different values of R, the optimal policy can be chosen. The algorithm is illustrated by a numerical example.

350

XI. Multi-Location Models

The multi-location models represent inventory systems where the items are stored at several locations which are interdependent by the applied inventory policy or by the costs. This group of models can be subdivided on the basis of the character of the connection among the locations; these subgroups can be further subdivided according to different aspects.

The models of the first subgroup assume the existence of a central storage location, from which inventories are distributed among the parallel locations of one or more lower level of the hierarchical system. The "outside", real demand appears at the lowest level. These models are different from the point of view of the deterministic or random type of demand, or according to the static or dynamic description of the system, or according to the cost factors considered.

The models of the second subgroup investigate the problem of the distribution of a given total inventory among parallel storage locations with a minimal total cost. Here, the models are distinguished according to the type of the inventory distribution, which can be static or dynamic. The models of the third subgroup have the joint feature that there are serially linked storage locations with hierarchical echelons. A typical example is the internal storage system of a production facility. These models can be subdivided according to the type of demand (deterministic or random) and according to the number of periods investigated (single-period or dynamic, multi-period systems).

The multi-item, multi-location models belong to the fourth subgroup. These are rather special models: there are distribution-type models (where the items are connected by some joint constraint, e.g., concerning the orders or shortage) and models dealing with an internal storage system of a production. A simulation model of a warehouse system is also included in this description.

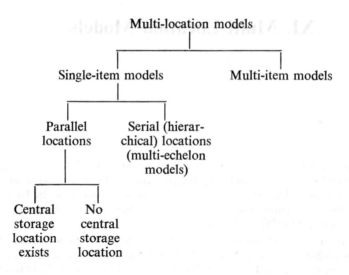

In the following, two variations of a model are described in more detail as they reflect many typical characteristics of the multi-location inventory models.

XI.1. Multi-Location Single-Item Models

XI.1.1. Several Parallel Locations

XI.1.1.1. Parallel Storage Locations with a Central Store

288. *Central inventory control of parallel stores*

Main codes:

$$1\ 8\ 0\ 1\ 0\ 1\ 1\ 1\ 1\ 0$$

Assumptions:

In this inventory system a single item is considered which is distributed by a central store among some parallel locations. Two different types of inventory policies are possible: 1) The various stores may have their independent ordering policies; 2) They centralize inventory decisions according to the joint optimum of the whole system. The second case has immediate advantages: by redistribution of the stocks among the parallel stores, shortage and surplus can be avoided. On the other hand, in the case of joint management the complicated control and inventory policy necessitates greater central administration. This type of model is described first for two parallel stores (Model 288), then it is generalized for n stores (Model 289).

For the two-store case the initial stock before ordering is I_i. A single period is considered, which has a random demand S_i at each store with probability density function $f_i(S_i)$ for $i=1$ and 2. The inventory holding or shortage cost is paid according to the amount of net inventory at the end of the period: the respective unit cost factors are denoted by r_i and v_i.

The unit purchasing cost is γ_i, the order level is denoted by x_i. Delivery takes place at the beginning of the order period at each store. During the order period there is a possibility of redistributing the stock among stores by direct transportation between the stores. The amount y transported is taken to be positive if it is transported in direction $2 \to 1$, and negative in the opposite case. The unit cost of transporting between the parallel depots is d, independent of the direction, where $d + \gamma_1 > \gamma_2$ and $d + \gamma_2 > \gamma_1$.

Objective:

The expected total cost of the system can be written as

$$C(x_1, x_2, y) = \gamma_1(x_1 - I_1) + \gamma_2(x_2 - I_2) + d|y| +$$

$$+ r_1 \int_0^{x_1+y} (x_1+y-S)f_1(S)\,dS + v_1 \int_{x_1+y}^{\infty} (S-x_1-y)f_1(S)\,dS +$$

$$+ r_2 \int_0^{x_2+y} (x_2-y-S)f_2(S)\,dS + v_2 \int_{x_2-y}^{\infty} (S-x_2+y)f_2(S)\,dS.$$

Solution:

Using the necessary optimality conditions

$$\partial C/\partial x_1 = 0 \quad \text{and} \quad \partial C/\partial x_2 = 0,$$

the following system is derived for the optimal x_1^* and x_2^*:

$$x_1 + y = x_1^* \quad (x_1 > I_1),$$
$$x_1 + y = I_1 + y \geqq x_1^* \quad (x_1 = I_1),$$

and

$$x_2 - y = x_2^* \quad (x_2 > I_2),$$
$$x_2 - y = I_2 - y \geqq x_2^* \quad (x_2 = I_2).$$

Depending on the initial stock level vector (I_1, I_2), four different cases have to be considered for the calculation of the optimal policy:

I. $x_1 > I_1$ and $x_2 > I_2$,

II. $x_1 > I_1$ and $x_2 = I_2$,

III. $x_1 = I_1$ and $x_2 > I_2$,

IV. $x_1 = I_1$ and $x_2 = I_2$.

For example, in case I, we have that

$$x_1 + y = x_1^* \quad (x_1 > I_1, y \geqq 0)$$
$$x_2 - y = x_2^* \quad (x_2 > I_2, y \leqq 0)$$

thus $y = 0$, $x_1 = x_1^*$ and $x_2 = x_2^*$; this is valid only when $I_1 > x_1^*$ and $I_2 > x_2^*$,

353

since $I_1 < x_1 = x_1^*$ and $I_2 < x_2 = x_2^*$. This implies the following optimal policy: both stores should order the quantity $x_i - I_i$ and there is no redistribution. The other cases can be treated similarly.

The outlined model can be generalized easily for more than 2 parallel stores, but in these cases the above procedure is very inefficient. An iterative procedure is illustrated for two depots which can be extended for the multi-store case in a more efficient way (Model 289).

All the variables are nonnegative if we introduce the notations $z_i = x_i - I_i$, furthermore, y_{12} is the transport in direction $1 \to 2$, and y_{21} in direction $2 \to 1$. The natural assumptions $0 \leq y_{12} \leq I_1$ and $0 \leq y_{21} \leq I_2$ are also introduced. The decision variables are collected in the set

$$H = \{x_1; x_2; x_3; x_4\}$$

where

$$x_1 = z_1$$

$$x_2 = z_2$$

$$x_3 = y_{21} \quad \text{and}$$

$$x_4 = y_{12}.$$

Let us define an iteration process in the following way. Assume that the values $x_i^{(m)}$ have resulted after the m-th step of iteration. Let U_i denote the upper bound for x_i, then $U_1 = U_2 = +\infty$; $U_3 = I_1$ and $U_4 = I_2$. The $(m+1)$-st iteration cycle consists of the following steps:

1. The partial derivatives of the cost function $\partial C / \partial x_i$ are calculated (using the new notations);
2. A subset is defined by the inequalities

$$\frac{\partial C}{\partial f_i} > 0 \quad \text{if} \quad x_i^{(m)} > 0 \quad \text{and}$$

$$\frac{\partial C}{\partial x_i} > 0 \quad \text{if} \quad x_i^{(m)} < U_i;$$

3. The partial derivative with largest absolute value is chosen; let us denote this by $\partial C / \partial x_{i_0}$;
4(a) If $\partial C / \partial x_{i_0} > 0$, the n x_{i_0} is decrease duntil $\partial C / \partial x_{i_0} = 0$ or $x_{i_0} = 0$ is reached;
 (b) If $\partial C / \partial x_{i_0} < 0$, then x_{i_0} is increased until $\partial C / \partial x_{i_0} = 0$ or $x_{i_0} = U_{i_0}$ is reached;
 (c) If the subset defined in (2) is empty, then the procedure is finished.

289. Parallel inventory locations with identical demands

Main codes:

$$1\ 8\ 0\ 0\ 0\ 1\ 4\ \ -1\ 0\ 0$$

Assumptions:

The system of a central and M-subordinated parallel stores is considered. The demands against the parallel stores are identical with known deterministic inten-

354

sity λ. The central store places orders of amount Q at fixed ordering intervals with length T, and these orders are distributed in lots of amount q among the parallel stores. $Q=kq$ and $M=rk$, where r is a positive integer. The fixed ordering cost is A at the central store and the distribution cost of a lot is "a" at each parallel store. The unit inventory holding cost is I. The value of k is chosen in such a way that $M \cdot A = k^2 a$ is valid. No shortage is allowed.

Objective:

The total cost of inventory holding, ordering and distribution,

$$C(Q, q) = M \frac{q}{2} I + M \frac{\lambda}{q} a + \frac{M\lambda}{Q} A$$

is to be minimized.

Solution:

The optimal amount of central order and distribution lot sizes are

$$Q = \sqrt{\frac{2M\lambda A}{I}} \quad \text{and} \quad q = \sqrt{\frac{2\lambda a}{I}}.$$

The optimal length of the central ordering interval is

$$T = \sqrt{\frac{2(A+ma)}{IM\lambda}}.$$

290. Parallel inventory locations with different demands

Main codes:

$$1\ 8\ 0\ 0\ 0\ 1\ 4\ -1\ 0\ 0$$

Assumptions:

This model is a generalized version of the previous one in the sense that demand at the different parallel stores may be different but has a constant rate λ_i for store i. The central order with ordering cycle T is distributed among the parallel stores. The i-th store receives a delivery at each time interval with length $N_i T$. where N_i is an integer. The central ordering cost is A, the distribution cost is a_i for depot i. The unit cost of inventory holding is denoted by I. No shortage is allowed.

Objective:

The total cost is

$$k(T, N_1, ..., N_M) = \frac{A}{T} + \sum_{i=1}^{M} \frac{a_i}{N_i T} + I \sum_{i=1}^{M} \frac{N_i \lambda_i T}{2}.$$

Solution:

An iterative procedure is suggested with starting values $N_i=1$. The iteration steps are the following:

1.

$$T_0 = \sqrt{\frac{2\left|A + \sum\limits_{i=1}^{M} \dfrac{a_i}{N_i}\right|}{I \sum\limits_{i=1}^{M} N_i \lambda_i}}.$$

2. Choose the largest integer N_i that satisfies the inequality

$$N_i(N_i - 1) \leqq \frac{2a_i}{T_0^2 I \lambda_i} \qquad i = 1, ..., M$$

and apply it in (1) (this yields a new value T_1 instead of T_0, etc.). The convergence of the above procedure is proved.

291. Optimal ordering frequency for central parallel stores

Main codes:

$$1\ 8\ 0\ 0\ 0\ 1\ 4\ -1\ 0\ 0$$

Assumptions:

Similarly to the previous model, the demand of M parallel stores is continuous and has the known rate λ_i for store i. The central store places orders periodically with the period length TN. The i-th store receives a delivery from the central store also periodically with a period length TN_i, where N_i and N/N_i are integers $(i=1, ..., M)$. The ordering costs are A and a_i for the parallel stores and A' for the central store. The unit cost of inventory holding is I_1 at the central store and I_2 at the parallel stores.

Objective:

The total inventory holding and shortage cost of the system,

$$K(T, N, N_1, ..., N_M) = \frac{1}{T}\left(\frac{A'}{N} + A + \sum_{i=1}^{M} \frac{a_i}{N_i}\right) +$$

$$+ \frac{T}{2}\left(\sum_{i=1}^{M}(N - N_i)\lambda_i I_1 + \sum_{i=1}^{M} N_i \lambda_i I_2\right)$$

has to be minimized.

Solution:

The optimal value is $T = \sqrt{2a/Q}$ with

$$a = \frac{A'}{N} + A + \sum_{i=1}^{M} \frac{a_i}{N_i}$$

and

$$Q = \sum_{i=1}^{M}(N - N_i)\lambda_i I_i + \sum_{i=1}^{M} N_i \lambda_i I_2.$$

The optimal values N and N_i $(i=1, ..., M)$ are the largest integers which satisfy

356

the following inequalities:

$$N(N-1) \leqq \frac{A'(\sum_{i=1}^{M} N_i \lambda_i)(I_2 - I_1)}{I_1(\sum_{i=1}^{M} \lambda_i) A + \sum_{i=1}^{M} \frac{a_i}{M_i}}$$

and

$$N_i(N_i-1) \leqq \frac{a_i[NI_1 \sum_{j=1}^{M} \lambda_j + \sum_{j \neq i} N_j \lambda_j (I_2 - I_1)]}{\left(\frac{A'}{N} + A + \sum_{j \neq i} \frac{a_j}{N_j}\right) \lambda_i (I_2 - I_1)}.$$

292. Stochastic inventory and transportation model for N stores

Main codes:

1 8 0 1 0 1 2 2 1 0

Assumptions:

The system consists of N centrally coordinated stores which have independent demands generated by Poisson processes at each store with different parameters. A central (s, q) policy, a normal (slow) and a fast (instant) redistribution among the parallel stores is considered. At any fixed moment only one store may receive inventory by redistribution. The following costs are considered:

— inventory holding cost: $c_1(q)$;
— shortage cost at the parallel stores with fixed and time-dependent factor: $\pi_j + \bar{\pi}_j \cdot t$;
— shortage cost when shortage appears in the whole system: $\pi + \bar{\pi} \cdot t$;
— transportation costs per year which depend on the mode of transportation: J;
— cost of the central ordering with fixed ordering cost c_3 and purchasing cost $K(q)$ for a unit item.

Objective:

The purpose of the model is to determine
— the time and amount of central ordering,
— the distribution policy of the central inventory among the parallel stores,
— the fast redistribution policy among the parallel stores.

The expected total cost of the system equals

$$C(s, q) = \frac{\lambda}{q} c_3 + J + \lambda K(q) + c_1(q)\left[\frac{q}{2} + s + \frac{1}{2}\right] + \pi E(s, q) + [\bar{\pi} + c_1(q)] B(s, q),$$

where $E(s, q)$ is the expected value of the shortage and $B(s, q)$ is the expected value of the demand satisfied with delay.

Solution:

The method of dynamic programming is used to determine the optimal policy.

293. *A two-echelon inventory model for low demand items*

Main codes:

$$1\ 8\ 0\ 1\ 1\ 3\ 5\ 2\ 1\ 1$$

Assumptions:

The inventory system of reparable spare parts is considered which consists of a central and of j local stores. Demand for spare parts occurs at the local stores according to a Poisson process with parameter λ_j at store j. The demand is instantly satisfied if it is possible, otherwise it is backordered and satisfied later according to a first-in-first-out rule. The failed items are examined in the local store and repaired if it is possible, with probability r_j. The time of reparaticn is R_j at the local store. The failed item is repaired in the central store with probability $(1-r_j)\varrho$, and $(1-r_j)(1-\varrho)$ is the probability that it is not reparable. When a broken item has been exchanged for a good one at a local store, immediately an order is placed to the central store which is delivered with a leadtime t_j, if there is no shortage in the central store. The time of reparation of a faulty item in the central store is R_0. The leadtime of the delivery for the central store is t_0 ($R_0 \leqq t_0$). The local stores are controlled by an (S_j-1, S_j) policy and the central store by an (s, S) policy.

Objective:

No explicit cost function is derived. The purpose is to analyze the stationary properties of the system for given decision parameters.

Solution:

The stationary probability distribution of the inventory level, of the number of spare parts under reparation and of the amount of shortage is determined for the central and local stores; the application possibilities of these characteristics are also outlined.

XI.1.1.2. Parallel Stores without a Central Store

294. *Multi-echelon system with deterministic demand*

Main codes:

$$1\ 8\ 0\ 0\ 0\ 1\ 7\ -1\ 0\ 0$$

Assumptions:

The model is constructed for the supply of a homogeneous item for a large region with a constant demand rate for each unit of territory. A multi-echelon inventory system is considered, where

— the stores of each echelon are uniformly distributed on the respective territory of the consumers;
— each store of an echelon serves a unique territory;
— the delivery between two locations is a linear function of the distance: $g=a+bd$ and is not discounted by the amount of delivery;
— the territory of service is a circle.

The system consists of three types of warehouses:

— a central warehouse which supplies the others within a territory sized S,
— n_2 is the number of class two warehouses;
— $n_1 \cdot n_2$ is the number of class one warehouses.

The model investigates the following three possible ways of the service organization:

— at each delivery the demand of a single store on a lower level is satisfied;
— the delivered amount is uniformly distributed among the stores of the lower level;
— the delivery is distributed among a group of stores according to some combination of the previous two policies.

Objective:

The total cost of the central store for a unit of time, with order interval length T, is equal to

$$L(T) = \frac{\mu s ST}{2} + \frac{q}{T},$$

where
   μ is the demand per unit of territory,
   s is the unit cost of inventory holding,
   S is the area of territory to be supplied.

Solution:

The optimal T is calculated from the equation $dL(T)dt=0$. This is a function of the delivery costs g, which is influenced by the type of service organization. The influence of these factors on the total cost is analyzed for the above three cases.

295. Distribution and redistribution for n stores with stochastic demand

Main codes:

$$1\,8\,0\,1\,0\,1\,2\,2\,1\,0$$

Assumptions:

A system of n parallel stores is considered, where the demand of store j is a random variable with probability density function $f_j(x)$. The initial inventory is distributed among the stores so that each store has an inventory z_j. Later on, there is a possibility for redistribution of the inventory among the stores. The amount q_{ij} is transported from store i to store j with unit transportation cost c_{ij}. The set of store indices which increase the stock at the redistribution is denoted by M^-, while the set of indices M^+ denotes the decrease of stocks.

Objective:

The distribution of the inventory is to be determined in such a way that the expected total cost of inventory redistribution and shortage

$$L = \sum_{j \in M^-} p_j \int\limits_{z_j + \sum\limits_{i \in M^+} q_{ij}}^{\infty} \left(x - z_j - \sum_{i \in M^+} q_{ij}\right) f_i(x)\, dx +$$

$$+ \sum_{i \in M^+} p_i \int\limits_{z_i - \sum\limits_{j \in M^-} q_{ij}}^{\infty} \left(x - z_i + \sum_{j \in M^-} q_{ij}\right) f_i(x)\, dx + \sum_{j \in M^-} \sum_{i \in M^+} c_{ij} q_{ij}.$$

is minimal, where p_i denotes the unit cost of shortage in store i.

Solution:

The algorithm is based on the solution of equations concerning the derivatives of the objective, while an inner transportation problem is also to be solved (concerning the redistribution for the determination of the sets M^- and M^+).

296. *Inventory redistribution by simulation*

Main codes:

$$1\ 8\ 0\ 1\ 1\ 1\ 4\ 1\ 1\ 1$$

Assumptions:

The initial stock of two parallel stores is denoted by S_1 and S_2 (the model can be generalized for an arbitrary number of stores). The time until the next central delivery is subdivided into t periods. At the beginning of each period a redistribution is possible between the parallel stores. The demand is generated by a Poisson distribution at each store.

Objective:

If the amount R is transported from store 2 to store 1, its cost for the t-th period is

$$C(R, t) = C_1(S_1 + R, t) + C_2(S_2 + R, t) + \alpha_t(kR + K),$$

where $kR + K$ is the cost of transportation and administration, C_1 and C_2 are the shortage cost functions of the stores and α_t is a discount factor depending on the period of the redistribution. The optimal R for each period t can be easily calculated for given α_t. The purpose of the model is to determine the optimal value of α_t.

Solution:

A simulation method has been developed for determining the optimal α_t which provides a joint optimum of the individually determined optimal values R of each period.

360

297. Stock allocation model for two stores with joint demand

Main codes:

1 8 0 1 1 1 5 1 1 0

Assumptions:

A certain item is stored in two parallel locations. At equidistant points in time, random demands occur at each location, which are stochastically dependent and have a joint probability density function $\varphi(\xi)$. For each period in which an order is placed, the vector of initial stocks is x, and the vector of inventory levels after ordering is y. The leadtime is zero. The joint inventory of the two stores is bounded by the capacity R, the occupation rate is $r=(r_1, r_2)$ per unit amount. Let c denote the unit cost of purchasing. The total inventory holding and shortage cost is described by the function $L(y)$ which is supposed to be strictly convex, twice continuously differentiable and can be separated for the two stores.

Objective:

The expected discounted total cost of n periods is to be minimized. Using the principle of dynamic programming, this can be formulated by the following functional equation:

$$C_n(\underline{x}) = \min_{\substack{y \geq x \\ R \geq ry}} \left\{ c(\underline{y}-\underline{x}) + L(\underline{y}) + \alpha \int_{\underline{\xi}>0} C_{n-1}(\underline{y}-\underline{\xi}) \varphi(\underline{\xi}) \, d\underline{\xi} \right\}$$

$$C_0(\underline{x}) = 0,$$

where α is the discount factor.

Solution:

The optimal cost function can be derived recursively, then the critical vector x_n can be calculated for all n which determines the optimal reorder point s_n and order level S_n in each period. In the case of $C_1/r_1 = C_2/r_2$, the critical vector x_n is identical for each n and can be calculated in a simpler way.

298. Optimal policy for a dynamic multi-echelon system

Main codes:

1 8 0 1 1 1 5 1 1 0

Assumptions:

The model considers an inventory system with n facilities, each of which carries stocks of a single commodity. The random demand for the item at facility j in period i is D_{ij}, and Φ_i denotes its joint distribution. At the beginning of period i, $x_i = (x_{ij})$ denote the inventories on hand at the facilities. A negative x_{ij} means a backlogged demand.

An order can be placed at the beginning of each period, and delivery takes place immediately. The inventories on hand after order are denoted by $y_i = (y_{ij})$. Thus $y_i - x_i$ is the vector of order quantities placed in period i at each of the n facilities. y_i must be chosen from a set Y_i of n-dimensional vectors. $S_i(y_i, D_i)$

361

is the supply policy, which specifies the amount of stock on hand after the demand occurs in period i.

The ordering cost in period i is c_i and other costs (storage, shortage, redistribution cost) are included in the g_i function.

Objective:

The expected total cost is to be minimized. The ordering, inventory holding, shortage and inventory redistribution costs are considered in the cost function:

$$W_i(y, t) = c_i y + g_i(y, t) - c_{i+1} s_i(y, t).$$

Solution:

The problem of determining the optimal policy is solved by a minimization of an n-dimensional problem. The optimal policy is to order up to the base stock level $\bar{y}_i$ in period i. The authors make specific assumptions about Y_i, s_i, g_i and characterize the properties of the model.

XI.1.2. Multi-Echelon Systems with Hierarchical Locations

299. *Central store–local store system with deterministic demand*

Main codes:

$$1\ 8\ 0\ 0\ 1\ 1\ 4\ -1\ 0\ 0$$

Assumptions:

The central store satisfies the order, for an amount Q on each occasion, of a local store, when the inventory level of the local store decreases to zero. The orders are delivered immediately. When the inventory level of either the central store or of the local store decreases to zero, central ordering can be initiated.

The inventory holding cost of the central store is h_0 per unit time and unit of item, the local unit cost of inventory holding is h with the same dimension. The fixed cost of a central order is K_0 and of a local order is K. No shortage is allowed.

Objective:

The cost of ordering and inventory holding of the two stores for a unit of time is

$$C(n, Q) = \frac{(K_0 + nK)D}{nQ} + Q\frac{nh_0 + h}{2},$$

where D is the rate of demand at the local store and n is the number of local orders during a central ordering cycle.

Solution:

The optimal policy is determined by the parameters which satisfy

$$Q^* = \sqrt{\frac{2(K_0 + n^* K)D}{(n^* h_0 + h) n^*}}$$

and

$$n^*(n^*-1) \leq \frac{K_0 h}{K h_0} \leq n^*(n^*+1).$$

300. *Optimal lot-sizes for a production line*

Main codes:

$$1\ 8\ 0\ 0\ 0\ 1\ 3\ -1\ 0\ 0$$

Assumptions:

The inventories in a production line are considered. The production is serial, the level i has the production lot sizes $Q_i=n_iQ_1$ (with an integer value of n_i). This is the amount of order of the level $(i-1)$ which is produced at a constant rate. It is stored in the production line for a certain time, then it is used up for the production of the level $(i-1)$ in lot sizes $Q_{i-1}=n_{i-1}Q_1$. The usage has also a constant rate. Between two series of production there is no usage. The production of the higher level provides for the supply of the lower level. The lot size of the final product is Q_1.

At each level of the production the following costs are considered:

— the setup cost of a production lot (S);

— the production cost (P);

— the unit cost of inventory holding and of capital invested (c_1).

The unit production cost is a monotonously decreasing exponential function of the lot-size. The inventory holding cost depends on the time of delivery and on the setup of a production lot.

Objective:

The total of the above costs in the production line equals

$$C(Q_i) = \sum_i \left\{ \frac{S}{Q_i} + \frac{P}{Q_i} + c_1\left[t_{2,1} - \frac{t_{1,i}}{2} + \frac{t_{1,i-1}}{2} + \frac{T_{i-1}}{2}\left(\frac{n_i}{n_{i-1}} - 1 \right) \right] \right\},$$

where $T_i=\frac{Q_i}{R}$, while $t_{1,i}$ is the time of delivery and $t_{2,i}$ is the time of setup of a production lot for the i-th level.

Solution:

The cost function cannot be minimized in a direct (analytical) way. An iterative procedure is suggested, starting with the optimal parameters of the individual stores; in this way, the joint optimum can be approximated with arbitrary accuracy.

301. *Chain of echelons with stochastic demand*

Main codes:

1 8 0 1 0 1 1 0 1 0

Assumptions:

The chain of echelons is a hierarchical sequence of storage locations where each subsequent store satisfies the demand of the previous store. Thus, each supplier is connected with a single customer. A three-echelon system of stores of a production process is considered. The first level is the finished-product store, the second one is an intermediate store and the third one is the raw material store. Only the demand of the finished product has to be considered, since the demand for lower-level inventories can be derived from this. A period with length T is considered which has a random demand with a probability density function $f(x)$. In the case of a shortage at store i the unsatisfied demand is transferred to the subsequent level $i+1$. The inventory is measured in units of completed goods at each level. The unit cost factors of shortage $p_{T,i}$ and the unit cost factors of inventory holding $s_{T,i}$ depend on the level of inventory Y_i: they increase together with the completion rate. Their dimension is [\$]/[amount]/[time].

Objective:

For given order levels Y_i ($i=1, 2, 3$) the total expected inventory holding and the shortage cost of the three levels is equal to

$$L = (s_{T,1}Y_1 + s_{T,2}Y_2 + s_{T,3}Y_3) \int_0^{Y_1} f(x)\, dx + s_{T,1} \int_0^{Y_1} x f(x)\, dx +$$

$$+ [s_{T,2}(Y_1+Y_2) + s_{T,3}Y_3] \int_{Y_1}^{Y_1+Y_2} f(x)\, dx - s_{T,2} \int_{Y_1}^{Y_1+Y_2} x f(x)\, dx +$$

$$+ s_{T,3}(Y_1+Y_2+Y_3) \int_{Y_1+Y_2}^{Y_1+Y_2+Y_3} f(x)\, dx + s_{T,3} \int_{Y_1+Y_2}^{Y_1+Y_2+Y_3} x f(x)\, dx -$$

$$- p_{T,1}Y_1 \left[1 - \int_0^{Y_1} f(x)\, dx\right] - p_{T,2}(Y_1+Y_2) \left[1 - \int_0^{Y_1+Y_2} f(x)\, dx\right] -$$

$$- p_{T,3}(Y_1-Y_2-Y_3) \left[1 - \int_0^{Y_1+Y_2+Y_3} f(x)\, dx\right] + p_{T,1} \int_{Y_1}^{\infty} x f(x)\, dx +$$

$$+ p_{T,2} \int_{Y_1+Y_2}^{\infty} x f(x)\, dx + p_{T,3} \int_{Y_1+Y_2+Y_3}^{\infty} x f(x)\, dx.$$

Solution:

The optimal order levels can be calculated from the system of equations

$$\partial L/\partial Y_i = 0 \quad i = 1, 2, 3.$$

364

302. *Two-echelon model with demand forecasting*

Main codes:

$$1\ 8\ 0\ 1\ 1\ 1\ 5\ 1\ 1\ 0$$

Assumptions:

Two stores are connected, the first echelon supplies the second one which supplies directly the demand of the customers. The estimation of the demand of the next period is η. The conditional expectation of the demand (ξ) on conditioning the estimation η is denoted by $E_{\xi|\eta}$, the expectation of the forecasting by E_η. The unit purchasing costs are denoted by c_1 and c_2. The expected cost of inventory holding and shortage of the first store is a convex function $L(y_1)$ of the order level y_1; at the second store, $L_2(y_2|\eta)$ is also a convex function of y_2 for any forecasting η, where y_2 means the order level of the second store.

Objective:

The cost optimum of n periods is denoted by $f_n(x_1, x_2)$ for an initial stock vector (x_1, x_2), the same cost optimum by forecasting η is denoted by $g_n(x_1, x_2, \eta)$. For these cost functions the following functional equations are valid (by applying a discount factor α):

$$g_n(x_1, x_2, \eta) = \min_{x_1 \leq y_2 \leq x_1} \{c_2(y_2 - x_2) + L_2(y_2|\eta) - \alpha E_{\xi|\eta}[f_{n-1}(x_1 - \xi, y_2 - \xi)]\},$$

$$f_n(x_1, x_2) = \min_{y_1 \geq x_1} \{c_1(y_1 - x_1) + L_1(Y_1) + E_\eta[g_n(y_1, x_2, \eta)]\}.$$

Solution:

The optimal cost functions f_n and g_n can be separated under the conditions given above. The consequence of this fact is that the optimal ordering policy of the first store is of the (s, S) type and the optimal order amount for the second store can be determined as follows:

$$\min\{\bar{x}_n(\eta) - x_2, S_n - x_2\} \quad \text{if} \quad x_1 \leq s_n \quad \text{and} \quad x_2 \leq \bar{x}_n(\eta),$$
$$\min\{\bar{x}_n(\eta) - x_2, x_1 - x_2\} \quad \text{if} \quad x_1 > s_n \quad \text{and} \quad x_2 \leq \bar{x}_n(\eta),$$
$$0 \quad \text{if} \quad x_2 > \bar{x}_n(\eta).$$

The values of s_n, S_n and $\bar{x}_n(\eta)$ can be calculated from respective single-variable cost functions using a recursive relation.

XI.2. Multi-Item Multi-Location Models

303. *Supply policy in a two-level multi-item multi-store system*

Main codes:

$$8\ 8\ 0\ 0\ 0\ 1\ 7\ -1\ 0\ 0$$

Assumptions:

A central store and m local stores are considered. The stores of the lower level are connected with each other only through the central store. The number of suppliers is n and each type of item can be ordered only from a single supplier. The model deals with the organization of supply which ensures a minimal replenishment and inventory holding cost. There are two different ways of organizing supply:

1. Each supplier delivers immediately to the store of the lower level (with cost L_1):
2. The supply of the lower level takes place through the central store: in this case, the delivery cost from the central to the local stores is L_2 and the cost of supplying the central store is L_3.

Objective:

The costs L_1, L_2, L_3 are expressed using the following notations:

M_i: the set of items purchased from supplier i,
Q_j: the set of items delivered to the local store j,
μ_{rj}: demand for item r at store j, $r \in Q_j$,
g_{ij}: cost of delivery from supplier i to store j,
g_i: cost of delivery from supplier i to the central store,
g_{0i}: cost of delivery from supplier i to store j,
g_{0j}: cost of delivery from the central store to store j,
c_r: the unit cost of inventory holding for item r,
k_{rj}: the rate of the item r ordered at store j in a given period relative to the set of items Q_j $(0 \leq k_{rj} \leq 1)$,
γ: the fixed cost factor of each delivery (independent of r, i and j).
k_r: the rate of the item r ordered at the central store

$$L_1 = \sum_{j=1}^{m} \sum_{i=1}^{n} \sqrt{ 2 g_{ij} \left(\gamma \sum_{r \in Q_j \cap M_i} \frac{1}{k_{rj}} + 1 \right) \sum_{r \in Q_j \cap M} \mu_{rj} c_r k_{rj} },$$

$$L_2 = \sum_{j=1}^{m} \sqrt{ 2 g_{0j} \left(\gamma \sum_{r \in Q_j} \frac{1}{k_{rj}} + 1 \right) \sum_{r \in Q_j} \mu_{rj} c_r k_{rj} },$$

$$L_3 = \sum_{i=1}^{n} \sqrt{ 2 g_i \left(\gamma \sum_{r \in M_i} \frac{1}{k_r} + 1 \right) \sum_{r \in M_i} c_r K_r \sum_{j=1}^{m} \mu_{rj} }.$$

Solution:

The choice of the optimal supply system may be made directly by comparing the costs calculated using the above expressions.

304. *Multi-item multi-location model with redistribution and random demands*

Main codes:

8 8 0 1 0 1 4 0 1 0

Assumptions:

A multi-item model with parallel locations and a (t, q) inventory policy is considered. There are two main tasks: to solve first the system of cost-optimal supply at the beginning of each period, then—during the periods—to optimize the use of possible redistributions among the parallel stores depending on the realization of the random demands. If the shortage cost of each item is summarized, then the problem can be separated for the different items. The single-item Model 294 can be used for all of the items, independently of each other.

If the shortage cost is calculated in each store on the basis of the largest shortage of items, then the redistribution has to be determined considering the inventory level of all the items. In this case, a "key-item" exists, which is in maximal shortage. It is most probably the item with a maximal shortage probability:

$$\max_r \bar{p}_r \int_{Z_{\delta,r}}^{\infty} f_{\delta,r}(x)\, dx,$$

where $\bar{p}_r$ is the mean of the shortage costs for all of the stores, $Z_{\delta,r}$ is the total inventory of item r in the system, and $f_{\delta,r}$ is the probability density function of the total demand for item r.

Objective:

No explicit cost function is formulated. The main goal is to give the optimal redistribution policy.

Solution:

The following three main steps are suggested:

a) to choose the "key-item",
b) to derive the optimal redistribution according to the key-item using the algorithm of Model 294,
c) to give a redistribution for the other items: this has to ensure that the shortage cost of any item do not exceed the shortage cost of the key-item and the delivery costs are minimal.

305. *Dynamic lot-size model for a multi-stage assembly system*

Main codes:

$$8\ 8\ 0\ 0\ 1\ 1\ 4\ -1\ 0\ 0$$

Assumptions:

The considered multi-stage assembly system has the specific assumption that any of the stages of assembling may have more than one preceding stage, but it may have only a single stage of continuation. There are N stages F_n, and the finishing stage is denoted by F_N.

T periods of the system operation are considered. Each period has a known demand of finished product which may be different in different periods. A dynamic lot-size system is investigated. The production cost is supposed to be concave, the inventory holding cost is linear and the two cost components can be separated.

The following notations are introduced:

$q_{n,t}$:   the amount produced in period t in stage n,
$y_{n,t}$:   the inventory of period t and stage n,
$C_{n,t}(q)$: production cost of period t and stage n of amount q,

$H_{n,t}$:   the unit cost of inventory holding in the stage n at the end of period t,
r_t: the demand for finished product in period t.
$b(n)$:   the set of indices of stages preceding stage n,
$a(n)$:   the index of stages following n.

The following inequalities are supposed to hold

$$C_{n,t}(q) \leqq C_{n,t-1}(q) \quad \text{for} \quad n = 1, ..., N, \quad t = 2, ..., T \quad \text{and}$$

$$H_{n,t} \geqq \sum_{m \in b(n)} H_{m,t} \quad \text{for} \quad n = 1, ..., N, \quad t = 1, ..., T.$$

Objective:

The total production and inventory holding cost in the system is equal to

$$\sum_{n=1}^{N} \sum_{t=1}^{T} [C_{n,t}(q_{n,t}) + H_{n,t} y_{n,t}],$$

where

$$y_{n,t} = y_{n,t-1} + q_{n,t} - q_{a(n),t}$$

for $n = 1, ..., N-1; \quad t = 1, ..., T$

and

$$y_{N,t} = y_{N,t-1} + q_{N,t} - r_t; \quad q_{n,t} \geqq 0, \; y_{n,t} \geqq 0$$

$$y_{n,0} = 0$$

for $t = 1, ..., T, \; n = 1, ..., N$.

Solution:

The optimal solution can be determined by dynamic programming. The solution obtained has the following properties:

$$y_{n,t-1} q_{n,t} = 0; \; q_{a(n),t} = 0 \quad \text{implies that} \quad q_{n,t} = 0;$$

$$q_{n,t} > 0 \quad \text{implies that} \quad q_{a(n),t} > 0, \quad t = 1, ..., T, \quad n = 1, ..., N.$$

306. Experience concerning the forecasting and inventory control of a warehouse system

Main codes:

$$8\ 8\ 0\ 1\ 1\ 3\ 4\ 1\ 2\ 1$$

Assumptions:

In a warehouse-system an (s, q) policy is applied for inventory control. Forecasting is used to determine the new values of the reorder point and order quantity. The forecasting and control parameters are compared with the actual inventory levels for each item to correct the order levels for each item.

368

Objective:

No explicit cost function is derived. The purpose of the simulation study is to provide an efficient control of the system.

Solution:

A simulation method has been developed. First, forecasting is made for the expected sales. The slow moving items are listed. In the case of frequent shortage, the decision parameters are corrected.

The paper describes the results of the practical application of the inventory control system: a simultaneous decrease of the inventory level and time of shortages.

XII. Reliability-Type Inventory Models

The joint characteristic of this family of models is that they focus on the reliability of the inventory system on the service level. Generally, they do not contain
an explicit cost function to be minimized (or a utility function to be maximized)
but the basic criterion of decision making is the fulfillment of some reliability
(service level) constraint. The major part of the models consider random delivery
processes, in which case the undisturbed satisfaction of demand is especially
difficult and important.

For constructing subgroups, the criteria of classification are the number of
items and the form of the inventory policy. In the models of the first subgroup,
the inventory review is periodical and at each review point an order is placed.
The amount of order is delivered after a known or random time interval (leadtime)
in one lot. The demand during the leadtime is random and has a known distribution type (e.g. a normal or gamma distribution) with unknown parameters
which are estimated (e.g. based on past demand characteristics). The decision
variable (which is either the amount of order or the order level) is to be determined by considering the random demand in such a way that the continuous supply
is ensured on a given service level. The service level may be prescribed by the
maximal permissible probability or expectation of shortage. The solution is
usually simple: after estimating the parameters of the demand distribution, the
inverse of this distribution has to be calculated for a given argument. (This may
be realized by numerical methods or by tabulating the distribution function.)

The inventory review and ordering are periodical also in the second subgroup,
but the delivery of a demand is realized not on one occasion but at multiple
occasions which are not known in advance. The supplier undertakes to deliver
the whole amount before the end of the period, but the times and amounts
of the transports within this period are not specified in advance. The delivery
process, which is considered as a random process at the time of ordering, has
different patterns. The delivery may occur at random points with different distributions while the transported lots may be known, or also random amounts with
different distributions. The purpose is to determine the initial inventory level in
such a way that the known or random demand should be continuously satisfied
with a prescribed probability. Depending on the pattern of delivery and demand,
the solution can be calculated by a simple formula, by a numerical procedure, or
by the use of simulation techniques. The inventory level at the end of a period
(as the initial stock for the next period) has to ensure the prescribed service level.
The appropriate level of this inventory can be attained by the order for the previous period.

In the models of the third subgroup, continuous reviewing is considered, and orders may be placed at an arbitrary time. The instants of ordering are determined by the reorder point: it has to be fixed in such a way that the probability of shortage may not exceed a prescribed level during the leadtime. The demand of the leadtime is random with some known distribution (normal, gamma, Weibull distribution), where the parameters may have known or estimated values. These models usually do not deal with the optimization of the lot-size, its quantity is determined independently of the reorder point. The economic order quantity (Wilson formula) or its different generalized versions are usually applied in practice. The methods for finding the reorder point are simple: similarly to the models of the second subgroup, they consist of the estimation of the parameters and of calculating the inverse of the distribution function.

The models of the fourth subgroup jointly determine the inventory level of several items considered. The purpose is to ensure an uninterrupted supply on the highest possible level under financial or storage capacity restrictions. The objective may be the maximization of the continuous supply probability (reliability maximization) or the minimization of the expected maximal shortage (decreasing the bottle-neck). Some models have the objective to minimize the capital invested in inventories by a prescribed minimal level of the probability of continuous supply. The determination of the exact joint optimum of the inventory levels usually leads to a rather difficult nonlinear programming problem, where the function values may be determined only by Monte-Carlo techniques. In practice, simple approximate formulas and approximate solution methods (based on Lagrange multipliers) are frequently derived.

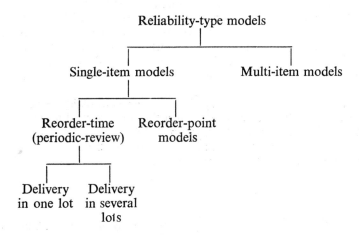

In the following a typical reliability-type model (belonging to subgroup XII.1.1.2.) is described in detail.

307. Reliability-type model for random demand and delivery

Main codes:

$$1\ 1\ 1\ 1\ 0\ 2\ 1\ 1\ 7\ 4$$

Assumptions:

The inventory is replenished with a fixed ordering frequency. The length of the order period T is prescribed. For the beginning of each period an initial stock is planned. It serves as the safety stock of the period for eliminating interruptions in demand satisfaction caused by random factors. Continuous supply is to be provided on a prescribed service level. (Service level is often measured by the probability of continuous supply.)

In the general model, both demand and delivery are random processes. This supposition is often necessary, since the information available at the time of decision making (ordering) is not sufficient for a deterministic description of any of the two processes. Consider the ordering period $[0, T]$. The cumulated demand of the period $[0, t]$ with $t \leq T$ is $\xi(t)$, and the cumulated delivery of the same period is $\eta(t)$, both being random amounts. The initial stock of the period is denoted by M. There is no shortage in the period investigated if the inequality

$$\xi(t) \leq M + \eta(t)$$

is valid for each $0 \leq t \leq T$. Due to the random nature of the processes, the fulfillment of the above conditions can be guaranteed usually only with a given probability. This is the value of the service level $1-\varepsilon$, prescribed in practice in the range $0.7 \leq 1-\varepsilon \leq 0.99$, depending on the importance of continuous supply.

In the reliability-type models, different demand and delivery processes are considered. One of the most important cases is when the deliveries of an order occur at n random instants in the period $[0, T]$.

If, at the time of ordering, no *a priori* information is available about the distribution of the time instants of deliveries, then they are assumed to occur at any moment of the period $[0, T]$ with the same probability. Thus the elements of an ordered sample taken from the uniform probability distribution in $[0, T]$ are considered as a realization of the delivery instants:

$$0 \leq t_1^* < t_2^* < \ldots < t_n^* \leq T.$$

(As the events $t_i^* = t_j^*$ (for some pair i, j $i \neq j$) are of zero probability, their possible occurrence is neglected here, and also in the sequel.)

The amounts delivered may have a random character, too. Suppose that the minimal amount δ, which arrives at a delivery with certainty, can be estimated. From the amount R ordered the remaining part $R - n\delta$ is randomly subdivided among the lots delivered by $n-1$ random points $\tau_1^* < \tau_2^* < \ldots < \tau_{n-1}^*$, which are uniformly distributed on the interval $[0, R - n\delta]$. The respective amounts (defined by the random subintervals of $[0, R - n\delta]$) are assigned to the delivery points and together with the fixed amount δ represent the amounts of deliveries.

372

This model of a delivery process can be described by the function

$$F_n(t, \lambda) = \begin{cases} 0 & \text{if} \quad 0 \leq t \leq t_1^* \\ \lambda R \dfrac{k}{n} + \tau_k^* & \text{if} \quad t_k^* < t \leq t_{k+1}^* \quad k = 1, ..., n-1, \\ R & \text{if} \quad t_n^* < t \leq T, \end{cases}$$

where the factor $\lambda = n\delta/R$ is the ratio of the deterministic and the total amount of deliveries.

The demand process is constructed similarly. The total demand of the period $[0, T]$ occurs at m random points uniformly distributed in time. The minimal amount demand is denoted by γ, and furthermore, $\mu = m\gamma$. The total amount of demand is known and is equal to the amount of order R. The demand $R - m\gamma$ is randomly subdivided among the m instants of demand, according to a uniform distribution, similar to the delivery process.

The initial stock of a period can be influenced by the ordering for the previous period. Thus, the planning of the appropriate stock must be performed one period ahead. These types of models control according to the (t_p, S) policy, where the order level S is determined as the sum of demand and the necessary closing stock of the following ordering period. The necessary closing stock is M, which is the required initial stock of the subsequent period.

Objective:

The purpose of the reliability-type inventory model is to determine the minimal level of the initial stock M which assures continuous supply of the ordering period on a prefixed probability level $1 - \varepsilon$. It means the solution of the following constrained optimization problem:
Find the minimum of M subject to

$$P\big(\xi(t) \leq M + \eta(t), \; 0 \leq t \leq T\big) \geq 1 - \varepsilon.$$

If the above probability is a strictly monotonous and continuous function of M, then the solution of this problem is equivalent to the solution of the following equation:

$$P\big(\xi(t) \leq M + \eta(t), \; 0 \leq t \leq T\big) = 1 - \varepsilon.$$

The latter is referred to as the "reliability equation" and it is usually described in the equivalent form

$$P\big(\sup_{0 \leq t \leq T} \{\xi(t) - \eta(t)\} \leq M \big) = 1 - \varepsilon.$$

Solution:

The exact solution of the reliability equation in the general case is a difficult task, but for practical applications it is often sufficient to derive an approximation of the optimal solution. For the processes $\xi(t)$ and $\eta(t)$ described above, the following approximation can be given:

$$P\left(\sup_{0\leq t\leq T}\{\xi(t)-\eta(t)\}\leq M\right)\approx$$

$$\approx 1-\exp\left\{-\frac{2M^2}{R^2}\sqrt{\frac{mn}{m+n+m(1-\lambda)^2+n(1-\mu)^2}}\right\},$$

which results in an approximate solution of the reliability equation in the explicit form

$$M\approx R\sqrt{\frac{1+(1-\lambda)^2}{n}+\frac{1+(1-\mu)^2}{m}}\cdot\sqrt{\frac{1}{2}\ln\frac{1}{\varepsilon}}.$$

This approximation is valid for sufficiently large m, n (for $m, n \geq 10$). The exact solution has not yet been derived.

XII.1. Reliability-Type Inventory Models for a Single Item

XII.1.1. Models with Fixed Ordering Period

XII.1.1.1. Delivery in a Single Lot

308. *Prescribed relative service level*

Main codes:

$$1\ 1\ 0\ 1\ 0\ 2\ 1\ 1\ 2\ 7$$

Assumptions:

The demand of a period is a random, normally distributed amount with an expected value M and standard deviation D. The order level is determined by $S=M+\lambda D$, where λ denotes the safety factor to be optimized. By definition, the relative service level is γ, if the ratio of the satisfied demand exceeding the average demand over the total demand has the expectation γ. The value γ is a prescribed constant.

Objective:

The value of the safety factor λ is determined as a function of the prescribed service level γ, as the solution of the equation

$$\gamma = \Phi(\lambda)+\frac{\lambda}{2\sqrt{\pi}}\int_\lambda^\infty\frac{1}{v}e^{-v^2/2}\,dv,$$

where Φ denotes the standard normal probability distribution function.

Solution:

The values of γ can be tabulated as a function of λ. Simple numerical procedure can also easily be derived for the solution of the reliability equation.

309. *Reliability model for the lost sales case*

Main codes:

$$1\ 1\ 0\ 1\ 1\ 2\ 4\ 1\ 2\ 1$$

Assumptions:

The length of the order period is prescribed. The leadtime has the length of n ordering periods. The demand x_k of the k-th ordering period is a random amount with a normal distribution, and the expectation and standard deviation are denoted by m and δ. The demands of different periods are independent. At shortage, the demand is lost.

Objective:

The probability of a shortage occurrence can be expressed by $n+1$ linear inequalities which contain the partial sums of the random amounts x_k ($k=0, 1, ..., n$). The reliability constraint is expressed in terms of the probability of continuous supply:

$$P(x_{k+1}+...+x_n \leq q_k+q_{k+1}+...+q_n) = 1-\varepsilon \quad k = 0, 1, ..., n,$$

where q_0 denotes the initial stock and q_k is the amount to be ordered in period k.

Solution:

By the transformation of the reliability criteria, the determination of q_k is reduced to a calculation of the values of an $n+1$-dimensional standard, but correlated, normal distribution function. The solution is defined by the inverse function which can be determined in two dimensions by using the tables of the function values. In higher dimensions, the Monte-Carlo method can be applied.

310. *Reliability model for a gamma distributed demand*

Main codes:

$$1\ 1\ 0\ 1\ 0\ 2\ 1\ 1\ 7\ 1$$

Assumptions:

The demand per unit time has a gamma distribution with parameters k and μ_λ. The leadtime L is constant. The length of the order interval is prescribed. The leadtime demand has the density function

$$f(x) = (\alpha^k x^{k-1} e^{-\alpha x})/\Gamma(k) \quad \text{with} \quad \alpha = k/\mu.$$

$\Gamma(k)$ denotes the complete gamma function defined as

$$\Gamma(k) = \int_0^\infty \alpha^k x^{k-1} e^{-\alpha x} \, dx.$$

Objective:

Two different reliability constraints are considered:

1. the probability of shortage occurrence $P = \int_R^\infty f(x)\,dx$;

2. the expected value of the amount short

$$L_s = \int\limits_R^\infty (x-R)f(x)\,dx.$$

Solution:

For the reliability constraint of type 1, the reorder level R is determined by an iterative process based on Newton–Raphson iteration. The initial value of the iteration can be selected by using the tables of the gamma distribution. If the value of the parameter k is greater than 50, an approximate formula can be applied (based on the approximation of a gamma distribution by normal distribution).

In the case of a reliability constraint of type 2, the order level can be determined using the tables of the incomplete gamma function. Approximating the demand by a normal distribution, an explicit formula is derived for the order level which approximates the required service level. The numerical approximation errors are tabulated.

311. *Reliability model for slow-moving items*

Main codes:

$$1\ 1\ 0\ 1\ 0\ 2\ 1\ 1\ 2\ 0$$

Assumptions:

The length of the reviewing period is prescribed. In the majority of the periods no demand occurs. The probability of demand in a period is $1/p$, thus, on the average, after every p periods a demand appears. The quantity of demand is also random, having a normal distribution with parameters μ and σ. At the end of each period, when demand occurs, an order is placed which is delivered at the beginning of the next period. At shortage, the demand is lost. L is the unit cost of the item, $100\,h$ is the percentage cost of stockholding per review interval, and C_n is the replenishment cost.

Objective:

The order level S is to be determined in such a way that the probability of shortage does not exceed a prescribed bound, which depends on the cost factors.

$$k = \{2\log_e [C_n/\sqrt{2\pi}\sigma Lhp]\}^{1/2}.$$

Solution:

The demand of the next reviewing period is forecasted by a modified version of the exponential smoothing method which takes into account the specifics of slow moving items. The order level at the end of the t-th reviewing period is expressed in the form

$$R = \mu + k\sigma.$$

The values of k for different probabilities of shortage occurrence are listed in a table.

376

312. *Reliability model for trend and seasonality in demand*

Main codes:

$$1\ 1\ 1\ 1\ 0\ 2\ 1\ 1\ 1\ 3$$

Assumptions:

Demand is characterized by trend and seasonality. It is forecasted by a second-order exponential smoothing method. The absolute error of the forecasting is denoted by E_j and the seasonal indices by s_j. The forecasting error is assumed to have a normal distribution. The length of the ordering period is prescribed. The order is delivered with a leadtime of k periods, where k is not necessarily an integer. The amount delivered is not necessarily equal to the amount ordered: The absolute value of their difference in period j is denoted by F_j.

Objective:

The probability of continuous supply is a prescribed value p. For an inventory holding cost factor c_1 and for a shortage cost factor c_2, the value $p=1-c_1/c_2$ is suggested as the required service level. The safety stock satisfying the reliability constraint is given in the form

$$K_{j+n}(p) = 1.25\,u(p)\,\sqrt{E_j^2\left[\sum_{i=1}^{j+n} s_i^2+(k-n-1)^2\,s_{j+n+1}^2\right]+(n+2)\,F_j^2}.$$

Solution:

The value of the safety factor $u(p)$ is determined by the equation $\Phi(u)=p$, where Φ means the standard normal probability distribution function, while E_j, F_j and s_i are forecasted by exponential smoothing.

The order quantity is defined by

$$Q_{j+n} = P_{j,k}+K_{j+n},$$

where $P_{j,k}$ is the expected demand of k periods, forecasted at the beginning of period j.

313. *Internal storage in a production line*

Main codes:

$$1\ 8\ 1\ 1\ 0\ 2\ 1\ 1\ 1\ 4$$

Assumptions:

The material supply of three production equipments, connected to each other, is investigated. The material arrives at the first equipment in fixed lot-sizes, but at random instants I_j. The number of deliveries of material is also fixed. The distribution of the random instants of delivery is approximated by a multi-dimensional normal distribution. The parameters (expectation, variances and covariances) are given by the production controller.

The first equipment supplies two parallel working equipments. The processing times $(a_i, b_i^{(1)}, b_i^{(2)})$ of the different lots are also random amounts, approximated by multi-dimensional normal distribution. The lots after processing get to the joint store of the parallel working equipments, thus the input occurs at random times in fixed lot-sizes similarly to the first equipment. The demand appears also at random times estimated by the production control. The storage capacities are bounded.

Objective:

The initial (minimal) stock level t_i ($i = 0, 1, 2$) is planned by the model in such a way that the material supply of the parallel working equipments is guaranteed on a prescribed probability level $1 - \varepsilon$:

$$P\left(\max t_0 + A_{u_i - 1}, I_{k_i} + a_{u_i} \leq t_j + \sum_{s=1}^{i-1} b_s^{(j)}, \ j = 1, 2\right) \geq 1 - \varepsilon,$$

where u_i and k_i are the indices of serial numbers, and $A_{u_i - 1}$ is the end time of processing the lot $u_i - 1$ on the first equipment measured from time t_0.

Solution:

The problem is formulated in a stochastic programming model which is proved to be a convex programming problem under the above assumptions. The values of the multi-dimensional normal distribution function can be determined by the Monte-Carlo method.

XII.1.1.2. Delivery in Several Lots

314. *Random delivery in identical lots*

Main codes:

$$1\ 1\ 1\ 0\ 0\ 2\ 1\ 1\ 7\ 4$$

Assumptions:

There is a continuous demand with a known constant rate r. The delivery of an order occurs in n parts, and in identical lot-sizes. The delivery instants are random points of the order interval $[0, T]$. The deliveries are assumed to occur at any moment of the period $[0, T]$ with the same probability. Thus the elements of an ordered sample taken from the uniform distribution in $[0, T]$ can be considered as the realization of the delivery instants: $t_1 < t_2 < ... < t_n$.

The amount ordered is equal to the total demand rT of the order period. The cumulated amount of delivery in the period $[0, T]$ is denoted by $F_n(t)$ for $0 \leq t \leq T$. According to the characteristics of the deliveries we have

$$F_n(t) = \begin{cases} 0 & \text{if } 0 \leq t \leq t_1, \\ cT \dfrac{k}{n} & \text{if } t_k < t \leq t_{k+1} \quad k = 1, ..., n-1, \\ cT & \text{if } t_n < t \leq T; \end{cases}$$

which is basically equivalent to the empirical probability distribution function of the uniform distribution on $[0, T]$.

Objective:

An initial stock M has to be determined which serves as a safety stock for the subsequent period as protection against random disturbances in delivery. In other words, the minimal level of the safety stock M is to be determined, which ensures a continuous supply during the whole period $[0, T]$ on a prescribed service level $1-\varepsilon$. For the optimal M, the probability of a continuous supply satisfies the equation

$$P\left(\sup_{0 \leq t \leq T} \{rt - F_n(t)\} \leq M\right) = 1 - \varepsilon.$$

The optimal amount of an order is $S = M + rT$, where M is the solution of the above equation.

Solution:

For the exact solution, the equation

$$1 - y \sum_{i=0}^{[n(1-y)]} \binom{n}{i} \left(1 - y - \frac{i}{n}\right)^{n-i} \left(y + \frac{i}{n}\right)^{i-1} = 1 - \varepsilon$$

is to be solved for y, then the optimal amount of the initial stock is $M = rTy$. An iterative procedure is suggested for the solution of the equation, starting with $y_0 = \sqrt{\dfrac{1}{2n} \ln \dfrac{1}{\varepsilon}}$.

The approximately optimal M can be expressed in explicit form

$$M \approx rT \sqrt{\frac{1}{2n} \ln \frac{1}{\varepsilon}}.$$

It is based on an asymptotic distribution, its error decreases with the increase of n. For $n > 10$, the approximation is good enough for most practical applications.

315. *Random delivery and random intensity of demand*

Main codes:

$$1\ 1\ 1\ 1\ 0\ 2\ 1\ 1\ 7\ 4$$

Assumptions:

The order is delivered in n identical lots at random instants of the period $[0, T]$. The random delivery process has the same description as in the previous model. The demand is constant during the interval $[0, T]$, but the rate of demand is not known exactly. Thus the rate of the total demand and total delivery is a random variable denoted by α.

Objective:

The initial stock M at time $t = 0$ has to be determined in such a way that the random demand is continuously satisfied with given probability $1 - \varepsilon$ during the interval $[0, T]$:

$$P\left(\sup_{0 \leq t \leq T} \{\alpha t - F_n(t)\} \leq M\right) = 1 - \varepsilon.$$

Solution:

The approximate value of the minimal necessary stock M is the solution of the equation

$$1 - e^{-2nM(M+1)} E[e^{2nM\alpha}] = 1 - \varepsilon$$

with the notations of the previous model and with E as a sign of expectation. If α has a normal distribution with an expectation m and standard deviation s, then the solution is expressed in an explicit form:

$$M = \frac{m-1}{2(1-ns)^2} + \sqrt{\left[\frac{m-1}{2(1-ns^2)}\right]^2 + \frac{\ln\frac{1}{\varepsilon}}{2n(1-ns^2)}}$$

for $s < 1/\sqrt{n}$. Here, usually $m=1$, i.e., the expected amount of the total demand equals the total amount of order, which has been chosen to be a unit quantity.

316. Random delivery and demand in identical lots

Main codes:

$$1\ 1\ 1\ 1\ 0\ 2\ 1\ 1\ 7\ 4$$

Assumptions:

The order is delivered in n identical lots at random instants of the order period $[0, T]$, similar to previous models. The total demand of this period is a known quantity rT. The demand is also assumed to occur in m identical lots at random times of the period $[0, T]$. The amount of a demand is $\frac{rt}{m}$. The pattern of the random times of demands is similar to the model of delivery times: they are considered as the elements of an ordered sample of size m taken from the uniform distribution on $[0, T]$, independently of the times of deliveries. The cumulative demand of period $[0, T]$ can be expressed by the empirical distribution function $G_m(t)$ similar to $F_m(t)$ described at the previous model $(0 \leq t \leq T)$.

Objective:

The initial stock M provides a continuous supply during the period $[0, T]$ with a prescribed probability $1 - \varepsilon$, if the following equation holds

$$P\left(\sup_{0 \leq t \leq T} \{G_m(t) - F_m(t)\} \leq M\right) = 1 - \varepsilon.$$

The order level is determined by $S = M + rT$.

Solution:

The exact distribution of the probability of continuous supply is not known for the above patterns of the demand and delivery process. The approximate solution of the above equation is derived from the asymptotic distribution in the form

$$M \approx rt \sqrt{\frac{n+m}{nm} \ln \frac{1}{\varepsilon}}.$$

The associated error of the formula decreases with an increase of n and m.

380

317. *Random delivery instants and uniformly distributed random delivery amounts*

Main codes:

$$1\ 1\ 1\ 0\ 0\ 0\ 2\ 1\ 7\ 4$$

Assumptions:

A deterministic demand is considered with constant rate r. The order is delivered at n random times which are uniformly distributed in the order period $[0, T]$, similar to the previous models. The amount delivered at the i-th instant is $\lambda r t/n + \beta_i$ $(i=1, ..., n)$, where the first component is the deterministic part of the delivery with $0 \leqq \lambda \leqq 1$. The random amount β_i is defined in the following way: the interval $[0, (1-\lambda)rt]$ is subdivided by $n-1$ independent, uniformly distributed random points to n parts, the lengths of these subintervals are $\beta_1, ..., \beta_n$.

Objective:

The minimal level of the initial stock M is to be determined, which ensures a continuous supply during the interval $[0, T]$ on a prescribed probability level $1-\varepsilon$.

Solution:

The probability of continuous supply is equal to $1-\varepsilon$ at the optimal M. First, the equation

$$(1-y)^n - y \sum_{k=1}^{r} k \binom{n}{k}\binom{n-1}{k} \int_0^{a_k} R_k^{k-1}(1-R_k)^{n-k}x^{k-1}(1-x)^{n-k-1}\, dx = \varepsilon$$

has to be solved for y, where

$$R_k = y + (1-\lambda)x + \lambda\frac{k}{n},$$

$$a_k = \min\left\{\frac{1-M-\lambda\dfrac{k}{n}}{1-\lambda}, 1\right\} \quad \text{and} \quad r = \min\left\{\frac{n}{\lambda}(1-M), n-1\right\}.$$

The optimal M is defined by $M = r\ Ty$ and the order level by $S = M + rT$. The solution of the above equation can be derived by an iterative process starting with

$$y_0 = \sqrt{1+(1-\lambda)^2}\cdot\sqrt{\frac{1}{2n}\ln\frac{1}{\varepsilon}}.$$

An approximate solution for M is given in the form

$$M = rT\sqrt{1+(1-\lambda)^2}\sqrt{\frac{1}{2n}\ln\frac{1}{\varepsilon}}$$

based on the asymptotic distribution of the probability of continuous supply.

381

318. Random delivery instants and entirely random, uniformly distributed delivery amounts

Main codes:

$$1\ 1\ 1\ 0\ 0\ 2\ 1\ 1\ 7\ 4$$

Assumptions:

This model is a special version of the previous model, in the sense that the delivery amounts are entirely random. It means that at the time of ordering no estimation is available about the minimal amount delivered at a single occasion, i.e., it may be arbitrarily small. With the notations of the previous model, $\lambda=0$ by this specification. All the other assumptions are the same as in the previous model.

Objective:

The minimal level of the initial stock is to be determined which ensures a continuous supply with probability $1-\varepsilon$.

Solution:

For this special case, a simpler, analytical solution is given. It is the root of the equation

$$\left(1-\frac{M}{rt}\right)^n \left(1+\frac{M}{rT}\right)^{n-1} = \varepsilon,$$

which can be solved by an iteration method starting with the initial value

$$M = rT\sqrt[n]{1-\sqrt{\varepsilon(1+y_0)}}, \quad \text{where} \quad y_0 = \sqrt{\frac{1}{n}\ln\frac{1}{\varepsilon}}.$$

This quantity is a good approximation for large values of n.

319. Random delivery instants and random delivery amounts with arbitrary distribution

Main codes:

$$1\ 1\ 1\ 0\ 0\ 2\ 1\ 1\ 7\ 4$$

Assumptions:

A deterministic demand is considered during the interval $[0, T]$ with a constant rate r. The order placed for the replenishment of stocks for this interval is delivered at n random instants which are uniformly distributed in $[0, T]$, similarly to the previous models. The amount delivered at the i-th instant is a random variable with the same distribution for each $i=1, ..., n$.

Objective:

The minimal level of the initial stock M is to be determined which ensures continuous supply with a prescribed probability $1-\varepsilon$.

Solution:

The minimal necessary stock is the solution of the equation

$$\left(1-\frac{M}{rT}\right)^n - \frac{M}{rT}\sum_{k=1}^{n}\binom{n}{k}^{n(1-(M/rT))}\int_{0}^{\;}\left(\frac{M}{rT}+\frac{x}{n}\right)^{k-1}\left(1-\frac{M}{rT}-\frac{x}{n}\right)^{n-k}h_k(x)\,dx = \varepsilon,$$

where $h_k(x)$ is the probability density function of the sum of the first k delivery amounts.

320. Random deliveries not uniformly distributed during the order interval

Main codes:

$$1\ 1\ 1\ 0\ 2\ 1\ 1\ 7\ 4$$

Assumptions:

A continuous demand with constant rate r is considered in the interval $[0, T]$. The replenishment of this period is completed by a delivery process which is a generalized version of the random delivery described by Models 316 and 317. The delivery of an order is realized in n random parts at random instants. The minimal time between two consecutive deliveries is γ. The interval $[0, T-n\gamma]$ is subdivided by L random points which have a uniform distribution on this interval. The deliveries in the order period may be cumulated at any part-period in a random way. Let $x_1^* < x_2^* < ... < x_L^*$ denote the ordered sample formed from L random points. Out of this ordered sample we select those which have subscripts $k_1 < k_2 < ... < k_{n-1}$ according to the cumulation of the deliveries. The choice of the indices L and k_i can be based on fitting them to the statistical data of earlier observations concerning the delivery instants. According to the model, the time points of deliveries are

$$x_{j_1}^* \quad \text{and} \quad i\gamma + x_{j_i}^* - x_{j_{i-1}}^* \quad \text{for} \quad i = 2, 3, ..., n.$$

The minimal amount delivered at one occasion is denoted by δ. The random part of the deliveries is subdivided among the delivery time points on a similar way. The interval $[0, rT-n\delta]$ is subdivided by N random points uniformly distributed on this interval, the ordered sample is $0 \leq y_1^* < y_2^* < ... < y_N^* \leq rT-n\delta$. Then the lots of deliveries are $\delta + y_{k_1}^*, \delta + y_{k_2}^* - y_{k_1}^*, ..., \delta + rT - y_{k_{n-1}}^*$ where the integers $1 \leq k_1 < k_2 < k_{n-1} \leq N$ are chosen according to the distribution of the random parts of deliveries (based on statistical data).

Objective:

The initial stock M is to be determined which provides for the continuous supply during the period $[0, T]$ with a prescribed probability level $1-\varepsilon$.

Solution:

The reliability criteria can be formulated using an n-dimensional probability in the form

$$P(\xi_i \leq M+(i-1)\delta-i\gamma, \quad i = 1, ..., n) = 1-\varepsilon,$$

where $\xi_1 = x_{j_1}^*$ and $\xi_i = x_{j_i}^* - y_{j_{i-1}}^*$ for $i = 2, 3, ..., n$.

The equation can be solved by an iterative procedure calculating the probability of the continuous supply by a Monte-Carlo method.

321. Multi-period reliability model

Main codes:

$$1\ 1\ 1\ 1\ 0\ 2\ 1\ 1\ 1\ 4$$

Assumptions:

D subsequent time periods are considered with a fixed length T of a reviewing period. The delivery and demand processes are both random, i.e., the numbers of deliveries and demand occurrences follow some (possibly unknown) joint discrete probability distribution; further, for any fixed numbers m, n of deliveries and demands, they can be described similarly to that which was applied in Model 306.

Objective:

The initial stock M of the first period is to be determined in such a way that continuous supply is ensured during D periods with a prescribed probabllity $1-\varepsilon_D$:

$$P\left(\max_{1\leq d\leq D}\left[\sup_{0\leq t\leq T}(\xi_d(t)-\eta_d(t))\right]\leq M\right)\geq 1-\varepsilon_D,$$

where

$$P(\xi_d(t)=G_m(t,\mu),\eta_d(t)=F_n(t,\lambda))=p_{mn}\sum_{m,n}p_{mn}=1,\quad d=1,\ldots,D.$$

Solution:

An approximate solution is given by the formula

$$M_D\approx\frac{-\ln\ln\dfrac{1}{1-\varepsilon_D}-B_D}{A_D},$$

where the constants A_D, B_D are estimated (on the basis of earlier observations) by graphical or numerical methods. This approximation is based on an asymptotic distribution, thus its error decreases with an increase in the number of demands and deliveries. (The approximation is acceptable in practice, if $\min_{m,n}\{m,n\}\geq 10$ and $D\geq 2$.)

322. Reliability model for continuous random demand and delivery

Main codes:

$$1\ 1\ 1\ 1\ 0\ 2\ 1\ 1\ 7\ 3$$

Assumptions:

The random demand of the time interval $[0, t]$ has the expected value rt and the standard deviation $\sigma_1\sqrt{t}$ for $0\leq t\leq T$; it is assumed to have a normal distribution for $0\leq t\leq T$ with the above parameter values and it has independent increments. This model of the demand process ξ_t is the so-called homogeneous

384

Wiener process. The delivery is also continuous and has, similarly, the pattern of a Wiener process η_t with parameters v and σ_2. It means that both delivery and demand have a mean intensity rate which is disturbed by random influences having a normal distribution.

Objective:

The initial stock M of the order period $[0, T]$ has to be determined in such a way that a continuous supply is ensured for the whole period at a given probability level $1-\varepsilon$.

Solution:

The above reliability constraint is expressed by the relation

$$P\left(\sup_{0 \leq t \leq T} \{\xi_t - \eta_t\} \geq M\right) = 1 - \varepsilon,$$

which is equivalent to the following equation (to be solved for M):

$$\Phi\left(\frac{M - (r - v)T}{\sigma \sqrt{T}}\right) - \exp\left(\frac{2M(r - v)}{\sigma^2}\right) \Phi\left(\frac{-M - (r - v)t}{\sigma \sqrt{T}}\right) = 1 - \varepsilon.$$

Here $\sigma = \sqrt{\sigma_1^2 + \sigma_2^2}$ and Φ denotes the standard normal distribution function. The above equation can be solved by numerical methods. In the case, when demand and delivery have the same mean rates $(r = v)$, then the solution is equal to

$$M = \sqrt{\sigma T} \Phi^{-1}\left(1 - \frac{\varepsilon}{2}\right),$$

where Φ^{-1} denotes the inverse of the distribution function Φ.

323. Reduction of the on-hand inventory at a given service level

Main codes:

$$1\ 1\ 1\ 1\ 0\ 2\ 1\ 1\ 1\ 1$$

Assumptions:

The quantity demanded in the time interval $[0, t]$ is a random amount with probability density $h(x)$. Orders are placed periodically with the fixed length T of a period. Orders placed at $t = 0$ arrive during the interval $(\tau, \tau + T)$ in n random parts at n random instants. The model of this process is similar to the delivery process described in Model 306. The demand is a continuous random process assumed to have stationary and independent increments. The order level is denoted by S. The inventory holding cost factor is IC, the shortage cost factor is p and $\beta = p/(IC + p)$.

Objective:

The purpose is to investigate what changes in the parameters of the demand or delivery processes make possible the reduction of the on-hand inventory

$D(S) = \int\limits_0^S (S-x)h(x)dx$ without increasing the expected shortage $B(S) =$

$= \int\limits_S^\infty (x-S)h(x)dx.$

Solution:

The optimal $S = S(\beta)$ is determined from the equation

$$\int\limits_0^S h(x)\, dx = \beta.$$

A general condition is derived in the form of an inequality concerning the mixture of the demand and delivery distributions: $h_2(S_2(\beta)) \geq h_1(S_1(\beta))$. If this inequality holds, the inventory reduction is possible for the new system with other parameters. For approximation by a normal or gamma distribution, the decrease of variance is a sufficient condition which can be fulfilled by changing different parameters of the demand or delivery.

XII.1.2. Reorder-Point Models

324. *Approximate formula for the safety factor*

Main codes:

1 1 7 1 0 2 2 2 7 7

Assumptions:

The demand during the leadtime is a random variable with an expected value $\bar{x}$ and standard deviation σ. The leadtime may also be random. The order is placed at a single instant. Continuous reviewing is assumed. The amount of order is determined independent of the reorder point: it is not specified, so the economic order quantity may be selected.

Objective:

A reorder point must be specified which keeps the shortage probability under a prescribed value q. It is expressed in the form $s = \bar{x} + k\sigma$, where k is the safety factor depending on p.

Solution:

If the value of $\bar{x}$ is much larger than σ, then the leadtime demand can be approximated by a normal distribution. The required value of k is determined by the inverse of the standard normal distribution function. This inverse function is approximated by Tukey's lambda distribution. It results in the following explicit expression for the approximation of the safety factor

$$k = 5.0633\,([1-p]^{0.135} - p^{0.135}] \quad 0 \leq p \leq 1.$$

325. *Reliability model with random leadtime and* (s, q) *policy*

Main codes:

$$1\ 1\ 1\ 1\ 0\ 2\ 2\ 2\ 1\ 3$$

Assumptions:

Unit amounts are demanded at random instants. The demand process is stationary with a mean rate M. The leadtime is also random. The demand distribution during the leadtime is characterized by the probability density function $p(x)$. The mean demand in leadtime is M_L. The holding and ordering cost is denoted by h and K, respectively.

Objective:

The expected cost of inventory holding and ordering per unit time $\left(\frac{Q}{2}+s-M_L\right)h+\frac{hI_2}{2Q}+\frac{KM}{Q}$ is minimized, where Q is the economic order quantity. No shortage cost is included. Instead, the reliability constraint $I_1/q=B$ is considered, where I_1 means the expected value of shortage and B is a prescribed constant.

Solution:

The above constrained minimization problem is reduced to the solution of a system of two equations, for which different numerical procedures are available. No specific procedure is suggested.

326. (s, q) *policy with shortage constraints applied for spare parts inventory*

Main codes:

$$1\ 1\ 0\ 1\ 0\ 2\ 2\ 1\ 1\ 1$$

Assumptions:

Spare parts are demanded in single units at random instants. The total demand during a reviewing period is approximated by a normal distribution with an expected value d and standard deviation σ. Orders may be placed at each reviewing time. The leadtime equals the length of L reviewing periods. The demands of the different reviewing periods are independent random amounts.

The model was applied for the inventory control of spare parts of machines in a mine. The reviewing period has the length of a month. The maximal number of the annual orders is restricted to W.

Objective:

The safety stock is determined by the prescribed limit S on the probability of the shortage: $\sum_{D=0}^{E(D)+M} P(D)=S$, where D is the demand in leadtime, $E(D)$ is its expectation, $P(D)$ is its probability and M is its lower limit.

Solution:

The safety stock is expressed in the form $K\sqrt{L}*S+0.1\,dL$, where the safety factor K is determined by the shortage constraint. A table is constructed, based on the normal distribution, which contains the values of K for different service levels of continuous supply. The reorder point s is determined as the sum of the safety stock and of the expected demand during leadtime:

$$s = K\sqrt{L} * S + 1.1\, dL.$$

The order quantity is expressed by means of the ordering cost P and the maximal number of annual orders W in the form $q = c\sqrt{d/P}$, where $c = 12dP/W$.

327. Determination of the reorder level under a shortage constraint

Main codes:

$$1\ 1\ 1\ 1\ 0\ 2\ 2\ 2\ 1\ 3$$

Assumptions:

Spare parts are demanded by units at random instants according to a Poisson process. The leadtime is also a random variable with a gamma distribution. Thus the leadtime demand has negative binomial distribution. The number of annual orders may not exceed the value W.

Objective:

The safety stock is to be determined in such a way which ensures a continuous supply with a prescribed probability. The amount of order is determined by the minimization of ordering cost under the constraint on the maximal number of annual orders.

Solution:

The leadtime demand is approximated by a normal distribution with an expected value dL and standard deviation $dL + d^2 r(L)$ where d means the average demand of an ordering period, L is the expected length and $r(L)$ is the standard deviation of the leadtime. The safety stock is expressed by $K\sqrt{Ld + d^2 r(L)}$, where the safety factor K depends on the prescribed probability of continuous supply and can be expressed using the inverse of the standard normal distribution. The amount of order is equal to

$$\frac{12d\sqrt{dp}}{w}$$

where p denotes the cost of an order.

328. Reorder point for a Weibull distributed leadtime demand

Main codes:

$$1\ 1\ 1\ 1\ 0\ 2\ 2\ 2\ 7\ 3$$

Assumptions:

At continuous reviewing, an order is placed at the reorder point s. The leadtime is a random variable: the case of a gamma and a normal distribution is considered. The demand for a time-interval with a fixed length L is also a random variable which may have a Poisson, gamma or normal distribution.

388

Objective:

The reorder point has to be determined which fulfills a reliability criterion. Two possible measures of reliability are investigated:

1. the probability of shortage is constrained,
2. the expected value of shortage is constrained.

Solution:

Under the above conditions, the total demand during the leadtime can be described by a Weibull distribution with parameters b and c for which an estimation method is derived. The reorder point for the reliability constraint type 1 is expressed in the explicit form $s = b[-\ln(1-p)]^{1/c}$, where p means the prescribed probability level. For the constraint type 2, the solution can be expressed using the tables of the incomplete gamma function. The order quantity is not considered by the model.

329. Bayesian determination of the reorder point

Main codes:

$$1\ 1\ 7\ 7\ 0\ 2\ 2\ 2\ 2\ 9$$

Assumptions:

The demand of the leadtime is a random variable with an unknown discrete distribution function with possible values $0, 1, ..., N$. Under continuous reviewing, the value of the reorder point j has to be determined which fulfills a reliability criterion. This is one of the following possible measures of the service level:

1. the probability of shortage may not exceed a prescribed level,
2. a prescribed rate of demand has to be fulfilled without delay (demand is lost at shortage),
3. a prescribed rate of demand has to be fulfilled instantaneously, the other part of the demand is fulfilled with delay (backorders case).

Objective:

Case 1: The minimal integer $j = j_\alpha$ of the reorder level is to be determined for which the probability of shortage is bounded by α

$$\sum_{k=j+1}^{N} \frac{m_k}{\sum_{i=0}^{N} m_i} \leq \alpha,$$

where α is the prescribed probability level and m_k is the number of leadtimes when the amount k has been demanded.

Case 2: The minimal integer $j = j_\beta$ is to be found for which there holds

$$\frac{\sum_{k=j+1}^{N} (k-j) m_k}{\sum_{i=0}^{N} m_i} \leq \frac{(1-\beta)q}{\beta},$$

where the left-hand side expresses the expected demand lost due to shortage, q is the amount of the order and β is the prescribed rate of satisfied demand.

Case 3: $j=j_\gamma$ is the minimal integer for which we have

$$\frac{\sum\limits_{k=j+1}^{N} (k-j)m_k}{\sum\limits_{i=0}^{N} m_i} \leq (1-\gamma)\,q,$$

where γ is the prescribed service level according to criterion 3.

Solution:

The algorithm is constructed on the basis of the Bayesian principle in the following steps for each of the above three types of service level:

1. The probability of a leadtime demand $z=j$ is denoted by p_j. The *a priori* estimation is $\bar{p}_j$. The likelihood function $P(E|p_0, p_1, \ldots, p_N)=kp_0^{n_0}p_1^{n_1}\ldots p_N^{n}$ has a multi-nomial distribution, where E denotes the event that the leadtime demand is observed n times and, of these, n_i times the demand for amount i was detected

$$\left(\sum_{i=1}^{N} n_i = n \right).$$

The corresponding *a priori* distribution is a multi-dimensional beta distribution of the $p_i's$ ($i=1, \ldots, N$) for which

$$P(z = k) = \frac{m_k}{\sum\limits_{i=0}^{N} m_i},$$

where the values of m_i are the theoretical equivalents of the empirical values n_i.

2. Knowing the values $\bar{p}_j$ $j=1, \ldots, N$ and using the above relation, the values of m_k are determined unambiguously, if an arbitrary additional constraint is considered.

3. The required type of service level is chosen from the three possible alternatives.

4. The parameters are modified using the new observations according to the additive rule $m_i''=m_i'+u_i$, where m_i' is the *a priori* parameter and m_i'' is the *a posteriori* parameter. After this, return to the step 3.

330. Stochastic $(S,-1, S)$ policy with a reliability constraint for waiting time

Main codes:

1 1 1 1 0 2 5 2 1 3

Assumptions:

The demand is random with discrete values. The time between two consecutive demand occurrences has an exponential distribution with the expected value

$$\frac{1}{\lambda(1-\psi)}$$

for a unit time. The amount demanded at a given occasion has a geometric distribution with mean $\frac{1}{1-\psi}$:

$$P(K = k) = (1-\psi)\psi^{k-1} \quad (0 < \psi < 1; \; k = 1, 2, ...).$$

Thus the demand of a unit of time has a compound geometric Poisson distribution. There is a continuous reviewing. As a demand occurs, an order is placed which increases the inventory level to the order level S. The units ordered are delivered after an independent exponentially distributed leadtime with mean $1/\mu$. Thus the units, ordered at the same instant, arrive usually at different instants. The demand is waiting in the case of stockout. The waiting time W is also a random variable.

Objective:

The order level S is to be determined considering the constraint on the waiting time $P(W \leq w) \geq \alpha$, where w and α are prescribed values.

Solution:

The steady-state probabilities of the waiting time are expressed by

$$F(w) = P(W \leq w) = \sum_{n=0}^{\infty} P(N = n) \sum_{k=1}^{\infty} P(k = k) \quad P(W = w | N = n, K = k),$$

where N is the number of items ordered and so far not delivered. The random variables N and K are assumed to be independent. The explicit form of the probability distribution $F(w)$ is derived for Poisson demand and the compound geometric Poisson demand described above. The probabilities are calculated for some parameter combinations.

XII.2. Multi-Item Reliability Models

331. *Reliability model for the ordering and transporting frequency*

Main codes:
$$8\;1\;0\;1\;0\;2\;0\;1\;0\;1$$

Assumptions:

The order has to be placed simultaneously for all of the items considered. The length of order period T is subject to control. The delivery of the order begins after a leadtime L and is realized in N identical parts at equidistant time points. The transportation cost of each item at each transport occasion is denoted by c_4. The unit cost factor of inventory holding is c_1. The fixed cost of an ordering is c_3. There is a continuous, random demand with mean rate r.

Objective:

The expected cost of inventory holding, ordering and transportation for a unit of time is to be minimized under a reliability constraint according to which the probability of shortage may not exceed the prescribed limit ε.

Solution:

First the constant λ is to be determined, for which the probability of a demand rate larger than λr is less than ε. Having obtained this value, the optimal length of the order period is expressed in the form

$$T_0 = \sqrt{W\sqrt{L+\sqrt{W\sqrt{L+\sqrt{W\sqrt{L+...}}}}}}, \quad \text{where} \quad W = \frac{2c_3}{c_1\lambda r}.$$

The optimal order level calculated item by item is

$$S_0 = r(L+T_0) + \lambda r\sqrt{L+T_0}$$

and the optimal number of transportations of an order is the integer part of

$$N_0 = T_0\sqrt{\frac{c_1 r}{2c_4}}.$$

332. *Multi-item model with constrained conditional expectation of shortage*

Main codes:

$$8\,1\,1\,0\,0\,2\,1\,1\,1\,4$$

Assumptions:

K different items are demanded continuously with constant rates in a given period $[0, T]$. The order placed for refilling the stock of item i in this period is delivered in n_i parts. The time points of these deliveries are random points of the interval $[0, T]$. The minimal time interval between two consecutive deliveries is denoted by γ_i. The interval $(0, T-n_i\gamma_i)$ is subdivided by n_i-1 random points which are chosen as the appropriate elements of an ordered sample of independent, uniformly distributed random points on the interval $[0, T-n_i\gamma_i]$. The yielded subintervals represent the random parts in the interarrival times of deliveries similar to Model 319. The amount delivered at one occasion consists of a fixed part δ_i and a random part which is modelled similarly to Model 319.

Objective:

An initial stock $M^{(i)}$ is planned for each item to assure continuous supply. This needs an extra capital investment $d_i M^{(i)}$. The initial stock of the K item has to be determined. With the minimal possible amount of total capital invested in initial stock,

$$\sum_{i=1}^{K} d_i M^{(i)}$$

a prescribed service level of the joint continuous supply has to be ensured. The service level is measured by the probability of shortage and by the conditional expectation of shortage (under the condition that shortage occurs in the considered period).

392

Solution:

The determination of the optimal level of the initial stocks under the above constraints means the solution of a stochastic programming problem. It is proved to be convex programming problem which has been solved by the SUMT method with logarithmic penalty function. The probability and conditional expectation of the shortage occurrence were determined by Monte-Carlo techniques.

333. Simulation model for the maximization of reliability under capacity constraints

Main codes:

$$8\ 1\ 1\ 1\ 0\ 2\ 1\ 1\ 7\ 1$$

Assumptions:

The demand of item i is described in the period $0 \leq t \leq T$ by a stochastic process η_t^i. The delivery during the same time interval $[0, t]$ is also a random variable denoted by ξ_t^i. The distribution of the random variables are known. They are continunous and the logarithm of their joint density function is a concave function (e.g. multi-dimensional normal distribution). The total storage capacity is limited, but constraints may be prescribed also for the maximal inventory of a group of items.

Objective:

The amount of the initial stocks M_i has to be determined under the capacity constraints of the store, and a way that the possible maximum of the service level should be reached. The service level is measured by the n dimensional joint probability of the continuous supply expressed in the form

$$P\left[\inf_{0 \leq t \leq T} (M_i + \xi_t^i - \eta_t^i) \geq 0,\ i = 1, ..., n\right].$$

Solution:

The service level can be calculated for an arbitrarily given vector of initial stocks $(M_1, ..., M_n)$ by simulation. The objective function is a convex function of the initial stocks, thus not having available gradients it can be minimized step by single variables until the change of the objective function decreases under step by a given value. The convergence of the procedure is proven.

334. Analytical model for the maximization of reliability under investment constraint

Main codes:

$$8\ 1\ 1\ 0\ 0\ 2\ 1\ 1\ 1\ 4$$

Assumptions:

A deterministic demand with constant rate r_i is considered in the time period $[0, T]$ for item i $(i = 1, 2, ..., k)$. The order placed for the refilling the stock of item i in this period is delivered in n_i parts of identical amounts. The deliveries occur at random instants of period $[0, T]$. The minimal time between two con-

393

secutive deliveries is γ_i. The time of the j-th delivery is $t_j = j \cdot \lambda_i + u_j^*$, where u_j^* denotes the j-th element of the ordered random sample of size n_i taken from the uniform distribution on $[0, T - n_i \gamma_i]$. The capital invested in the initial stocks of the items is $\sum_{i=1}^{k} d_i M_i$, which amount is bounded from above.

Objective:

The amount of the initial stocks is determined under the investment constraint. The purpose is to maximize the reliability of the system which is measured by the probability of the continuous supply during the interval $[0, T]$:

$$\prod_{i=1}^{k} P\left(\sup_{0 \leq t \leq T} \{r_i t - b_i(t)\} \leq M_i \right),$$

where $b_i(t)$ denotes the cumulative amount of delivery during the interval $[0, T]$ according to the random delivery process defined above.

Solution:

The probability of the continuous supply is approximated by a simple analytical formula based on asymptotic distribution theory. The constrained optimization problem is reduced by the Lagrange multiplier method to the solution of a single equation with one variable. The other decision variables can be expressed by this value. This simple solution method provides an approximation of the optimum, sufficiently close for practical applications.

335. Multi-item model under two different shortage criteria

Main codes:

$$8\ 1\ 0\ 1\ 0\ 2\ 4\ 1\ 1\ 0$$

Assumptions:

A single period is considered in which a random demand appears for item k with a probability distribution F_k ($k = 1, 2, ..., n$). The different items are demanded independently of each other. The amount q_k is purchased for unit price c_k at the beginning of the period. The total invested capital is bounded:

$$\sum_{k=1}^{n} c_k q_k \leq C.$$

Objective:

The total capital invested is to be divided among purchases of the items in such a way that a shortage criteria is optimally met. A usual criterion is the expected value of the total shortage for all of the items:

$$\min_{q_k} \sum_{k=1}^{n} \int_0^\infty [1 - F_k(q_k + y)]\, dy.$$

Another criterion is also investigated. Here the purpose is to minimize the expected shortage for the item which has the maximal shortages (it has the meaning

394

of elimination of the bottle-neck). The form of this objective function is

$$\min_{q_k} \int\limits_0^\infty \left[1 - \prod_{k=1}^n F_k(q_k+y)\right] dy.$$

Solution:

The optimality conditions can be described for the above two different criteria by the Lagrange multiplier method in the following forms:
For the first criteria

$$\int\limits_0^\infty f_i(x_i+y)\,dy + \tau_2 c_i + \tau_{2,i} = 0 \quad i = 1, 2, ..., n$$

and for the second criteria

$$\int\limits_0^\infty \sum_{k\ne i} F_k(q_k+y) f_i(q_i+y)\,dy + \tau_i c_i + \tau_{1,i} = 0 \quad i = 1, 2, ..., n$$

with the Lagrange multipliers τ_i, $\tau_{1,i}$, $\tau_{2,i}$ $i=1, ..., n$.

336. *Continuous-review multi-item model with shortage criterion*

Main codes:

$$8\ 1\ 0\ 1\ 0\ 2\ 2\ 2\ 1\ 3$$

Assumptions:

An (s, q) policy system is considered for N items under continuous reviewing and random leadtime with an expected value τ_i. The leadtime demand is normally distributed with an expected value λ_i and standard deviation σ_i for item i. The unit cost factor of inventory holding is denoted by c_1, the fixed cost of an ordering is c_3. The reliability of the continuous supply can be constrained item by item or jointly. Two different measures of this reliability are considered:

1. The probability that no shortage occurs during an order interval

$$\frac{\left[\displaystyle\sum_{i=1}^N \frac{\lambda_i}{q_i} \Phi(t_i)\right]}{\left[\displaystyle\sum_{i=1}^N \frac{\lambda_i}{q_i}\right]}$$

is constrained from below;

or

2. The expected rate of shortage and amount of order

$$\frac{\displaystyle\sum_{i=1}^N \left[\frac{\lambda_i \sigma_i}{q_i} \varphi(t_i) - t_i \Phi(t_i)\right]}{\displaystyle\sum_{i=1}^N \lambda_i}$$

395

is constrained from above, where t_i denotes the quotient $\dfrac{s_i}{\tau_i}$, while Φ and φ denotes the probability distribution and probability density function of the standard normal distribution.

Objective:

The sum of the expected cost of inventory holding and ordering,

$$\sum_{i=1}^{N} K_i = c_3 \sum_{i=1}^{N} (\lambda_i/q_i) + c_1 \sum_{i=1}^{N} \left(\frac{q_i}{2} + t_i \sigma_i \right)$$

is minimized under the reliability constraint of type 1 or 2. The optimal value of s_i is determined by the reliability constraint.

Solution:

For both constraints the Lagrange multiplier method can be applied and a system of nonlinear equations is derived for the optimal values of s_i and q_i which can be solved by iterative procedures. An approximate formula for finding the optimal solution is also derived.

Appendix

I. Sources of Models

1. Naddor (1966)
2. Hadley—Whitin (1963)
3. Naddor (1966)
4. Naddor (1966)
5. Naddor (1966)
6. Schussel (1968)
7. Buchan—Koenigsberg (1967)
8. Naddor (1966)
9. Naddor (1966)
10. Naddor (1966)
11. Naddor (1966)
12. Naddor (1966)
13. Naddor (1966)
14. Barbosa—Friedman (1978)
15. Schussel (1968)
16. Naddor (1966)
17. Naddor (1966)
18. Schwartz (1970)
19. Simon (1952)
20. Simon (1952)
21. Naddor (1966)
22. Hadley—Whitin (1963)
23. Naddor (1966)
24. Hadley—Whitin (1963)
25. Tate (1964)
26. Hadley (1964)
27. Naddor (1966)
28. Naddor (1966)
29. Рыжиков (1969)
30. Naddor (1966)
31. Naddor (1966)
32. Naddor (1966)
33. Naddor (1966)
34. Henery (1979)
35. Donaldson (1977)
36. Donaldson (1977)
37. Resh—Friedman—Barbosa (1976)
38. Ледин—Ермаков (1978)
39. Peckelman (1974)
40. Elmaghraby—Bawle (1972)
41. Kunreuther—Schrange (1973)
42. Schussel (1978)
43. Pierskalla—Roach (1972)
44. Goyal (1977)
45. Dave (1979)
46. Falkner (1970)
47. Hadley—Whitin (1963)
48. Hadley—Whitin (1963)
49. Hadley—Whitin (1963)
50. Hadley—Whitin (1963)
51. Hadley—Whitin (1963)
52. Buchan—Koenigsberg (1967)
53. Buchan—Koenigsberg (1967)
54. Buchan—Koenigsberg (1967)
55. Buchan—Koenigsberg (1967)
56. Рыжиков (1969)
57. Рыжиков (1969)
58. Grinold (1967)
59. Daniel (1963)
60. Suddenth (1965)
61. Prékopa (1973)
62. Nahmias (1975)
63. Nahmias (1975)
64. Fries (1975)
65. Wagner (1968)
66. Naddor (1966)
67. Naddor (1966)
68. Naddor (1966)
69. Naddor (1966)
70. Naddor (1966)
71. Gross (1963)
72. Jagannathan (1978)
73. Jagannathan (1978)
74. Iglehart—Jaquette (1969)
75. Schneeweiss (1974)
76. Kao (1975)
77. Gonedes—Lieber (1974)
78. Рыжиков (1969)
79. Buchan—Koenigsberg (1963)
80. Hochstädter (1969)
81. Jagannathan (1978)
82. Naddor (1966)
83. Naddor (1966)
84. Naddor (1966)
85. Naddor (1966)
86. Naddor (1966)
87. Hadley—Whitin (1963)
88. Hadley—Whitin (1963)
89. Hadley—Whitin (1963)
90. Prékopa (1972)
91. Hochstädter (1969)
92. Hadley—Whitin (1963)
93. Hochstädter (1969)
94. Naddor (1966)

95. Klemm—Mikut (1972)
96. Klemm—Mikut (1972)
97. Hausman—Thomas (1972)
98. Naddor (1966)
99. Naddor (1966)
100. Naddor (1966)
101. Naddor (1966)
102. Рыжиков (1969)
103. Psoinos (1976)
104. Donaldson (1974)
105. Gerencsér (1970)
106. Hadley—Whitin (1963)
107. Hadley—Whitin (1963)
108. Лаврениенко (1972)
109. Рыжиков (1972)
110. Wagner (1968)
111. Klemm (1973)
112. Hadley—Whitin (1963)
113. Hadley—Whitin (1963)
114. Nahmias—Wang (1979)
115. Naddor (1966)
116. Naddor (1966)
117. Snyder (1974)
118. Schwartz (1970)
119. Hausman—Thomas (1972)
120. Naddor (1966)
121. Arrow—Karlin—Scarf (1958)
122. Valisalo—Sivazlian—Mailott (1972)
123. Hochstädter (1969)
124. Naddor (1966)
125. Naddor (1966)
126. Girlich (1971)
127. Girlich (1971)
128. Hochstädter (1969)
129. Naddor (1966)
130. Croston (1974)
131. Rosenshine—Obee (1976)
132. Geisler (1964)
133. Naddor (1966)
134. Greenberg (1965)
135. Klemm (1973)
136. Geisler (1964)
137. Schneider (1978)
138. Schneeweiss (1971)
139. Rose (1972)
140. Gross—Harris (1971)
141. Gross—Harris (1971)
142. Wessels (1973)
143. Naddor (1966)
144. Sivazlian
145. Snyder (1974)
146. Gerencsér (1973)
147. Henin (1972)
148. Iglehart—Morey (1972)
149. Cawdery (1976)
150. Hadley—Whitin (1963)
151. Das (1975)
152. Gebhardt (1972)
153. Herron (1967)

154. Burgin (1970)
155. Richards (1976)
156. Mann (1973)
157. Рыжиков (1969)
158. Kulcsár (1979)
159. Wagner (1968)
160. Das (1975)
161. Ziermann (1953)
162. Рыжиков (1969)
163. Рыжиков (1969)
164. Рыжиков (1969)
165. Рыжиков (1969)
166. Рыжиков (1969)
167. Рыжиков (1969)
168. Рыжиков (1969)
169. Рыжиков (1969)
170. Рыжиков (1969)
171. Рыжиков (1969)
172. Рыжиков (1969)
173. Рыжиков (1969)
174. Рыжиков (1969)
175. Buchan—Koenigsberg (1967)
176. Prékopa (1972)
177. Gerencsér (1972)
178. Agin (1966)
179. Buchan—Koenigsberg (1967)
180. Smith (1977)
181. Das (1977)
182. Feeney—Sherbrooke (1966)
183. Loo—Ghossal (1971)
184. Mumford (1977)
185. Hochstädter (1969)
186. Hochstädter (1969)
187. Рыжиков (1969)
188. Denardo (1968)
189. Iwaniec (1979)
190. Iglehart—Jaquette (1969)
191. Kao (1975)
192. Veinott (1963)
193. Thomas (1974)
194. Iglehardt—Jaquette (1969)
195. Stidham (1977)
196. Wijngaard (1973)
197. Porteus (1971)
198. Porteus (1972)
199. Hochstädter (1969)
200. Tijms (1974)
201. Snyder (1975)
202. Veinott (1963)
203. Hadley—Whitin (1963)
204. Iglehart—Jaquette (1969)
205. Iglehart—Jaquette (1969)
206. Brown—Corcoran—Lloyd (1971)
207. Schneeweiss (1971)
208. Tur (1972)
209. Howe (1974)
210. Wijngaard (1973)
211. Veinott (1965)
212. Simon (1952)
213. Zangwill (1966)

214. Рыжиков (1969)
215. Hadley—Whitin (1963)
216. Rao (1976)
217. Wagner (1968)
218. Goyal (1977)
219. Hadley—Whitin (1963)
220. Blackburn—Kunreuther (1972)
221. Blackburn—Kunreuther (1972)
222. Liberatore (1977)
223. Hochstädter (1969)
224. Hochstädter (1969)
225. Karlin (1960)
226. Scarf (1960)
227. Hadley—Whitin (1963)
228. Wagner 1968)
229. Sobel (1971)
230. Bessler—Zehna (1967)
231. Cohen (1976)
232. Nahmias (1977)
233. Hochstädter (1969)
234. Hadley—Whitin (1963)
235. Hadley—Whitin (1963)
236. Riis (1965)
237. Naddor (1966)
238. Naddor (1966)
239. Naddor (1966)
240. Naddor (1966)
241. Goyal (1974)
242. Рыжиков (1969)
243. Tate (1964)
244. Рыжиков (1969)
245. Рыжиков (1969)
246. Рыжиков (1969)
246. Рыжиков (1969)
247. Рыжиков (1969)
248. Рыжиков (1969)
249. Рыжиков (1969)
250. Polyzos—Xirokostas (1976)
251. Рыжиков (1972)
252. Silver (1965)
253. Hochstädter (1973)
254. Hochstädter (1973)
255. Veinott (1965)
256. Veinott (1965)
257. Dékány (1974)
258. Rényi—Ziermann (1963)
259. Hadley—Whitin (1963)
260. Hadley—Whitin 1963)
261. Buchan—Koenigsberg (1967)
262. Buchan—Koenigsberg (1967)
263. Gerson—Brown (1967)
264. Gerson—Brown (1967)
265. Gerson—Brown (1967)
266. Kelle (1977)
267. Prékopa (1973)
268. Wharton (1975)
269. Garman (1976)
270. Hadley—Whitin (1963)
271. Page—Paul (1976)
272. Naddor (1966)
273. Рыжиков (1969)
274. Zoller (1977)
275. Hadley—Whitin (1963)
276. Рыжиков (1969)
277. Prékopa—Kelle (1976)
278. Bomberger (1966)
279. Hodgson (1972)
280. Hodgson (1972)
281. Dorsey—Hodgson—Ratliff (1974)
282. Simmons (1972)
283. Simmons (1972)
284. Goyal (1973)
285. Pentico (1976)
286. Kleindorfer—Newson (1975)
287. Wolfson (1965)
288. Gross (1963)
289. Whitin (1973)
290. Whitin (1973)
291. Whitin (1973)
292. Hadley—Whitin (1963)
293. Simon (1971)
294. Рыжиков (1969)
295. Рыжиков (1969)
296. Berman (1961)
297. Iglehart—Lalchandani (1967)
298. Bessler—Veinott (1966)
299. Schwarz (1973)
300. Schussel (1968)
301. Рыжиков (1969)
302. Iglehart—Morey (1971)
303. Рыжиков (1969)
304. Рыжиков (1969)
305. Crowston—Wagner—Williams (1972)
306. Bishop (1974)
307. Prékopa (1963)
308. Rényi—Ziermann (1961)
309. Yaspan (1972)
310. Burgin (1975)
311. Croston (1972)
312. Mann (1971)
313. Kelle (1978)
314. Ziermann (1963)
315. Kelle (1984)
316. Ziermann (1963)
317. Prékopa (1963)
318. László (1973)
319. Kelle (1984)
320. Prékopa (1973)
321. Pintér (1978)
322. Németh (1971)
323. Gerencsér (1971)
324. Silver (1977)
325. Cawdery (1976)
326. Magson (1979)
327. Magson (1979)
328. Tadikamalla (1978)
329. Silver (1965)
330. Feyerherm—Machado (1975)

331. Polyzos—Xirokostas (1976)
332. Prékopa—Kelle (1976)
333. Pintér (1973)

334. Kelle (1977)
335. Miller (1971)
336. Herron (1967)

II. Sources of Models and Numbers of Models

Agin, N. [1966]: A Min-Max Inventory Model. Management Science, Vol. 12, No. 7, pp. 517—529.
Model 178.

Arrow, K. J., Karlin, S. and *Scarf, H. [1958]:* Studies in the Mathematical Theory of Inventory and Production. Stanford University Press.
Model 121.

Barbosa, L. C. and *Friedman, M. [1978]:* Deterministic Inventory Lot Size Models — A General Root Law. Management Science, Vol. 24, No. 8, pp. 819—926.
Model 14.

Berman, E. B. [1961]: Monte-Carlo Determination of Stock Redistribution. Operations Research, pp. 500—506.
Model 296.

Bessler, S. A. and *Veinott, A. F. [1966]:* Optimal Policy for a Dynamic Multi-Echelon Inventory Model. Naval Research Logistics Quarterly, Vol. 13, No. 4, pp. 355—389.
Model 298.

Bessler, S. A. and *Zehna, P. W. [1967]:* An Application of Servomechanism to Inventory. Naval Research Logistics Quarterly, pp. 157—168.
Model 230.

Bishop, J. L. [1974]: Experience with a Successful System for Forecasting and Inventory Control. Operations Research, Vol. 22, No. 6, pp. 1224—1231.
Model 306.

Blackburn, J. and *Kunreuther, H. [1972]:* Planning and Forecast Horizons for the Dynamic Lot Size Model with Backlogging. University of Chicago, Research Report.
Models 220, 221.

Bomberger, E. E. [1966]: A Dynamic Programming Approach to a Lot Size Scheduling Problem. Management Science, Vol. 12, No. 11, pp. 778—784.
Model 278.

Brown, G. F., Corcoran, T. M. and *Lloyd, R. M. [1971]:* Inventory Models with Forecasting and Dependent Demand. Management Science, Vol. 17, No. 7, pp. 498—499.
Model 206.

Buchan, J. and *Koenigsberg, E. [1967]:* Nautschnoye upravleniye zapasami. Nauka, Moscow (Scientific Inventory Management. Englewood Cliffs, N.J., Prentice-Hall [1963])
Models 7, 52, 53, 54, 55, 79, 175, 179, 261, 262.

Burgin, T. A. [1975]: The Gamma Distribution and Inventory Control. Operational Research Quarterly, Vol. 26, pp. 507—525.
Model 310.

Burgin, T. A. [1970]: Back Ordering in Inventory Control. Operational Research Quarterly, Vol. 21, No. 4, pp. 453—461.
Model 154.

Cawdery, M. N. [1976]: The Lead Time and Inventory Control Problem. Operational Research Quarterly, Vol. 27, No. 4, pp. 971—982.
Models 149, 325.

Cohen, M. A. [1976]: Analysis of Single Critical Number Ordering Policies for Perishable Inventories. Operations Research, Vol. 24, No. 4, pp. 726—741.
Model 231.

Croston, J. D. [1972]: Forecasting and Stock Control for Intermittent Demands. Operational Research Quarterly, Vol. 23, No. 3, pp. 289—303.
Model 311.

Croston, J. D. [1974]: Stock Levels for Slow-Moving Items. Operational Research Quarterly, Vol. 25, No. 1, pp. 123—130.
Model 130.

Crowston, W. B., Wagner, H. M. and *Williams, J. F. [1973]:* Economic Lot Size Determination in Multi-Stage Assembly Systems. Management Science, Vol. 19, No. 5, pp. 517—527.
Model 305.

Daniel, K. H. [1963]: A Delivery-Lag Inventory Model with Emergency, in: *Scarf, H. E., Gilford, D. M.* and *Shelly, M. W.:* Multistage Inventory Models and Techniques. Stanford University Press, pp. 32—46.
Model 59.

Das, C. [1975]: Effect of Lead Time on Inventory: A Static Analysis. Operational Research Quarterly, Vol. 26, No. 2, pp. 273—282.
Models 151, 160.

Das, C. [1977]: The $(S-1, S)$ Inventory Model under Time Limit on Backorders. Operations Research, Vol. 25, No. 5, pp. 835—850.
Model 181.

Dave, U. [1979]: On a Discrete-In-Time Order-Level Inventory Model for Deteriorating Items. Journal of the Operational Research Society, Vol. 30, No. 4, pp. 349—354.
Model 45.

Dékány, I. [1974]: Numerical Solution of a Stochastic Control Problem Derived from Bnusoussan—Lions' Inventory Model, in: *Prékopa, A.* (ed.): Progress in Operations Research. Eger, pp. 231—241.
Model 257.

Denardo, E. V. [1968]: Separable Markovian Decision Problems. Management Science, Vol. 14, No. 7, pp. 451—462.
Model 188.

Donaldson, W. A. [1977]: Inventory Replenishment Policy for a Linear Trend in Demand—An Analytical Solution. Operational Research Quarterly, Vol. 28, No. 3, pp. 663—670.
Models 35, 36.

Donaldson, W. A. [1974]: The Allocation of Inventory Items to Lot Size/Reorder Level (Q, r) and Periodic Review (T, Z) Control Systems. Operational Research Quarterly, Vol. 25, No. 3, pp. 481—485.
Model 104.

Dorsey, R. C., Hodgson, T. J. and *Ratliff, H. D. [1974]:* A Production-Scheduling Problem with Batch Processing. Operations Research, Vol. 22, No. 6, pp. 1271—1279.
Model 281.

Elmaghraby, S. E. and *Bawle, V. Y. [1972]:* Optimalization of Batch Ordering under Deterministic Variable Demand. Management Science, Vol. 18, No. 9, pp. 508—517.
Model 40.

Falkner, C. F. [1970]: Jointly Optimal Deterministic Inventory and Replacement Policies. Management Science, Vol. 16, No. 9, pp. 622—635.
Model 46.

Feeney, G. J. and *Sherbrooke, C. C. [1966]:* The $(s-1, s)$ Inventory Policy under Compound Poisson Demand. Management Science, Vol. 12, No. 5, pp. 391—411.
Model 182.

Fries, B. E. [1975]: Optimal Ordering Policy for a Perishable Commodity with Fixed Lifetime. Operations Research, Vol. 23, No. 1, pp. 46—61.
Model 64.

Garman, M. B. [1976]: Multi-Product Economic Order Quantity Analysis under Minimum Inventory Valuation Constraints. Operational Research Quarterly, Vol. 27, No. 4, pp. 983—989.
Model 269.

Gebhardt, D. [1973]: Zur Bestimmung der Parameter der (s, q)—Bestellpolitik bei laufender Lagerüberwachung. Zeitschrift für Operations Research, Band 17, Serie B. 83—95.
Model 152.

Geisler, M. A. [1964]: The Sizes of Similation Samples Required to Compute Certain Inventory Characteristics with Stated Precision and Confidence. Management Science, Vol. 10, No. 2, pp. 261—286.
Models 132, 136.

Gerencsér, L. [1972]: Egy folyamatos nyilvántartású sztochasztikus készletgazdálkodási modell ismertetése és érzékenységi vizsgálata). (A Continuous Review Stochastic Inventory Model and its Sensitivity Analysis). MTA SZTAKI Közlemények 10. (Hungarian Academy of Sciences)
Model 105.

401

Gerencsér, L. [1972]: Az *S* szintre felrendelés elnevezésű készletmodellek költségfüggvényeinek összehasonlítása (Comparison of Cost Functions of *S*-Type Inventory Models). MTA SZTAKI Közlemények 9. (Hungarian Academy of Sciences)
Model 177.

Gerencsér, L. [1973]: Átlagkészlet csökkenése folyamatos nyilvántartású modell bevezetésével (Reduction of Average Inventories with Application of Continuous Review Model). MTA SZTAKI Közlemények 11. (Hungarian Academy of Sciences)
Model 146.

Gerencsér, L. [1973]: Reduction of the On-Hand Inventory on Given Level of Reliability, in: *Prékopa, A.* (ed.): Colloquia Math. Soc. J. Bolyai, 7. Inventory Control and Water Storage. Győr, 1971, North Holland.
Model 323.

Gerson, G. and *Brown, R. G. [1970]:* Decision Rules for Equal Shortage Policies. Naval Research Logistics Quarterly, Vol. 3, pp. 351—358.
Models 263, 264, 265.

Girlich, H. J. [1973]: Zur diskreten Roberts-approximation, in: *Prékopa, A.* (ed.): Colloqua Math. Soc. J. Bolyai, 7. Inventory Control and Water Storage, Győr, 1971. North Holland.
Model 126, 127.

Gonedes, N. J. and *Lieber, Z. [1974]:* Production Planning for a Stochastic Demand Process. Operations Research, pp. 771—787.
Model 77.

Goyal, S. K. [1973]: Economic Packaging Frequency for Items Jointly Replenished. Operations Research, pp. 644—647.
Model 284.

Goyal, S. K. [1974]: Optimum Ordering Policy for a Multi-Item Single Supplier System. Operational Research Quarterly, Vol. 25, No. 2, pp. 293—298.
Model 241.

Goyal, S. K. [1977]: Economic Packaging Frequency of Perishable Jointly Replenished Items. Operational Research Quarterly, Vol. 28, No. 1, pp. 215—219.
Model 44.

Goyal, S. K. [1977]: A Method for Improving the Solution of an Inventory Problem under Certainty. Operational Research Quarterly, Vol. 28, No. 1, pp. 69—78.
Model 218.

Greenberg, H. [1965]: Stock Level Distributions for (*s, S*) Inventory Problems. Naval Research Logistics Quarterly. pp. 343—349.
Model 134.

Grinold, R. C. [1976]: Manpower Planning with Uncertain Requirements. Operations Research, Vol. 24, No. 3, pp. 387—399.
Model 58.

Gross, D. [1963]: Centralized Inventory Control in Multilocation Supply Systems, in: *Scarf, H. E., Gilford, D. M.* and *Shelly, M. W.:* Multistage Inventory Models and Techniques. Stanford University Press.
Models 71, 288.

Gross, D. and *Harris, C. M. [1971]:* On One-For-One-Ordering Inventory Policies with State-Dependent Leadtimes. Operations Research, Vol. 19, No. 3, pp. 735—760.
Models 140, 141.

Hadley, G. [1964]: A Comparison of Order Quantities Computed Using the Average Annual Cost and the Discounted Cost. Management Science, Vol. 10, 472—476.
Model 26.

Hadley, G. and *Whitin, T. M. [1963]:* Analysis of Inventory Systems. Prentice-Hall.
Models: 2, 22, 24, 47, 48, 49, 50, 51, 87, 88, 89, 92, 106, 107, 112, 113, 150, 203, 215, 219, 227, 234, 235, 259, 260, 270, 275.

Hadley, G. and *Whitin, T. M. [1963]:* An Inventory-Transportation Model with *N* Locations, in: *Scarf, H. E., Gilford, D. M.* and *Shelly, M. W.:* Multistage Inventory Models and Techniques. Stanford University Press, pp. 116—142.
Model 292.

Hausman, W. H. and *Thomas, L. J. [1972]:* Inventory Control with Probabilistic Demand and Periodic Withdrawals. Management Science, Vol. 18, No. 5, pp. 265—275.
Models 97, 119.

Henery, R. J. [1979]: Inventory Replenishment Policy for Increasing Demand. Journal of the Operational Research Society, Vol. 30, No. 7, pp. 611—617.
Model 34.

Henin, C. [1972]: Optimal Replacement Policies for a Single Loaded Sliding Standby. Management Science, Vol. 18, No. 11, pp. 706—715.
Model 147.

Herron, D. P. [1967]: Inventory Management for Minimum Cost. Management Science, Vol. 14, No. 4, pp. 219—235.
Models 153, 336.

Higa, I., Feyerherm, A. M. and *Machado, A. L. [1975]:* Waiting Time in an $(S-1, S)$ Inventory System. Operations Research, Vol. 23, No. 4, pp. 647—680. Technical notes: *Sherbrooke, C. C.:* Waiting Time in and $(S-1, S)$ Inventory System-Constant Service Time Case. Operations Research, pp. 819—820.
Model 330.

Hochstädter, D. [1973]: The Stationary Solution of Multi-Product Inventory Models, in: *Prékopa A.* (ed.): Colloquia Math. Soc. J. Bolyai J., 7. Inventory Control and Water Storage, Győr, Hungary (1971). North Holland.
Models 253, 254.

Hochstädter, D. [1969]: Stochastische Lagerhaltungsmodelle. Springer-Verlag,
Models 80, 91, 93, 123, 128, 185, 186, 199, 223, 224, 233, 199.

Hodgson, T. J. [1972]: An Analytical Model of a Two-Products, One-Machine Production-Inventory System. University of Florida, Research report.
Models 279, 280.

Howe, W. G. [1974]: A New Look at Brown's Dynamic Inventory System. Operations Research, Vol. 22, No. 4, pp. 848—857.
Model 209.

Iglehart, D. L. and *Jaquette, S. C. [1969]:* Multi-Class Inventory Models with Demand a Function of Inventory Level. Naval Research Logistics Quarterly, Vol. 16, No. 4, pp. 495—502.
Models 74, 190.

Iglehart, D. L. and *Jaquette, S. C. [1969]:* Optimal Policies under the Shortage Probability Criterion for an Inventory Model with Unknown Dependent Demands. Naval Research Logistics Quarterly, Vol. 16, No. 4, pp. 485—493.
Models 194, 204, 205.

Iglehart, D. L. and *Lalchandani, A. P. [1967]:* An allocation model. SIAM Journal on Appl. Math. Vol. 15, No. 2, pp. 303—323.
Model 297.

Iglehart, D. L. and *Morey, R. C. [1972]:* Inventory Systems with Imperfect Asset Information. Management Science, Vol. 18, No. 8, pp. 388—394.
Model 148.

Iglehart, D. L. and *Morey, R. C. [1971]:* Optimal Policies for a Multi-Echelon Inventory System with Demand Forecasts. Naval Research Logistics Quarterly, Vol. 18, No. 1, pp. 115—118.
Model 302.

Iwaniec, K. [1979]: An Inventory Model with Full Load Ordering. Management Science, Vol. 25, No. 4, pp. 374—384.
Model 189.

Jagannathan, R. [1978]: A Minimax Ordering Policy for the Infinite Stage Dynamic Inventory Problem. Management Science, Vol. 24, No. 11, pp. 1138—1149.
Models 72, 73, 81.

Kao, E. P. C. [1975]: A Discrete Time Inventory Model with Arbitrary Interval and Quantity Distribution of Demand. Operations Research, Vol. 23, No. 6, pp. 1131—1142.
Models 76, 191.

Karlin, S. [1960]: Dynamic Inventory Policy with Varying Stochastic Demands. Management Science, Vol. 6, No. 3, pp. 231—258.
Model 225.

Kelle, P. [1977]: Stochastische Mehrproduktmodelle für die Sicherheitsbestände bei eines Serienfabrikation. Rostocker Betriebswirtschaftliche Manuscripte, 20. Teil II.
Models 334, 266.

Kelle, P. [1978]: Stochastische Optimierungsmodelle für die Produktionslager einer Walzwerkes. Mitteilungen der Mathematischen Gesellschaft der DDR, 78/1.
Model 313.

Kelle, P. [1984]: On the Safety Stock Problem for Random Delivery Process. European Journal of Operations Research, Vol. 17. 191—200.
Models 315, 319.

Kleindorfer, P. R. and *Newson, E. F. P. [1978]:* A Lower Bounding Structure for Lot-Size Scheduling Problems. Operations Research, Vol. 23, No. 2, pp. 299—311.
Model 286.

Klemm, H. [1973]: On the operating characteristic "service level", in: *A. Prékopa* (ed.): Colloqua Math. Soc. J. Bolyai 7. Inventory Control and Water Storage, Győr. North Holland
Models 135, 111.

Klemm, H. and *Mikut, M. [1972]:* Lagerhaltungsmodelle. Verlag Die Wirtschaft, Berlin,
Models 95, 96.

Kulcsár, T. [1979]: Egy sztochasztikus késési idős, folytonos készletellenőrzéses költségminimalizáló készletmodell (A Cost Minimizing Inventory Model with Stochastic Lead-Time). IX. Magyar Operációkutatási Konferencia, Győr, 1979.
Model 158.

Kunreuther, H. and *Schrange, L. [1973]:* Joint Pricing and Inventory Decisions for Constant Priced Items. Management Science, Vol. 19. No. 7, pp. 732—738.
Model 41.

László, Z. [1971]: Some recent results concerning reliability-type inventory models. in: *A. Prékopa* (ed.): Colloquia Math. Soc. J. Bolyai 7. Inventory Control and Water Storage, Győr, North Holland, 1973
Model 318.

Лавренченко, А. С. [1973]: Решение некоторых задач оптимального управления запасами методом усреднения уровня запаса. Экономика и математические методы, том IX, вып. 3, пп. 562—567.
Модел 108.

Ледин, М. И. анд *Ермаков, В. И. [1978]:* Об одной модели управления запасами с переменным спросом. Экономика и математические методы, том XIУ, вып. 3, пп. 531—538.
Модель 38.

Liberatore, M. J. [1977]: Planning Horizons for a Stochastic Lead-Time Inventory Model. Operations Research, Vol. 25, No. 6, pp. 977—988.
Model 222.

Loo, S. G. and *Ghosal, A. [1971]:* Optimization of the Capacity in a Storage System. Operations Research, Vol. 22, No. 2, pp. 544—548.
Model 183.

Magson, D. W. [1973]: Stock Control When the Lead-Time Cannot be Considered Constant. Journal of the Operational Research Society, Vol. 30, No. 4, pp. 317—322.
Models 326, 327.

Mann, Q. [1973]: Lagerhaltungsmodell für nicht Stationäre Nachfrage, in: *A. Prékopa* (ed.): Colloquia Math. Soc. J. Bolyai 7. Inventory Control and Water Storage, Győr. North Holland, 1973.
Models 312, 156.

Miller, B. L. [1971]: A Multi-Item Inventory Model with Joint Backorder Criterion. Operations Research, Vol. 22, No. 6, pp. 1467—1476.
Model 335.

Mumford, R. J. [1977]: The Numerical Generation of Lead-Time Demand Distributions for Inventory Models. Operational Research Quarterly Vol. 28, No. 1, pp. 79—85.
Model 184.

Naddor, E. [1966]: Inventory Systems. John Wiley and Sons.
Models: 1, 3, 4, 5, 8, 9, 10, 11, 12, 13, 16, 17, 21, 23, 27, 28, 30, 31, 32, 33, 66, 67, 68, 69, 70, 82, 83, 84, 85, 86, 94, 98, 99, 100, 101, 115, 116, 120, 124, 125, 129, 133, 143, 237, 238, 239, 240, 272.

Nahmias, S. [1975]: Optimal Ordering Policies for Perishable Inventory—II. Operations Research, Vol. 23, No. 4, pp. 735—749.
Models 62, 63.

Nahmias, S. and *Wang, S. S. [1979]:* A Heuristic Lot Size Reorder Point Model for Decaying Inventories. Management Science, Vol. 25, No. 1, pp. 90—97.
Model 114.

Nahmias, S. [1977]: Higher — Order Approximations for the Perishable — Inventory Problem. Operations Research Vol. 25, No. 4, pp. 630—640.
Model 232.

404

Németh, Gy. [1971]: Sztochasztikus készletmodellekkel kapcsolatos vizsgálatok (A Study of Some Stochastic Models). MTA III. Osztály Közleményei, Vol. 20, pp. 133—155.
Model 322.

Page, E. and *Paul, R. J. [1976]:* Multi-Product Inventory Situations with One Restriction. Operational Research Quarterly, Vol. 27, No. 4, pp. 815—834.
Model 271.

Peckelman, D. [1974]: Simultaneous Price-Production Decisions. Operations Research, IV. 788—794.
Model 39.

Pentico, D. W. [1976]: The Assortment Problem with Nonlinear Cost Functions. Operations Research. Vol. 24, No. 6, pp. 1129—1142.
Model 285.

Pierskalla, W. P. and *Roach, C. D. [1972]:* Optimal Issuing Policies for Perishable Inventory. Management Science, Vol. 18, No. 11, pp. 603—614.
Model 43.

Pintér, J. [1975]: Empirikus eloszlásfüggvény-sorozatok maximális eltérésének vizsgálata, alkalmazás egy többperiódusú megbízhatósági készletmodellre (Investigations on the Maximal Distance of Empirical Distribution Function Series). Alkalmazott Matematikai Lapok, Vol. 1, pp. 189—195. (Journal of Appl. Mathematics)
Model 321.

Pintér, J. [1973]: Egy sztochasztikus célfüggvényű irányítási feladat optimális megoldása (Optimal Solution of Control problem with Stochastic Objective). Egyetemi Számítóközpont Közleményei.
Model 333.

Polyzos, P. S. and *Xirokostas, R. A. [1976]:* An Inventory Control System with Partial Deliveries. Operational Research Quarterly, Vol. 27, No. 3. pp. 683—695.
Models 250, 331.

Porteus, E. L. [1971]: On the optimality of Generalized (s, S) policies, Management Science, Vol. 17, No. 7. pp. 411—426.
Model 197.

Porteus, E. L. [1972]: The Optimality of Generalized (s, S) Policies Under Uniform Demand Densities. Management Science, Vol. 18, No. 11, pp. 644—646.
Model 198.

Prékopa, A. [1963]: Reliability Equation for an Inventory Problem and Its Asymptotic Solutions, in: *A. Prékopa* (ed.): Colloquium on the Application of Mathematics to Economics. Akadémiai Kiadó, Budapest, pp. 317—327.
Models 317, 307.

Prékopa, A. [1972]: Az „S szintre felrendelés" elnevezésű sztochasztikus készletmodell és kiterjesztése intervallumszerű érkezések esetére (The "Order up to S" Stochastic Inventory Model and Its Extension for the Case of Interval Arrivals). Számológép, 1972/2.
Models 90, 176.

Prékopa, A. [1973]: Stochastic Programming Models for Inventory Control and Water Storage Problems, in: *A. Prékopa* (ed.): Colloquia Math. Soc. J. Bolyai 7. Inventory Control and Water Storage, Győr. North Holland.
Models 320, 267, 61.

Prékopa, A. and *Kelle, P. [1976]:* Sztochasztikus programozáson alapuló megbízhatósági jellegű készletmodell. (Reliability-Type Inventory Models Based on Stochastic Programming) Alkalmazott Matematikai Lapok, Vol. 2, pp. 1—16.
Models 277, 332.

Psoinos, D. [1976]: On the Joint Calculation of Safety Stocks and Replenishment Order Quantities. Operational Research Quarterly, Vol. 25, No. 1, pp. 173—177.
Model 103.

Rao, M. R. [1976]: Optimal Capacity Expansion with Inventory. Operations Research, Vol. 24, No. 2, pp. 291—300.
Model 216.

Rényi, A. and *Ziermann, M.:* Üzletek áruellátásával kapcsolatos szélsőérték feladatok (Tasks Relating to Extreme Values in Connection with the Goods Supply of Retail Outlets). MTA Alkalmazott Matematikai Intézet Közleményei.
Models 258, 308.

Resh, M., Friedman, M. and *Barbosa, L. [1976]:* On a General Solution of the Deterministic

27 Chikán

Lot Size Problem with Time-Proportional Demand. Operations Research, Vol. 24, No. 4, pp. 718—725.
Model 37.

Richards, F. R. [1976]: A Stochastic Model of a Repairable-Item Inventory System with. Attrition and Random Lead Times. Operations Research, Vol. 24, No. 1, pp. 118—130.
Model 155.

Riis, J. O. [1965]: Discounted Markov Programming in a Periodic Process. Operations Research, Vol. 13, No. 4, pp. 925—929.
Model 236.

Рыжиков, Ю, И. [1969]: Управление запасами (Inventory Control). Наука, Москва.
Models 29, 56, 57, 78, 102, 157, 162, 163, 164, 165, 166, 167, 168, 169, 170, 171, 172, 173, 174, 187, 214, 242, 244, 245, 246, 247, 248, 249, 273, 276, 294, 295, 301, 303, 304.

Рыжиков, Ю. И. [1972]: К расчёту пороговых строений управления запасами. Кибернетика, *П.* 4. пп. 50—55. Models 109, 251.

Rose, M. [1972]: The (s, S) Inventory Model with Arbitrary Backordered Demand and Constant Delivery Times. Operations Research, Vol. 20, No. 5, pp. 1020—1032.
Model 139.

Rosenshine, M. and *Obee, D. [1976]:* Analysis of a Standing Order Inventory System with Emergency Orders. Operations Research, Vol. 24, No. 6, pp. 1143—1156.
Model 131.

Scarf, H. [1960]: Some Remarks on Bayes Solutions to the Inventory Problem. Naval Research Logistic Quarterly, Vol. 7. pp. 591—596.
Model 226.

Schneeweiss, C. A. [1971]: Smoothing Production by Inventory — on Application of the Wiener Filtering Theory. Management Science, Vol. 17, No. 7, pp. 472—483.
Models 138, 207.

Schneeweiss, C. A. [1974]: Optimal Production Smoothing and Safety Inventory. Management Science, Vol. 20, No. 7, pp. 1122—1130.
Model 75.

Schneider, H. [1978]: Die Einhaltung einer Servicegrades bei (s, S) — Lagerhaltungspolitiken — eine Simulationsstudie. Zeitschrift für Operations Research, Band 22, Heft 4.
Model 137.

Schussel, G. [1968]: Job-Shop Lot Release Sizes. Management Science, Vol. 14, No. 8, pp. 449—472.
Models 6, 15.

Schussel, G. B. [1978]: Lagerhaltung bei Preisehöhrung. Zeitschrift für Operations Research, Band 22. Heft 3.
Models 42, 300.

Schwartz, B. L. [1970]: Optimal Inventory Policies in Perturbed Demand Models. Management Science, Vol. 16. No. 8, pp. 509—517.
Models 18, 118.

Schwartz, B. L. [1973]: A Simple Continuous Review Deterministic. One-Warehouse N-Retailer Inventory Problem. Management Science, Vol. 19, No. 5, pp. 555—566.
Model 299.

Silver, E. A. [1965]: Some Characteristics of a Special Joint-Order Inventory Model. Operations Research, Vol. 13, No. 2, pp. 319—322.
Model 252.

Silver, E. A. [1965]: Bayesian Determination of the Reorder Point of a Slow Moving Item. Operations Research, Vol. 13, No. 6, pp. 989—998.
Model 329.

Silver, E. A. [1977]: A Safety Factor Approximation Based on Tukey's Lambda Distribution. Operational Research Quarterly Vol. 28, No. 3, pp. 743—746.
Model 324.

Simmons, D. M. [1972]: Optimal Inventory Policies under Hierarchy of Setup Costs. Management Science, Vol. 18, No. 10, pp. 591—599.
Models 282, 283.

Simon, H. A. [1952]: On the Application of Servomechanism Theory in the Study of Production Control. Econometrica, Vol. 20, pp. 247—268.
Models 19, 20, 212.

Simon, R. M. [1971]: Stationary Properties of a Two-Echelon Inventory Model for Low Demand Items. Operations Research, Vol. 19, No. 3, pp. 761—773.
Model 293.

Sivazlian, B. D. [1974]: Continuous Review (s, S) Inventory System with Arbitrary Inter. arrival Distribution between Unit Demand. Operations Research, Vol. 20, No. 1, pp. 65—71-
Model 144.

Smith, S. A. [1977]: Optimal Inventories for an $(S-1, S)$ System with no Backorders. Management Science, Vol. 23, No. 5, pp. 522—528.
Model 180.

Snyder, R. D. [1974]: A Note on Fixed and Minimum Order Quantity Stock Systems. Operational Research Quarterly, Vol. 25, No. 4, pp. 635—639.
Models 117, 145.

Snyder, R. D. [1975]: A Dynamic Programming Formulation for Continuous Time Stock Control Systems. Operations Research, Vol. 25, pp. 383—385.
Model 201.

Sobel, M. J. [1971]: Production Smoothing with Stochastic Demand II: Infinite Horizon Case. Management Science, Vol. 17, No. 11, pp. 724—735.
Model 229.

Stidham, S. Jr. [1977]: Cost Models for Stochastic Clearing Systems. Operations Research, Vol. 25, No. 1, pp. 100—127.
Model 195.

Suddenth, W. D. [1965]: Another Formulation of the Inventory Problem. Operations Research, Vol. 13, No. 3, pp. 504—507.
Model 60.

Tadikamalla, P. R. [1978]: Application of the Weibull Distribution in Inventory Control. Journal of the Operational Research Society, Vol. 29, No. 1, pp. 77—83.
Model 328.

Tate, T. B. [1964]: In Defence of the Economic Batch Quantity. Operational Research Quarterly, Vol. 15, No. 4, pp. 329—339.
Models 25, 243.

Thomas, L. J. [1974]: Price and Production Decisions with Random Demand. Operations Research, Vol. 20, No. 3, pp. 513—518.
Model 193.

Tijms, H. C. [1974]: An Iterative Method for Approximating Average Cost Optimal (s, S) Inventory Policies. Zeitschrift für Operations Research, Vol. 18, pp. 215—223.
Model 200.

Тур. Л. П. [1972]: Об одном подходе к задаче управления запесами. Кибернетика, (On an Inventory Management Approach). No. 6.
Model 208.

Valisalo, P. E., Sivazlian, B. D. and *Mailott, J. F. [1972]:* Experimental Results on a New Computer Method for Generating Optimal Policy Variables in (s, S) Inventory Control Problems. Proceedings of the Spring Joint Computer Conference, University of Florida.
Model 122.

Veinott, A. F. Jr. [1965]: Optimal Policy for a Multi-Product Dynamic, Nonstationary Inventory Problem. Management Science, Vol. 12, No. 3, pp. 206—222.
Models 255, 256.

Veinott, A. F. Jr. [1965]: The Optimal Inventory Policy for Batch Ordering. Operations Research, Vol. 13, No. 3, pp. 424—432.
Model 211.

Veinott, A. F. Jr. [1963]: Optimal Stockage Policies with Non-Stationary Stochastic Demands, in: *Scarf, M. E., Gilford, D. M.* and *Shelley, M. W.:* Multistage Inventory Models and Techniques. Stanford University Press, Stanford, pp. 85—115.
Models 192, 202.

Wagner, H. M. [1968]: Principles of Operations Research. Prentice-Hall, Englewood Cliffs, N. J.
Models 65, 110, 159, 217, 228.

Wessels, J. [1973]: Inventory Control with Unknown Demand Distribution: A Discrete Time-Discrete Level Case, in: *A. Prékopa* (ed.): Colloquia Math. Soc. J. Bolyai 7. Inventory Control and Water Storage, Győr. North Holland.
Model 142.

27*

Wharton, F.: On estimating Aggregate Inventory Characteristics. Operational Research Quarterly, Vol. 26, No. 3, pp. 543—551.

Whitin, T. M. [1973]: Simple Multi-Stage Inventory Models, in: *A. Prékopa* (ed.): Colloquia Math. Soc. J. Bolyai 7. Inventory Control and Water Storage, Győr. North Holland. Models 289, 290, 291.

Wijngaard, J. [1973]: An Inventory Problem with Constrained Order Capacity. „Spezialtagung": Mathematische Lagerhaltungsmodelle. Weimar. Models 196, 210.

Wolfson, M. L. [1965]: Selecting the Best Length to Stock. Operations Research, Vol. 13, No. 4, pp. 570—585. Model 287.

Yaspan, A. [1972]: Fixed-Stockout-Probability Order Quantities with Lost Sales and Time Lag. Operations Research, Vol. 18, No. 4, pp. 903—904. Model 309.

Zangwill, W. J. [1966]: A Deterministic Multi-Period Production Scheduling Model with Backlogging. Management Science, Vol. 13. No. 1, pp. 105—119. Model 213.

Ziermann, M. [1953]: A raktárkészlet pótlásáról. A készletpótló rendelés (On the Replenishment of Stocks). MTA Matematikai Kutatóintézetének Közleményei, 1953/2, 203—217. Model 161.

Ziermann, M. [1963]: A Szmirnov-tétel alkalmazása egy raktározási problémára (Applying the Smirnoff Theorem to an Inventory Problem). MTA Matematikai Kutatóintézetének Közleményei, Ser. B. pp. 509—518. Models 314, 316.

Zoller, K. [1977]: Deterministic Multi-Item Inventory Systems with Limited Capacity. Management Science, Vol. 24, No. 4, pp. 451—455. Model 274.

References

Ackoff, R. L. (1961): *Progress in Operations Research.* Wiley, New York.

Aggarwal, S. C. (1974): A Review of Current Inventory Theory and Its Applications. *International Journal of Production Engineering,* Vol. 12, 443—482.

Agin, N. (1966): A Min-Max Inventory Model. *Management Science,* Vol. 12, No. 7, 517—529.

Allen, S. (1958): Redistribution of Total Stock over Several User Locations. *Naval Research Logistics Quarterly,* Vol. 5, 337—345.

Allen, S. (1961): Redistribution Model with Set-up Charge. *Management Science,* Vol. 8, 99—108.

Allen, S. and d'Esopo, D. A. (1968): An Ordering Policy for Repairable Stock Items. *Operations Research,* Vol. 16, 669—674.

Andersson, J. (1981): A Framework for Distributed Computerised Inventory Control Systems, in: Chikán (ed.) (1981).

Andler, K. (1929): *Rationalisierung der Fabrikation und Optimale Losgrösse.* Oldenburg, Munich.

Arrow, K., Harris, T. and Marschak, J. (1951): Optimal Inventory Policy. *Econometrica,* Vol. 19, 250—272.

Arrow, K., Karlin, S. and Scarf, H. (1958): *Studies in the Mathematical Theory of Inventory and Production.* Stanford.

Baker, K. (1977): An Experimental Study of the Effectiveness of Rolling Schedules in Production Planning. *Decision Science,* Vol. 8.

Baker, K., Dixon, P., Magazine, M. J. and Silver, E. A. (1978): An Algorithm for the Dynamic Lot Size Problem with Time-Varying Production Capacity Constraints. *Management Science,* Vol. 24, 1710—1720.

Balintfy, J. L. (1964): On a Basic Class of Multi-Item Inventory Problems. *Management Science,* Vol. 10, 287—297.

Bankaiev, A. A., Kostina, N. I. and Jarowicki, N. W. (1974): *Algoritmische Modellierung ökonomischer Probleme,* Berlin.

Barancsi, É., Bánki, Gy., Borlói, R., Chikán, A., Kelle, P., Kulcsár, T. and Meszéna, Gy.: On a Research of Inventory Models, in: Chikán (ed.) (1981)

Barankin, E. W. (1961): A Delivery-Lag Inventory Model with an Emergency Provision. *Naval Research Logistics Quarterly,* Vol. 8, 285—311.

Barbosa, L. C., Friedman, M. (1979): Inventory Lot-Size Models with Vanishing Market. *Journal of Operational Research Society,* Vol. 30, 1129—1132.

Bartmann, D. (1975): *Optimierung Markovscher Entscheidungsprozesse.* Dissertation, TK. Munich.

Beckman, M. J. (1961): An Inventory Model for Arbitrary Interval and Quantity Distributions of Demand. *Management Science,* Vol. 8, 35—57.

Beckman, M. J. and Hochstädter, D. (1968): Berechnung optimaler Entscheidungsregeln für die Lagerhaltung. *Jahrbücher für Nationalökonomie und Statistic,* Vol. 182, 106—123.

Beer, S. (1959): Cybernetics and Management. *The English Universities Press,* London.

Beesack, P. R. (1967): A Finite Horizon Dynamic Inventory Model with a Stockout Constraint. *Management Science,* Vol. 13, 618—630.

Bellman, R. (1957): Optimal Inventory Equation, in: Bellman: Dynamic Programming, Princeton.

Bellman, R., Glicksberg, I. and Gross, O. (1955): On the Optimal Inventory Equation. *Management Science,* Vol. 2, 83—104.

Belsley, D. A. (1969): *Industry Production Behavior: The Order-Stock Distinction.* Amsterdam, North Holland.

Bensoussan, A., Hurst, E. G. and Naslund, B. (1974): *Management Applications of Modern Control Theory.* North Holland.

Berács, J. (1973): A készletezési mechanizmusok vizsgálata (On the Inventory Mechanisms). *Research Report,* Karl Marx University of Economics, Budapest (in Hungarian)

Berács, J. (1977): A normák szerepe a készletgazdálkodásban (The Role of Norms in Inventory Management). *Research Report,* Hungarian Academy of Sciences, Socialist Enterprise Research Project (in Hungarian).

Berman, E. B. (1961): "Monte-Carlo" Determination of Stock Redistribution. *Operation Research,* Vol. 10, 500—506.

Bertalanffy, L. von (1951): General System Theory: A New Approach to Unity of Science. *Human Biology,* Vol. 23, 303—361.

Bessler, S. A. and Veinott, A. F. (1966): Optimal Policy for a Dynamic Multiechelon Inventory Model. *Naval Research Logistics Quarterly,* Vol. 13, 355—389.

Bessler, S. A. and Zehna, P. W. (1967): An Application of Servo-Mechanism to Inventory. *Naval Research Logistics Quarterly,* Vol. 14, 157—168.

Blackburn, J. D. and Millen, R. A. (1979): Selecting a Lot-Sizing Method for a Single Level Assembly Process. *Production and Inventory Management,* Vol. 20.

Blackburn, J. D. and Millen, R. A. (1981): Guidelines for Lot-Sizing Selection in Multi-Echelon Requirement Planning Systems, in: Chikán (ed.) (1981)

Bomberger, E. E. (1966): A Dynamic Programming Approach to a Lot-Size Scheduling Problem. *Management Science,* Vol. 12, 778—784.

Boulding, K. J. (1956): General System Theory: The Skeleton of Science. *Management Science,* Vol. 2, 197—208.

Brown, R. G. (1967): *Decision Rules for Inventory Management.* Rinchart-Winston, New York.

Brown, R. G. (1959): *Statistical Forecasting for Inventory Control.* McGraw Hill, New York.

Buchan, J. and Koenigsberg, E. (1963): *Scientific Inventory Management.* Prentice Hall.

Bulinskaya, E. V. (1964): Some Results Concerning Optimum Inventory Policies. *Theory of Probability and Its Applications,* Vol. 9, 389—403.

Burgin, T. A. (1970): Back Ordering in Inventory Control. *Operational Research Quarterly,* Vol. 21, 453—461.

Burgin, T. A. (1975): The Gamma Distribution and Inventory Control. *Operational Research Quarterly,* Vol. 26, 507—525.

Burgin, T. A. and Norman, J. M. (1976): A Table for Determining the Probability of a Stock out and Potential Lost Sales for a Gamma Distributed Demand. *Operational Research Quarterly,* Vol. 27, 621—631.

Bylka, S. (1981): Policies of (Multi-S)-Type in Dynamic Inventory Problem, in: Chikán (ed.) (1981)

Chikán, A. (1970): Többraktáras készletezési modellek és felhasználásuk kérdései a termelő-eszköz-kereskedelemben (Multilevel Inventory Models and Their Application in Trade). *Raktárgazdálkodás,* Vol. 1 (in Hungarian).

Chikán, A. (1973): Készletezési rendszerek matematikai modellezéséről (On Mathematical Modelling of Inventory Systems). *Research Report,* Hungarian Academy of Sciences, Socialist Enterprise Research Project (in Hungarian).

Chikán, A. (1977): Systems Approach to Inventory Management, in: Rose, J. and Bilciu, C. (eds): *Modern Trends in Cybernetics and Systems.* Proceedings of the Third International Congress of Cybernetics and Systems, Springer Verlag, Berlin–Heidelberg–New York.

Chikán, A. (ed.) (1983): *Készletezési modellek* (Inventory Models). Közgazdasági és Jogi Könyvkiadó, Budapest (in Hungarian).

Chikán, A. (1978): Some Comments on the Interpretation of the Lagrange-multiplier in Multi-product Inventory Models. *Mitteilungen der Mathematischen Gesellschaft der DDR,* Heft 1.

Chikán, A. (1980a): A vállalati készletgazdálkodás fő elvi és módszertani kérdései (Principles and Methods of Inventory Management) *Research Report,* Hungarian Academy of Sciences, Socialist Enterprise Research Project (in Hungarian).

Chikán, A. (1980b): The National Economy, as a Material Stock and Flow System, in: Rose (ed.) (1980): *Proceedings of the Fourth International Congress on General Systems and Cybernetics,* Springer-Verlag, Berlin.

Chikán, A. (1970): Többraktáras készletezési rendszerek matematikai modelljei (Mathematical Models of Multi-Level Inventory Systems). *Szigma,* Vol. 3, 43—67 (in Hungarian).

Chikán, A. and Meszéna, Gy. (1973): A Multi-Stage Stochastic Inventory Model and its Application, in: Prékopa (ed.) (1973).

Chikán, A. (1981): Market Disequilibrium and the Volume of Stocks, in: Chikán (ed.) (1981).

Chikán, A. (ed.) (1981): Economics and Management of Inventories, *Proceedings of the First International Symposium on Inventories.* Akadémiai Kiadó, Budapest, Elsevier Scientific Publishing Company, Amsterdam.

Chikán, A. and Nagy, M. (1978): *Készletgazdálkodás* (Inventory Management). Tankönyvkiadó, Budapest (in Hungarian).

Chikán, A., Fábri, E. and Nagy, M. (1978): *Készletek a gazdaságban* (Inventories in the National Economy). Közgazdasági és Jogi Könyvkiadó, Budapest (in Hungarian)

Churchman, C. W. (1968): *The Systems Approach.* Dell Publishing Co., New York.

Churchman, C. W., Ackoff, R. L. and Arnoff, E. L. (1957): Introduction to Operations Research. Wiley, New York.

Clark, A. J. (1958): A Dynamic Single-Item, Multi-Echelon Inventory Model. Rand Co.

Clark, A. J., and Scarf, H. (1960): Optimal Policies for a Multi-Echelon Inventory Problem. *Management Science,* Vol. 6, 475—490.

Cowdery, M. N. (1976): The Lead Time and Inventory Control Problem. *Operational Research Quarterly,* Vol. 27, 971—982.

Croston, J. D. (1972): Forecasting and Stock Control for Intermittent Demands. *Operational Research Quarterly,* Vol. 23, 289—303.

Croston, J. D. (1974): Stock Levels for Slow-moving Items. *Operational Research Quarterly,* Vol. 25, 123—130.

Daniel, K. H. (1963): A Delivery-Lag Inventory Model with Emergency, in: Scarf, Gilford, Shelly (1963)

Das, C. (1975): Effect of Lead Time on Inventory: A Static Analysis. *Operational Research Quarterly,* Vol. 26, 273—282.

Das, C. (1977): The $(S-1, S)$ Inventory Model under Time Limit on Backorders. *Operations Research,* Vol. 25, 835—850.

Denardo, E. V. (1968): Separable Markovian Decision Problems. *Management Science,* Vol. 14, 451—462.

Dietsch, V. (1977): *Dynamische Minimax-Eutscheidungsmodelle.* Dissertation, KMU. Leipzig.

Donaldson, W. A. (1977): Inventory Replenishment Policy for a Linear Trend in Demand —An Analytical Solution. *Operational Research Quarterly,* Vol. 28, 663—670.

Dvoretzky, A., Kiefer, J. and Wolfowitz, J. (1953): On the Optimal Character of the (s, S) —Policies in Inventory Theory. *Econometrica,* Vol. 21, 586—596.

Dzielinski, B. P. and Manne, A. S. (1961): Simulation of a Hypothetical Multi-Echelon Production and Inventory. *Journal of Industrial Engineering,* Vol. 12, 417—421.

Eilon, S. (1977): The Production Smoothing Problem. *The Production Engineer,* Vol. 123.

Evans, R. V. (1967): Inventory Control of a Multi-Product System with a Limited Production Resource. *Naval Research Logistics Quarterly,* Vol. 14, 173—184.

Evans, R. V. (1969): Inventory Control of By-Products. *Naval Research Logistics Quarterly,* Vol. 16, No. 1, 85—92.

Fábri E. (1981): Paradoxes of Macroeconomic Accumulation of Stocks, in: Chikán (ed.) (1981).

Farkas, K. (1976): Készlet és adaptáció (Inventory and Adaptation). *Research Report,* Hungarian Academy of Sciences, Institute of Economics (in Hungarian).

Freeney, G. F. and Sherbrooke, C. C. (1966): The $(S-1, S)$ Inventory Policy under Compound Poisson Demand: A Theory of Recoverable Item Storage. *Management Science,* Vol. 12, 391—411.

Fries, B. E. (1975): Optimal Ordering Policy for a Perishable Commodity with Fixed Lifetime. *Operations Research,* Vol. 23, 46—61.

Fukuda, Y. (1961): Optimal Disposal Policies. *Naval Research Logistics Quarterly,* Vol. 28, 221—227.

Fukuda, Y. (1964): Optimal Policies for the Inventory Problem with Negotiable Leadtime. *Management Science,* Vol. 10, 690—708.

Gelliher, H. P., Morse, P. M. and Simond, M. (1959): Dynamics of Two Classes of Continuous Review Inventory Systems. *Operations Research,* Vol. 7, 362—384.

Gebhardt, D. (1973): Zur Bestimmung der Parameter der (s, q) Bestellpolitik bei laufender Lagerüberwachung. *Zeitschrift für Operations Research,* Vol. 17, 83—95.

411

Geisler, M. A. (1964): The Sizes of Simulation Samples Required to Compute Certain Inventory Characteristics. *Management Science,* Vol. 10, 261—285.

Gerencsér, L. (1973): Egy folyamatos nyilvántartású készletmodell (A Continuous Inventory Model). *MTA SZTAKI Közlemények,* Vol. 10 (in Hungarian).

Gerencsér, L. (1972): Az S szintre felrendelés elnevezésű készletmodellek költségfüggvényeinek összehasonlítása (Cost Function Comparison in Order-up-to-S Inventory Models). *MTA SZTAKI Közlemények,* Vol. 9 (in Hungarian).

Gerencsér, L. (1973): Reduction of the On-hand Inventory on Given Level of Reliability, in: Prékopa (ed.) (1973).

Gerson, G. and Brown, R. G. (1970): Decision Rules for Equal Storage Policies. *Naval Research Logistics Quarterly,* Vol. 3, 351—358.

Ghali, M. A. (1981): Production Smoothing and Inventory Behaviour: A Simple Model, in: Chikán (ed.) (1981).

Girlich, H. J. (1977): *Bedienungstheoretische Behandlung von Lagerhaltungssystemen. Betriebsw.* Manuscripte, Rostock.

Girlich, H. J. (1973): Zur discreten Roberts Approximation, in: Prékopa (ed.) (1973).

Girlich, H. J. (1973): *Stochastische Entscheidungsprozesse.* Teubner, Leipzig.

Girlich, H. J. (1977): *Bedienungstheoretische Behandlung von Lagerhaltungssystemen.* Rostock, Universität Betriebswirtschaftliche Manuscripte, Vol. 20, 17—25.

Girlich, H. J. (1979): *Stochastische Lagerhaltungssysteme und die wirtschaftliche Losgrösse.* Akademie d. Wissenschaften d. DDR, Berlin.

Greenberg, H. (1964): Stock Level Distributions for (s, S) Inventory Problems. *Nav Researchal Logistics Quarterly,* Vol. 11, 343—349.

Griessbach, G. (1975): *Ein adaptiver Ansatz zur Bestimmung der optimalen Strategieparameter eines stationären LHM.* Dissertation, FSU Jena.

Goyal, S. K. and Belton, A. S. (1980): A Note on: A Simple Method of Determining Order Quantities in Joint Replenishments under Deterministic Demand. *Management Science.*

Günther, R. (1976): *Adaptive Vorhersage von stationären zufälligen Folgen mit rationaler Sprektraldichte.* Dissertation, FSU Jena.

Hadley, G. (1974): A Comparison of Order Quantities Computed Using the Average Cost and the Discounted Cost. *Management Science,* Vol. 10, 472—476.

Hadley, G. (1962): Generalizations of the Optimal Final Inventory Model. *Management Science,* Vol. 8, 454—457.

Hadley, G., and Whitin, T. M. (1963): *Analysis of Inventory Systems.* Prentice Hall.

Hajnal, A. (1976): A modellek modellje (Az interdiszciplináris szemlélet problémái) (The Model of Models. On the Problems of Interdisciplinarity), in: Kindler, J., Kiss, I. (1976).

Harris, F. (1915): *Operations and Cost.* A. W. Shaw Co., Chicago.

Hausmann, F. (1961): A Survey of Inventory Theory from the Operations Research Viewpoint, in: Ackoff (ed.) (1961)

Henery, R. J. (1979): Inventory Replenishment Policy for Increasing Demand. *Journal of Operational Research Society,* Vol. 30, 611—617.

Herron, D. P. (1967): Inventory Management for Minimum Cost. *Management Science,* Vol. 14, 219—235.

Higa, I., Feyerherm, A. M. and Machado, A. L. (1975): Waiting Time in an $(s-1, s)$ Inventory System. *Operations Research,* Vol. 23, 674—680.

Hochstädter, D. (1969): *Stochastische Lagerhaltungsmodelle.* Springer Verlag.

Hochstädter, D. (1973): The Stationary Solution of Multi Product Inventory Models, in: Prékopa (ed.) (1973).

Hochstädter, D. (1970): An Approximation of the Cost Function for Multi-Echelon Inventory Model. *Management Science,* Vol. 16, 716—727.

Hollier, R. and Vrat, P. (1976): A Review of Multi-Echelon Inventory Control Research and Application. *Research Report,* University of Birmingham.

Holmach, U. (1978): Nekatoriye modeli planirovaniya i operativnovo upravleniye potochnikami proizvodstvom (A Model of Production Planning and Scheduling). Dissertation, Leningrad, (in Russian).

Holt, C. C., Modigliani, F., Muth, J. F. and Simon, H. A. (1960): *Planning Production, Inventories and Work Force.* Prentice Hall, London.

Horowitz, G. and Samuelson, P. A. (eds) (1974): *Trade, Stability and Macroeconomics.* Academic Press, New York.

412

Howard, R. A. (1960): *Dynamic Programming and Markov Processes.* MIT Press, Cambridge Mass.

Hruckij, E. A., Szakovics, V. A. and Koloszov, Sz. P. (1977): *Optimizaciya hozyaystvennuh svyazey materialnuh zapasov* (Optimization of Inventories in Economic Cconnections). Ekonomika, Moscow, (in Russian).

Iglehart, D. L. (1963): Optimality of (s, S) Policies in the Infinite Horizon Dynamic Inventory Problem. *Management Science,* Vol. 9, 259—267.

Iglehart, D. L. (1965): Capital Accumulation and Production for the Firm: Optimal Dynamic Policies. *Management Science,* Vol. 12, 193—205.

Iglehart, D. L. and Lalchandani, A. P. (1967): An Allocation Model. *SIAM Journal,* Vol. 15, 303—323.

Iglehart, D. L. and Morey, R. C. (1971): Optimal Policies for a Multi-Echelon Inventory System with Demand Forecasting. *Naval Research Logistics Quarterly,* Vol. 18, 115—119.

Ignall, E. (1969): Optimal Continuous Review Policies for Two Product Inventory Systems with Joint Setup Costs. *Management Science,* Vol. 15, 278—283.

Ignall, E. and Veinott, A. (1969): Optimality of Myopic Inventory Policies for Several Substitute Products. *Management Science,* Vol. 15, 284—304.

Inderfurth, K. (1977): *Zur Grüte linearer Entscheidungsregeln in Produktions-Lagerhaltungs-Modellen.* Opladen.

Jennings, J. B. (1973): Blood Bank Inventory Control. *Management Science,* Vol. 19, 637—345.

Johnson, E. L. (1967): Optimality and Computation of (σ, S) Policies in the Multi-Item Infinite Horizon Inventory Problem. *Management Science,* Vol. 13, 475—491.

Johnson, E. L. (1968): On (s, S) Policies. *Management Science,* Vol. 15, 80—101.

Kalin, D. (1976): *Optimalität von verallgemeinerten (s, S) Politiken bei Mehrproduktlager-modellen.* Dissertation, Bonn University.

Kaminsky, F. C. (1966): Inventory Control with Random and Regular Replenishment: Constant Order Quantity Model. ORSA Tagung.

Kao, E. P. C. (1975): A Discrete Time Inventory Model with Arbitrary Interval and Quantity Distributions of Demand. *Operations Research,* Vol. 23, 1132—1142.

Kapitány, Zs. (1981): Dynamic Stochastic Systems Controlled by Stock and Order Signals, in: Chikán (ed.) (1981).

Kelle, P. (1977): Stochastische Mehrproduktmodelle für die Sicherheitsbestände bei einer Serienfabrikation. Rostocker Betriebswirtschaftliche Man., 20.

Kelle, P. (1978): Stochastische Optimierungsmodelle für die Produktionslager eines Walzwerkes. Mitteilungen der Math. Gesellschaft d. DDR, 1.

Kelle, P. (1980): Megbízhatósági készletmodellek és alkalmazásaik (Reliability Type Inventory Models and Their Application). *MTA SZTAKI Tanulmányok,* 107 (in Hungarian).

Kelle, P. (1979): Megbízhatósági készletmodellek és vállalati alkalmazásaik (Reliability Inventory Models and Their Applications at Firms). *Research Report,* Hungarian Academy of Sciences, Socialist Enterprise Research Project (in Hungarian).

Kiss, J. and Kindler, J. (eds) (1971): *Rendszerelmélet. Válogatott tanulmányok* (System Theory. Selected Papers). Közgazdasági és Jogi Könyvkiadó, Budapest (in Hungarian).

Kleijnen, J. P. C. (1976): *Computerized Inventory Management: A Critical Analysis of IBM's IMPACT System.* Reeks Ter Diskussic, 76.

Kleindorfer, P. R. and Newson, E. F. P. (1975): A Lower Bounding Structure for Lot Size Scheduling Problems. *Operations Research,* Vol. 23, 299—311.

Klemm, H and Mikut, M. (1972): *Lagerhaltungsmodelle.* Verlag Die Wirtschaft, Berlin.

Klemm, H. (1973): On the Operating Characteristic "Service Level", in: Prékopa (ed.) (1973)

Kocsondi, A. (1976): *A modell-módszer. A modellek szerepe a tudományos megismerésben* (The Model Method. The Role of Models in Scientific Research). Akadémiai Kiadó, Budapest (in Hungarian).

Kornai, J. (1971): *Anti-equilibrium.* Közgazdasági és Jogi Könyvkiadó, Budapest (in Hungarian).

Kornai, J. (1980): *The Economics of Shortage.* Elsevier Scientific Publishing Company, Amsterdam.

Kornai, J. and Martos, B. (1981): *Szabályozás árjelzések nélkül* (Economic Control without Prices). Akadémiai Kiadó, Budapest (in Hungarian).

Kornai, J. and Martos, B. (1971): Gazdasági rendszerek vegetatív működése (Vegetative Operation of Economic Systems). *Szigma,* Vol. 1, (in Hungarian).

413

Küenle, H. U. (1977): Ein neuer Optimalitätsbeweis für (*s*, *S*) Lagerhaltungsstrategien. *Mathematische Nachrichten, 77.*

László, Z. (1970): *Egy teljesen véletlen megbízhatósági jellegű készletmodell* (A Stochastic Reliability Type Inventory Model). Dissertation, University of Veszprém (in Hungarian).

László, Z.: Some Recent Results Concerning Reliability Type Inventory Models, in: Prékopa (ed.) (1973).

Lavratshenko, A. S. (1973): Reseniye nekatorih zadatsch optimalnovo upravleniya zapasami medotom usredniya urovna zapasa (Solution of Some Inventory Models with Method of Average Stock Level). *Ekonomika i Matematitscheskiye Metodi,* Vol. 9. 562—567.

Liberatore, M. J. (1977): Planning Horizons for a Stochastic Lead Time Inventory Model. *Operations Research,* Vol. 25, 977—988.

Ligeti, J. and Sivák, J. (1978): *Növekedés, szabályozás és stabilitás a gazdasági folyamatokban* (Growth, Regulation and Stability in Economic Processes). Közgazdasági és Jogi Könyvkiadó, Budapest (in Hungarian).

Limberg, H. (1968): Über die Optimalität von (*s*, *S*)-Strategien bei stochastischen Lagerhaltungsproblemen mit Lagerkostfunktionen, die eine Sprungstelle haben. *Operationsforschung und mathematische Statistik,* Vol. 1, 56—66.

Love, R. F. (1965): Convexity Properties of a Family of Stochastic Inventory Models. TR no. 13., Stanford.

Magson, D. W. (1979): Stock Control When the Lead Time Cannot be Considered Constant. *Journal of Operational Research Society,* Vol. 30, 317—382.

Mann, Q. (1973): Lagerhaltungsmodell für nichtstationäre Nachfrage, in: Prékopa (ed.) (1973)

Marchi, L. (1981): Role of Computers in Inventory Management. The Retailing Cage. in: Chikán (ed.) (1981).

Margansky, B. (1933): *Wirtschaftliche Lagerhaltung.* Berlin.

Martin, J. J. (1967): Bayesian Decision Problems and Markov Chains. New York.

Martin, W. (1980): Ermittlung der wesentlichen Faktoren und ihrer Wirkungsweise mit Hilfe von Modellexperimenten am Beispiel von TUL- Prozessen. The Fourth *Scientific Congress Mathematics and Cybernetics in the Economy,* Rostock, 1980.

Matthes, B. and Müller, J. A. (1977): *Ermittlung optimaler Parameter von stochastischen Systemen.* Betriebw. Man., Rostock.

Meier, R. C., Newell, W. T. and Paser, H. L. (1969): *Simulation in Business and Economics.* Englewood Cliffs, N. J.

Metzler, L. (1977): Factors Governing the Length of Inventory Cycles. *The Review of Economic Statistics,* Vol. 1, 1—15.

Miethe, M. (1978): Ein semi-markovsches Lagerhaltungsmodell. *Mathematische Operationsforschung und Statistik,* Vol. 9.

Miller, D. W. and Starr, M. K. (1969): *Executive Decision and Operations Research.* Prentice-Hall, Inc. Englewood Cliffs, N. J.

Moriguchi, Ch. (1967): *Business Cycles and Manufacturers' Short-Term Production Decisions.* North Holland, Amsterdam.

Móritz, A. (1978): Ein mathematische Modell für die Lagerhaltung im ungarischen Binnenhandel. *Zeitschrift für Operations Research,* Vol. 22, B 35—56.

Morse, P. M. (1959): Solutions of a Class of Discrete Time Inventory Problems. *Operation Research,* Vol. 7, 67—78.

Morse, P. M. and Kimbel, G. E. (1951): *Methods of Operations Research.* MIT Press.

Müller, J. A. (1981): About the Optimum Execution of Stock Keeping Processes by Means of Planned Model Tests. in: Chikán (ed.) (1981).

Müller, N. (1975): "Effectiveness", Reporting from Decision–Making Computer–Based Systems. *Production and Inventory Management,* No. 3.

Naddor, E. (1966): *Inventory Systems.* J. Wiley.

Nagy, M. (1975): A vállalati készletgazdálkodás szabályozása (Regulation of Company Inventory Management). *Research Report,* Hungarian Academy of Sciences, Socialist Enterprise Research Project (in Hungarian).

Nagy, M. (1975): A vállalati anyagellátás gazdaságosságának elemzése (Analysis of Profitability of Material Supply in Companies). *Research Report,* Hungarian Academy of Sciences, Socialist Enterprise Research Project (in Hungarian).

Nagy, M. (1981): Some Specialities of the Connection between Material Flows and Stocks in the Hungarian Economy, in: Chikán (ed.) (1981).

Nahmias, S. (1975): Optimal Ordering Policies for Perishable Inventory II. *Operations Research*, Vol. 23, 735—749.

Nahmias, S.: Queuing Models for Controlling Perishable Inventories, in: Chikán (ed.) (1981).

Nemény, V. (1973): *Gazdasági rendszerek irányítása* (Control in Economic Systems). Közgazdasági és Jogi Könyvkiadó, Budapest (in Hungarian).

Németh, Gy. (1971): Sztochasztikus készletmodellekkel kapcsolatos vizsgálatok (Analysis of Stochastic Inventory Models). *MTA Közleményei*, 133—155 (in Hungarian).

Neuts, M. F. (1964): An Inventory Model with an Optimal Time Lag. *SIAM Journal*, Vol. 12, 179—185.

O'Grady, P. J. and Bonney, M. C. (1981): Optimal Estimation of Inter-Machining Inventory Levels, in: Chikán (ed.) (1981).

Orliczky, J. (1975): *Material Requirements Planning*. McGraw Hill, New York.

Page, E. and Paul, P. J. (1976): Multi-Product Inventory Situations with One Restriction. *Operational Research Quarterly*, Vol. 27, 815—834.

Papathanassiou, B. (1981): Production Smoothing-Stock Level Middle-Term Planning of the Preconstruction Elements, in: Chikán (ed.) (1981).

Pawelzig, G. (1974): *Az objektiv rendszerek fejlődésének dialektikája* (Dialectics of Development of Objective Systems). Gondolat, Budapest (in Hungarian).

Pervoswanskij, A. A. (1975): *Matematitscheskiye modeli v upravlenii prauzvodstvom* (Mathematical Models in Inventory Management). Nauka, Moscow.

Pervoswansky, A. A. and Hollmach, U. (1978): Zu einigen Fragen der Modellierung von Prozessen der operativen Steuerung von Fliess bandproduktion. Frühjahrtagung d. IGr. Lagerhaltungsmodelle Weimar.

Pegels, C.: A Batch Size Model for the Hospital Pharmacy, in: Chikán (ed.) (1981).

Peterson, R. and Silver, E. A. (1979): *Decision Systems for Inventory Management and Production Planning*. John Wiley, New York.

Pintér, J. (1977): Egy sztochasztikus célfüggvényű irányítási feladat optimális megoldása (Optimal Solution of Stochastic Objective Function Control Problem). *Egyetemi Számítóközpont Közleményei*, Budapest (in Hungarian).

Pintér, J. (1975): Empirikus eloszlásfüggvény-sorozatok maximális eltérésének vizsgálata, alkalmazás egy többperiódusú megbízhatósági készletmodellre (Examination of Maximal Deviation of Empirical Distribution Function Series. Application for Multi-Period Reliability Type Inventory Models). *Alkalmazott Matematikai Lapok* 1, 189—195 (in Hungarian).

Popp, W. (1968): *Einführung in die Theorie der Lagerhaltung*. Springer Verlag.

Popplevel, K. and Bonney, M. C. (1977): *The Application of Discrete Linear Control Theory. Advances in OR*. Amsterdam.

Prékopa, A. (1975): Reliability Equation for an Inventory Problem and its Asymptotic Solutions. Colloquium on the Application of Mathematics to Economics. Akadémiai Kiadó, Budapest.

Prékopa, A. (1972): Az „S szintre való felrendelés" elnevezésű sztochasztikus készletmodell és annak kiterjesztése intervallumszerű érkezések esetére (The „Order-up-to-S" Stochastic Inventory Model and Its Extension for the Case of Interval Arrivals). *Számológép*, 2, 34—45 (in Hungarian).

Prékopa, A. (1973): Stochastic Programming Models for Inventory Control and Water Storage Problems. Colloquia Math. Soc. Bolyai J., Inventory Control and Water Storage, Győr, 1971, North Holland, Amsterdam, Akadémiai Kiadó, Budapest.

Prékopa, A. and Kelle, P. (1976): Sztochasztikus programozáson alapuló megbízhatósági jellegű készletmodellek (Reliability Type Inventory Models Based on Stochastic Programming) *Alkalmazott Matematikai Lapok*, 2, 1—16 (in Hungarian).

Psoinos, D. (1974): On the Joint Calculation of Safety Stocks and Replenishment Order Quantities. *Operational Research Quarterly*, Vol. 25, 173—177.

Rao, M. R. (1976): Capacity Expansions with Inventory. *Operations Research*, Vol. 24, 291—300.

Rempala, R. (1981): On the Multicommodity Arrow—Karlin's Inventory Model, in: Chikán (ed.) (1981).

Rényi, A. and Ziermann, M. (1960): Üzletek áruellátásával kapcsolatos szélsőérték feladatok (Supply Problems of Retailing System). *Alkalmazott Matematikai Intézet Közleményei*, Vol. 5B, No. 4, 495—505 (in Hungarian).

Richter, K. (1976): Untersuchungen eines linearen dynamischen Produktionsplanungsmodells. *Wiss. Zeitschrift TH Karl-Marx-Stadt*, 18, 399—402.

Richter, K. (1980): Equivalente Optimierungsaufgaben und Dekomposition. Teil I. *Mathematische Operationsforschung und Statistik*, Vol. 11, 1—14.

Riis, I. O. (1965): Discounted Markov Programming in Periodic Process. *Operations Research*, Vol. 13, 920—929.

Ritchie, E. (1981): Practical Inventory Policies for the Stocking of Spare Parts, in: Chikán (ed.) (1981).

Roberts, D. M.: Approximations to Optimal Policies in a Dynamic Inventory Model, in: Arrow, Karlin, Scarf (1962).

Robinson (1957): *Report on a System Simulation*. ORSA, New York.

Ryshikow, Jn. I. (1969): *Upravlenie zapasami* (Inventory Management). Nauka, Moscow,

Ryshikow, Jn. I. (1972): K rastschotu pragovüh strategiy upravleniya zapasami. *Kibernetika*, 4 50—55.

Sadowsky, V. N. (1976): *Az általános rendszerelmélet alapjai* (The Basic Features of General System Theory). Statisztikai Kiadó, Budapest (in Hungarian).

Sasieni, M., Yaspan, A., and Friedman, L. (1959): *Operations Research: Methods and Problems*. Wiley, New York.

Scarf, H. (1959): The Optimality of (s, S) Policies in the Dynamic Inventory Problem, in: Arrow, Karlin, Suppes (1959): Mathematical Methods in the Social Sciences.

Scarf, H., Gilford and Shelly (1963): *Multistage Inventory Models and Techniques*. Stanford.

Schneeweiss, C. H. (1977): Inventory Production Theory. Lecture Notes in Economics and Mathematical Systems, 151.

Schneider, H. (1978): Methods for Determining the Reorder Point of an (s, S) Ordering Policy when a Service Level is Specified. *Journal of Operational Research Society*, Vol. 29, 1181—1193.

Schneieder, H. (1981): On Obtaining a Required Service-Level in a Periodic Inventory Model, in: Chikán (ed.) (1981).

Schussel, G. (1968): Job-Shop Lot Release Sizes. *Management Science*, Vol. 14, 449—478.

Sherbrooke, C. C. (1968): Metric: A Multi-Echelon Technique for Recoverable Item Control. *Operations Research*, Vol. 16, 122—141.

Silver, E. A. (1965): Some Characteristics of a Special Joint-Order Inventory Model. *Operations Research*, Vol. 13, No. 2, 319—322.

Silver, E. A. (1976): A Simple Method of Determining Order Quantities in Joint Replenishment under Demand. *Management Science*, Vol. 22, 1351—1361.

Silver, E. A. and Massard, N. E. (1981): Setting of Parameter Values in Coordinated Inventory Control by Means of Graphical and Hand Calculator Methods, in: Chikán (ed.) (1981).

Sivazlian, B. D. (1974): A Continuous Review (s, S) Inventory System with Arbitrary Interarrival Distribution between Unit Demand. *Operations Research*, Vol. 23, 65—71.

Smith, S. (1977): Optimal Inventories for an (s − 1, s) System with Backorders. *Management Science*, Vol. 23, 522—528.

Snyder, R. D. (1974): Fixed and Minimum Order Quantity Stock Systems. *Operational Research Quarterly*, Vol. 25, 635—639.

Snyder, R. D. (1975): A Dynamic Programming Formulation for Continuous Time Stock Control Systems. *Operations Research*, Vol. 33, 383—385.

Steffanic-Allmayer, K. (1927): Die Güngstigste Bestellunge beim Einkauf. *Sparwirtschaft*, Vienna.

Stoff, V. (1973): *Modell és filozófia* (Model and Philosophy). Kossuth Kiadó, Budapest (in Hungarian).

Stohr, E. A. (1979): Information Systems for Oberving Inventory Levels. *Operations Besearch*, Vol. 27, 242—259.

Suddenth, W. D. (1965): Another Formulation of the Inventory Problem. *Operations Research*, Vol. 13, 504—507.

Tadikamalla, P. R. (1978): Application of the Weibull Distribution to Inventory Control *Journal of Operational Research Society*, Vol. 29, 77—83.

Tapiero, C. S. (1977): Optimization of Information Measurement with Inventory Applications. *INFOR* Vol. 15, 50—61.

Tijms, H. C. (1972): *Analysis of (s, S) Inventory Models*. Amsterdam: Mathematisch Centrum.

Tur, L. P. (1972): Ob obnom podhodie k zadatsch upravlenija zapasami. *Kibernetika* (in Russian).

Valisalo, P. E., Sivazlian, B. D., and Maillott, J. F. (1972): Experimental Results on a New Computer Method for Generating Optimal Policy Variables. Proceedings of the Joint Computer Conference, Florida.

416

Vander Ecken, J. (1981): Aggregate Inventory Management. A Case Study, in: Chikán (ed.) (1981).

Van Numen, J. (1976): Contracting Markov Decision Processes. Mathematics Centre Tracts 71, Amsterdam.

Veinott, A. F. (1966): On the Optimality of (s, S) Inventory Policy: New Conditions and a New Proof. *SIAM Journal*, Vol. 14, 1067—1083.

Veinott, A. F. and Wagner, H. M. (1965): Computing Optimal (s, S) Inventory Policies. *Management Science*, Vol. 11, 525—552.

Veinott, A. F. (1965): Optimal Policy for a Multi-Product, Dynamic, Nonstationary Inventory Problem. *Management Science*, Vol. 12, 206—222.

Ventcel, E. Sz. (1969): *A dinamikus programozás elemei* (The Elements of Dynamic Programming). Közgazdasági és Jogi Könyvkiadó, Budapest (in Hungarian).

Vrat, P. and Babu, A. S.: Optimal Policies for Spares in Multi-Echelon Repair Inventory Systems, in: Chikán (ed.) (1981).

Wagner, H. M. (1960): On the Optimality of Pure Strategies. *Management Science*, Vol. 6, 268—269.

Wagner, H. M. (1962): *Statistical Management of Inventory Systems*. John Wiley, New York.

Wagner, H. M. (1969): *Principles of Operations Research with Applications to Managerial Decisions*. Prentice Hall, N. J.

Wagner, H. M., O'Hagan, M., and Lundh, E. (1965): An Empirical Study of Exactly and Approximately Optimal Inventory Policies. *Management Science*, Vol. 11, 690—723.

Wagner, H. M.. and Whitin, T. M. (1958): Dynamic Version of the Economic Lot Size Model. *Management Science*, Vol. 5, 89—96.

Waldman, K. H. (1976): Stationäre Baysche Entscheidungsmodelle mit Anwendung in der Lagerhaltung. Dissertation, TH Darmstadt.

Whitin, T. M. (1957): *The Theory of Inventory Management*. Princeton University Press, Princeton, N. J.

Whitin, T. M. (1973): Simple Multi-Stage Inventory Models, in: Prékopa (ed.) (1973).

Wijngaard, J. (1973): An Inventory Problem with Constrained Order Capacity. *Mathematische Lagerhaltungsmodelle*, Weimar.

Wilson, R. W. (1934): A Scientific Routine for Stock Control. *Harvard Business Review*, Vol. 13, 116—128.

Wolfson, M. L. (1965): Selecting the Best Length to Stock. *Operations Research*, Vol. 23, 507—585.

Wolski, Z. S. (1981): Inventory Control of Spare Parts in Supplying Organizations, in: Chikán (ed.) (1981).

Wright, G. P. (1968): Optimal Policies for a Multi-Product Inventory System with Negotiable Lead Times. *Naval Research Logistics Quarterly*, Vol. 15, 375—401.

Yaspan, A. (1972): Fixed-Stockout-Probability Order Quantities with Lost Sales and Time Lag. *Operations Research*, Vol. 18, No. 3, 903—904.

Zabel, E. (1962): A Note on the Optimality of (s, S) Policies in Inventory Theory. *Management Science*, Vol. 9, 123—125.

Zangwill, W. I. (1966): A Dynamic Multi-Product, Multi-Facility Production and Inventory Model. *Operations Research*, Vol. 14, 486—507.

Ziermann, M. (1953): A raktárkészlet pótlásáról II (On the Replenishment of Stocks). *Alkalmazott Matematikai Intézet Közleményei*, Vol. 2, 203—216 (in Hungarian).

Ziermann, M. (1963): A Szmirnov tétel alkalmazása egy raktározási problémára (The Application of the Smirnoff Thesis in an Inventory Problem). *Matematikai Kutató Intézet Közleményei*, Vol. 8, 509—516 (in Hungarian).

Zoller, K. (1977): Deterministic Multi-Item Inventory Systems with Limited Capacity. *Management Science*, Vol. 24, 451—455.

417

Theory and Decision Library

Series B: Mathematical and Statistical Methods

Already published:

Robustness of Statistical Methods and Nonparametric Statistics
edited by Dieter Rasch and Moti Lal Tiku
ISBN 90—277—2076—2

Stochastic Optimization and Economic Models
by Jati K. Sengupta
ISBN 90—277—2301—X

A Short Course on Functional Equations
by J. Aczel
ISBN 90—277—2376—1

Optimization Models Using Fuzzy Sets and Possibility Theory
edited by J. Kacprzyk and S. A. Orlovski
ISBN 90—277—2492—X

Advances in Multivariate Statistical Analysis
edited by A. K. Gupta
ISBN 90—277—2531—4

Statistics with Vague Data
edited by Rudolf Kruse and Klaus Dieter Meyer
ISBN 90—277—2562—4

Applied Mathematics for Economics
by Jati K. Sengupta
ISBN 90—277—2588—8

Multivariate Statistical Modelling and Data Analysis
edited by H. Bozdogan and A. K. Gupta
ISBN 90—277—2592—6

Risk, Decision and Rationality
edited by Bertrand R. Munier
ISBN 90—277—2624—8

Multiple Criteria Decision Analysis in Regional Planning
by Fukimo Seo and Masatoshi Sakawa
ISBN 90—277—2641—8

418

Theory of Statistical Inference and Information

by Igor Vajda
ISBN 90—277—2781—3

Decision Theory and Decision Behaviour

by Anatol Rapoport
ISBN 0—7923—0297—4